The Psychology of Traffic

This book offers a comprehensive exploration of the intricate relationship between human behaviour and transportation systems. It examines how individual psychology intersects with mobility patterns, revealing the complex factors that drive our daily travel decisions and interactions on the road.

The authors explore key topics like the multidimensional nature of quality of life in relation to transportation, the psychological impact of driving fear and active transport benefits, the symbolic meaning of vehicles, and the influence of both formal regulations and informal social rules. It examines communication patterns among road users, decision-making processes, and how policy interventions and infrastructure design can psychologically influence behaviour towards more sustainable and safer mobility choices.

The Psychology of Traffic is primarily aimed at academic scholars and advanced students in transport studies and applied psychology. It can be useful for policymakers and practitioners seeking to understand and influence road user behaviour to create more sustainable, safe, and socially beneficial transportation systems that enhance overall quality of life.

Ralf Risser is a psychologist and sociologist (University of Vienna). Owned a private research institute FACTUM OG 1988–2019. Part-time work at Institute of Technology & Society, TU Lund, Sweden 1988–2013, 2006–2013 Adjunct Professor. 2011–2017 President of International Cooperation on Theories and Concepts in Traffic safety (ICTCT). Key researcher in projects such as EU-funded CONSOL, investigating case of the elderly in connection with traffic safety and mobility. Currently, Visiting Professor at the Palacký University Olomouc, Czechia, Department of Psychology. Elderly road users, vulnerable road users, and qualitative methods are in the focus of his research.

Matúš Šucha (26 April 1980, Bratislava) is a Czech traffic psychologist and a professor of psychology at Palacký University, Olomouc, Czechia. His main area of professional interest is traffic psychology, with a specific focus on mobility and traffic safety. He is a former vice-president of International Cooperation on Theories and Concepts in Traffic safety (ICTCT).

Routledge Studies in Transport Analysis

22. **Strategies for Sustainable Air Services Development**
An Airport-Destination Collaborative Approach
Edited by Chrystal Zhang and Kareem Yarde

23. **Artificial Intelligence in Shipping**
The State of Digital Innovation
Maria Lambrou

24. **Railway Safety Management**
Systems, Practices, and Emerging Trends
Katarzyna Chruzik and Marzena Graboń-Chałupczak

25. **Port Management Strategy for Enhancing Cybersecurity and Resilience**
Learning from Case Studies
Chalermpong Senarak

26. **Electric Vehicle Supply Chain Management**
Edited by Muhammad Shujaat Mubarik, Jyri Vilko, and Sajid Nazir

27. **Airline Pilots in the Age of AI**
A Global Study of Work, Safety and Technology
Geraint Harvey

28. **Electric Vehicle Supply Chain Management**
Edited by Muhammad Shujaat Mubarik, Jyri Vilko, Sajid Nazir

29. **The Psychology of Traffic**
Edited by Ralf Risser and Matúš Šucha

For more information about this series, please visit: www.routledge.com/Routledge-Studies-in-Transport-Analysis/book-series/RSTA

The Psychology of Traffic

Ralf Risser and Matúš Šucha

Routledge
Taylor & Francis Group
LONDON AND NEW YORK

First published 2026
by Routledge
4 Park Square, Milton Park, Abingdon, Oxon OX14 4RN

and by Routledge
605 Third Avenue, New York, NY 10158

Routledge is an imprint of the Taylor & Francis Group, an informa business

British Library Cataloguing-in-Publication Data
A catalogue record for this book is available from the British Library

ISBN: 978-1-041-00151-5 (hbk)
ISBN: 978-1-041-00152-2 (pbk)
ISBN: 978-1-003-60839-4 (ebk)

DOI: 10.4324/9781003608394

Typeset in Sabon
by Apex CoVantage, LLC

This publication was made possible thanks to targeted funding provided by the Faculty of Arts, Palacký University Olomouc, project No. FPVC2026/9

Contents

Contributors

Eva Adamovská is a psychologist specialising in traffic psychology and human behaviour in transport. Her professional focus includes psychological assessment, road safety, and applied traffic-psychological practice. She is engaged in interdisciplinary cooperation linking psychology, mobility, and prevention. Her work is grounded in evidence-based approaches aimed at improving safety and quality of transport systems.

Jiří Ambros is a psychologist working in the area of traffic psychology and road safety. His professional activities focus on understanding behaviour of drivers and other road users in real traffic conditions. He is engaged in applied psychological practice and expert work related to assessment, prevention, and safety interventions. His approach is grounded in empirical evidence and aims to support safer and more sustainable mobility systems.

Kateřina Bucsuházy is a researcher specialising in traffic psychology and human behaviour in mobility systems. Her work addresses behavioural mechanisms related to road safety, risk perception, and transport environments. She is involved in empirical research and interdisciplinary collaboration in transport studies. Her professional focus supports data-driven improvements in traffic safety and sustainable mobility.

Dana Černochová is a psychologist specialising in traffic psychology and road user behaviour. Her professional activities focus on psychological assessment, prevention, and safety in transport contexts. She contributes to applied practice and interdisciplinary cooperation in the field of mobility and traffic safety. Her work is grounded in evidence-based psychological approaches to improving road safety and behaviour.

Elisabeta Drimlová is a psychologist and academic engaged in research and practice in psychology, including areas related to behaviour and risk. She has been involved in scholarly activities that span educational, clinical, and behavioural psychology, and her work appears in contexts where traffic-related psychological research and risk behaviour are discussed.

Vít Gabrhel is a psychologist whose work concentrates on traffic psychology and human behaviour in road transport. He focuses on psychological determinants of driver behaviour, risk perception, and traffic safety. His activities span applied assessment as well as participation in research and professional projects. His approach is grounded in analytical, evidence-informed psychological practice within transport contexts.

Lucie Hartmannová is a psychologist affiliated with Palacký University Olomouc, Laboratory of Traffic Psychology. She specialises in traffic psychology, with a focus on driver behaviour, road safety, and psychological assessment. Her work combines applied research with professional practice in transport psychology. She contributes to evidence-based approaches to improving safety and behaviour in traffic systems.

Darina Havlíčková is a therapist affiliated with the Therapeutic Centre Nové Elysium in Brno. She works within the field of mental health and psychosocial support, focusing on therapeutic care for clients. Her professional practice is grounded in evidence-based therapeutic approaches and clinical experience. She contributes to multidisciplinary cooperation in the provision of comprehensive mental health services.

Kristýna Josrová is a researcher focusing on traffic psychology and human factors in transport. Her work centres on empirical research into road user behaviour, safety, and psychological determinants of mobility. She is involved in data-driven analyses and applied research projects related to traffic safety and prevention. Her professional activities emphasize evidence-based approaches to improving transport systems and road safety.

Martin Kořán is a clinical psychologist working in a hospital setting. His professional expertise includes traffic psychology, particularly psychological factors influencing driver behaviour and fitness to drive. He combines clinical assessment with applied psychological evaluation in transport-related contexts. His work is oriented toward evidence-based practice at the interface of health psychology and road safety.

Václav Linkov is a psychologist affiliated with Masaryk University. He specialises in traffic psychology, with a focus on driver behaviour, road safety, and psychological assessment. His work combines applied research with expert practice in the field of transport psychology. He contributes to evidence-based approaches to improving traffic safety and mobility systems.

Anton Pashkevich is a researcher at the Cracow University of Technology within the Chair of Transportation and Logistics Systems. His academic discipline aligns with *civil engineering, geodesy and transport systems*, with a focus on transport behaviour and dynamics. He contributes to empirical and applied research related to traffic, mobility, and system performance. His work supports evidence-based understanding of transport behaviour and safety.

Vlasta Rehnová is a psychologist specialising in traffic psychology and road user behaviour. Her professional focus includes psychological assessment of drivers, prevention, and traffic safety. She is engaged in applied psychological practice related to mobility and transport systems. Her work is grounded in evidence-based approaches aimed at improving road safety and responsible behaviour.

Jakub Štefec is a PhD student in psychology at Palacký University Olomouc. His doctoral research focuses on traffic psychology, particularly psychological factors influencing road user behaviour and safety. He is involved in empirical research and data-based analysis within transport and mobility contexts. His work contributes to the development of scientifically grounded approaches to traffic safety and prevention.

Mikuláš Toman is a researcher affiliated with Palacký University Olomouc. His work is situated within the academic environment of the university, with involvement in research and higher education activities. He contributes to scholarly work through research, collaboration, and academic dissemination. His professional activities reflect the university's emphasis on interdisciplinary and evidence-based research.

Petr Zámečník works in the field of psychology with a professional focus on traffic psychology and road safety. His activities address psychological aspects of driver behaviour, decision-making, and risk in traffic environments. He is involved in applied practice and/or expert work connected to assessment and prevention in transport. His approach is grounded in scientifically informed methods aimed at improving safety and responsible mobility.

Figures

Tables

1 Traffic psychology in everyday life

Ralf Risser and Matúš Šucha

Introduction

All of us need to be mobile in order to fulfil our daily needs and tasks. Thus, traffic is generated by us – by all individuals – acting and interacting in the public space in order to reach certain goals and fulfil those daily needs and tasks. We do this on the basis of our own personal features, such as attitudes, habits, and interests. However, what we do also depends on a given physical frame – the traffic infrastructure – and on using different types of transport: cars, bicycles, means of public transport, or simply walking. Finally, individual behaviour also has a major influence on societal features such as laws, informal rules, media reporting, and the social climate and vice versa.

Psychology, as the scientific discipline that deals with human behaviour, its understanding, and its control, seeks to explain the relationships between these preconditions and the resulting behaviour and to offer advice for improvements, if needed. The ultimate goal is to support the individual road user and to shape the physical and social environment, thereby respecting social, economic, and ecological criteria, in order to achieve a sustainable traffic system. Traffic safety is only one important feature of sustainable traffic.

What is the scope of traffic psychology?

In this respect, we try to define what the scope of traffic psychology could – or should – be. The discussion follows the Diamond Model (Figure 1.1); this model implies that our behaviour as road users is not only steered by our own attitudes, personalities, motives, or other individual traits. A strong influence also comes from outside our individual characteristics, specifically from the interaction with other people in the public space, from cultural and societal preconditions, from infrastructure features, and from the preconditions provided by the chosen – or available – mobility mode.

DOI: 10.4324/9781003608394-1

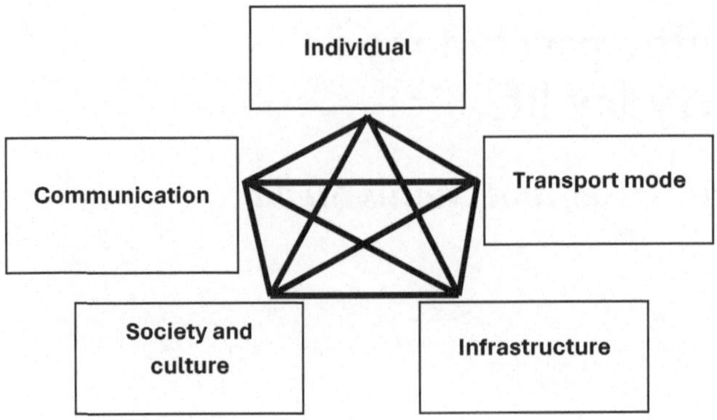

Figure 1.1 Diamond Model (Risser, 2006).

Individual – what lies behind behaviour?

The behaviour of an individual in traffic is a highly complex phenomenon shaped by a wide range of factors. As suggested earlier, these include factors related to the individual (such as personality, emotions, habits, and socio-demographic characteristics) as well as those linked to the situation (i.e. situational factors – both the micro and macro contexts). Arguably, one may add that one component of individual behaviour is to avoid disadvantages for oneself, or also to achieve advantages. To follow one line of thinking, one could say that quality of life (QoL) should not be influenced negatively by what we experience in the public space, or it should even improve.

QoL is understood as a combination of objective conditions and the perceived quality of an individual's daily life – that is the subjective assessment of the degree of well-being. This encompasses all the emotional, social, and physical aspects of a person's life. Thus, QoL may be seen as a multidimensional construct that reflects individual values (Snoek, 2000) and/or indicates whether a person's needs across various domains have been met.

QoL is often discussed as a construct that has three dimensions (Risser et al., 2006): physical (health), psychological (self-control, self-perceived competence, love, satisfaction, joy, morals, confidence, control over one's life, life expectation, beliefs, and desires), and social, divided into private (social network, support, income, education, and job) and public (community, social climate, social security, housing quality, environment, aesthetics of the environment, traffic, crime rate, equality, and justice).

As regards traffic, we claim that the following qualities are considered as relevant by most road users, while their relative weight may vary between different individuals and groups:

- mobility for all (availability, accessibility, and usability)
- a safe environment (safety)

- a comfortable environment (comfort)
- a secure environment (security)
- a clean environment (cleanliness and the absence of pollution)
- an appealing environment (aesthetics)
- a busy environment (availability of facilities)
- a lively and friendly environment (social life)
- public participation in decision-making.

QoL is important throughout the entire journey – from door to door – and thus for mobility in general. One of the relevant issues in traffic psychology that has a significant impact on QoL is *fear of driving*. It is estimated that up to 15% of the population experience symptoms of driving phobia (Taylor & Paki, 2008). A severe form of driving phobia (amaxophobia) affects approximately 1% of the population, predominantly young women. Fear of driving is also a concern among the older population (55+), as 25% of these drivers report at least mild fear (Deane et al., 2000). The causes and pathogenesis of driving phobia are not yet fully understood. Existing research (e.g. Deane et al., 2000) has mostly considered it a disorder resulting from motor or non-motor vehicle accidents. Indeed, some recent studies show that 18–77% of people who have experienced a serious accident subsequently develop the phobia (Taylor & Paki, 2008).

QoL is also an important aspect of professional driving. Wong et al. (2012) showed that professional drivers reported poorer physical and mental health-related QoL compared to the general population. Societal aspects that are important for this group are the preconditions provided by their employers in connection with an economic context and infrastructure circumstances.

Lack of physical activity presents a serious challenge in modern societies. A key point is that in the course of daily routines considerable portions of the daily recommended physical activity (as set by the World Health Organization [WHO]) could be achieved through walking, cycling, also combined with the use of public transport – without any additional effort, like visiting the gym. The fitness and health benefits associated with these modes are substantial, yet there appears to be limited awareness of this fact among road users. Psychology could – and should – contribute to the development of persuasive information regarding the fitness benefits of various transport modes. The habitual use of cars should be replaced by more rational behaviour, enabling individuals to choose other modes of transport when appropriate. To support this change, the conditions for using alternative modes must be made more attractive. This can scarcely be achieved without an understanding of road users' motives, wishes, and interests. It is widely recognised that tackling non-sustainable travel patterns requires a detailed understanding of travel behaviour and the reasons for preferring one mode of transport over another. Psychological factors need to be used to explain mode choice (Forward, 1994). Traffic psychologists focus on analysing the motives, attitudes, problem awareness, norms, and beliefs of road users in order to gain a better understanding of individual transport mode choices and to identify opportunities for shifting towards more sustainable transport options. The latest findings suggest that such a shift is achievable if it has a positive influence on QoL. As indicated earlier, QoL depends on how individuals evaluate their own living conditions, with the term "subjective well-being" often being used synonymously (Risser et al., 2006). It is a concept that encompasses numerous subjective elements. Hakamies-Blomqvist and Jutila (1997) found that people consider the following factors as relevant:

- *Social values*
  One issue – whether consciously or unconsciously experienced by non-motorised road users and residents – is the low social status associated with their modes of transport in comparison with driving a car. The car often symbolises concepts such as power, prestige, independence, freedom, and status. Infrastructure, too, assigns value in a certain sense: for example, traffic calming measures formally prioritise walking and cycling, thereby lending them greater importance. Moreover, the nature of interactions between different road users – not least displaying power relations – has a significant influence on individuals' sense of self-assurance. Traffic calming contributes to more equitable conditions for such interactions. Additionally, the perceived level of safety is strongly shaped by the nature of these interactions.
- *Health*
  Walking and cycling promote good health. Consequently, infrastructure that supports these modes of transport can have a positive impact on public health. Good health is also generally associated with a higher degree of subjective well-being. In contrast, public health issues connected to car use are, in the literature, mainly discussed in connection with traffic safety, but hardly ever in a more general way.
- *Comfort*
  The public road space serves not only for transportation but also for socialising, recreation, and relaxation. Specific amenities that enhance the comfort of those using the road space in this way include benches, litter bins, signposts, shelters, public toilets, cafés, kiosks, and restaurants. The mobility of non-motorised road users is strongly determined by pavement conditions. Negative experiences are often due to dirt or obstacles on the pavement, which should be avoided. Feelings of insecurity are commonly linked to poor surface conditions – such as potholes, broken pavements, litter, or shards of glass – and to very narrow pavements. Where such conditions are managed effectively, infrastructure can contribute significantly to users' comfort.

  At the same time, as a truism, one may say that everything is done to ensure the comfort of car drivers.

- *Safety*
  Perceived safety does not always correlate with objective safety. To enhance cyclists' sense of safety, physical separation from motor vehicles is important. Pedestrians, particularly older ones, may feel jeopardised by both cars and cyclists. Personal security can also be compromised by inadequate lighting, which is particularly concerning for women. Infrastructure that responds to the needs of non-motorised road users and offers people reassurance – such as parents feeling confident that their children are safe in calmed areas – can make a substantial contribution to subjective well-being.
  On the other hand, car drivers hardly feel a lack of safety, except those affected by driving phobia.
- *Mobility*
  This term refers to being able to move about in the public space without physical or psychological obstacles. Three key aspects of infrastructure influence mobility: space, traffic safety, and social security. For example, a lack of parking can deter people from visiting certain places by car, while narrow pavements or missing links in the pedestrian network can discourage walking. Traffic safety concerns influence mobility choices: older drivers tend to avoid challenging roads. Parents are reluctant to let children walk along routes where criminal acts could take place, and women tend to avoid walking through sparsely populated areas in the dark.
- *Aesthetics*
  Car drivers and passengers mainly keep their eyes on the road, noticing other road users, animals, or obstacles that could affect their progress. When they look beyond the road, their attention tends to focus on large landmarks, such as buildings. Pedestrians and cyclists, by contrast, move at a slower pace and have more time to observe their surroundings. An agreeable environment, especially a green one, is appreciated. Traffic noise and pollution are commonly experienced as both aesthetics- and health-related concerns.

Societal issues – social and cultural impacts

Rules and norms in traffic result from political power and political decisions. If society, for example, regards speeding or the infringement of pedestrians' right of way by car drivers as a serious problem, this will influence both formal (= laws) and informal *traffic rules*. The media play an important role in this respect, as they support specific viewpoints and shape public perception of traffic issues. Problems such as inappropriate choice of speed, an insufficient distance from vehicles ahead or to the side, or infringing the right of way may result from individual shortcomings with respect to both lack of skills and lack of will to respect rules. However, the *problem of informal norms* is equally significant. How do individuals assess their behaviour when driving above the speed limit alongside many others doing the same in a traffic stream? One might certainly try to avoid such behaviour, but doing so may be difficult or even impossible and could, in fact, endanger traffic safety. In many cases, the desired state is for all drivers to behave in a similar, homogeneous way. In this context, young drivers present a particular challenge. They tend to behave more "dynamically" or "energetically". It may be assumed that adhering to informal norms reflecting such dynamics – including exceeding the speed limit – is especially attractive to this group and such behaviour may even be exaggerated (Chaloupka-Risser et al., 2011). Problems can become particularly severe when peer influence promotes the acceptance of such informal rules (e.g. a high-speed trip home after a night out, with several passengers in the car). Such behaviour is often not due to a lack of ability but rather a lack of motivation to act as desired. Equally, it may stem from a lack of experience or simply from having "never learned anything else".

In many cases, social conditions encourage specific behaviours that conflict with adherence to rules. These include the working hours of professional drivers, the characteristics of youth, peer pressure among young drivers, and many others. What is defined as a "problem" is shaped in complex ways by society, and informal rules may differ from formal ones (laws). However, if the public reaches a consensus on what constitutes a problem (if these "issues" become widely recognised), then aberrant behaviour tends to diminish gradually, and the desired behaviour starts to prevail.

The behaviour of individuals in traffic is shaped through a socialisation process influenced by cultural dynamics. The traffic environment serves as a social context with very distinct rules and norms, which are passed on between road users over time and across generations. For example, according to Rosenbloom (2009), males are more likely to go through intersections than females when there is a red light. There are also situational differences that affect ignoring a red light; the tendency to proceed in contravention of the signal increases when fewer people are waiting at the kerb. Attitudes towards traffic behaviour are largely learned and shaped by various influences, including parents, peers, media, and other road users. Parents represent a significant source of cultural and normative influence on traffic behaviour, particularly for young people. Another key influence comes from the media, which, however, often tend to portray danger and risk as normal aspects of traffic behaviour rather than promoting safety.

Interaction between road users – communication

Communication between road users is a crucial issue (Risser et al., 2010). Not least, communication shapes our behaviour: it is through communication that we learn how one is expected to behave on the road, that is, how formal norms are followed and what informal norms should be adhered to. We acquire this understanding by observing others, which itself is a form of communication. Training and appropriate infrastructure design (such as paving and layout) can enhance communication and thereby improve safety. Communication between road users must be given the highest priority in infrastructure design, since accidents involving two or more participants (the vast majority of which occur in populated areas) should always be regarded as a breakdown in communication. It is estimated that between 60% and 80% of injury-related crashes involving more than one participant are the result of failed communication (depending on the type of road user, country, and reporting).

Thus, it is essential to understand how to interact effectively with others. A major problem is that we do not seem to learn systematically, particularly in driving schools, about the impact of our behaviour (especially as drivers) on other road users in our immediate surroundings. Moreover, road users are not taught how to express their needs and intentions in a clear yet respectful way. Thus, to a varying degree, we lack the ability to support others in ways that foster smooth traffic flow and increased safety, that is, outcomes that ultimately benefit everyone.

For road users inside motor vehicles, the room for meta-communication – that is, to give and receive social feedback – is significantly restricted, given the brevity of the interactions and physical separation. This often means acting solely according to one's own judgement and self-interest. For example, it is very difficult to signal to another road user that something done a moment earlier was not meant as an offence when one is enclosed in a car. Such situations are certainly highly problematic, as they contribute to stress and antagonism and ultimately worsen traffic safety (Risser et al., 1990).

From a safety perspective, communication is unquestionably one most important factor. Thus, one of the most pressing steps towards improved road safety – though not only for safety – is to enhance communication.

It is not easy to envision how communication between road users could be actively shaped. *Infrastructure measures* may be useful. For example, forcing reduced speeds in sensitive situations can help improve communication, as this provides more time to understand each other. For example, if drivers should slow down before a pedestrian crossing, speed humps may serve as a prompt to do so. This kind of measure helps reduce vehicle speed and consequently improves communication between, for instance, drivers and pedestrians. Law enforcement is another potential avenue, though it will only be effective if traffic checks are carried out consistently, and if clearly defined communicative misconduct is sanctioned systematically. Road users need to understand which forms of communication are problematic and must be made aware of these. In an ideal scenario, drivers would be fined every time they endangered pedestrians or failed to yield as required (in psychological terms: consistent reinforcement). Developing such a system appears difficult. Nonetheless, psychologists should contribute to finding viable solutions. Measures such as educational initiatives, awareness-raising campaigns, and infrastructural adaptations must be grounded in a robust understanding of human behaviour and interpersonal communication.

Part of the process of communication between road users results in more or less overt interpersonal conflicts. A wide variety of conflicts between two road users occur particularly at intersections and may lead to critical incidents or even accidents. For instance, Kim et al. (2007) found that over 50% of collisions involving a bicycle and another vehicle – most commonly a car (in 70% of cases) – took place at intersections.

However, communication between road users generally plays an important role in allowing road traffic to progress smoothly, or at all. In fact, road traffic would not function at all without the continuous process of communication among those participating in it. Not all movements and interactions can be regulated to the same extent as in, for example, air transport. In many situations, road users must negotiate manoeuvres with one another in real time. This is a domain in which psychology should be of assistance. This fact has become an explicit research topic in connection with car automation towards autonomous driving. The question how autonomous cars should be programmed in order to interact effectively with all non-equipped or non-automated road users has become a central research topic (Reyes-Muñoz & Guerrero-Ibáñez, 2022).

Infrastructure

There is no doubt that infrastructure influences the behaviour of road users. The difficulty lies in understanding the specific mechanisms through which this influence occurs. Infrastructure presents certain limits and can also unintentionally hinder safe traffic behaviour. For example, an unrealistically low speed limit on a straight, wide road with minimal traffic may lead drivers to disregard it. On the contrary, wide and straight roads encourage higher speeds, suggest (psychologically) priority at crossings, draw focal attention forwards, and thereby impair peripheral vision in the immediate vicinity. Smooth and well-maintained vehicle lane surfaces may make the choice of appropriate speed difficult. Similarly, pedestrians are likely to ignore a red traffic light if there is no vehicular traffic and the waiting time is perceived as excessive (Theeuwes, 2021).

Road administration probably knows why certain regulations are necessary, but road users may not understand them. Therefore, efforts need to be made to design the road environment in such a way that its functions and roles are understood by the road users, while such understanding should systematically result in wished-for behaviour (Kuiken et al., 2012; Theeuwes & Van der Horst, 2017). Infrastructure concepts that take advantage of these behavioural influences include self-explaining roads, traffic calming measures, and shared spaces, among others. All of these approaches are characterised by attempts to influence road user behaviour through the design of road space and accompanying support measures (e.g. information provision).

One could say that traffic calming measures were "invented" in densely populated areas with a specific goal in mind: the aim was not to calm traffic in general, but specifically motor vehicle traffic. The reason for this was that motor vehicle traffic was increasingly perceived as disruptive to the many other functions that roads in urban areas are also meant to fulfil. In order to support these other functions, the speed of motor vehicles needs to be reduced through infrastructural measures, smooth driving should be encouraged, and part of the motor vehicle traffic should be replaced by walking and cycling (in combination with public transport). The results of various EU projects suggest that such measures have great potential: a study conducted within the EU WALCYNG project (Hydén et al., 1998) found that 50% of all trips in Europe are shorter than 5 km (equivalent to 10–15 minutes by bike) and 15%–20% are shorter than 1 km (10–15 minutes on foot). More recent British research (Le Vine et al., 2009) has confirmed these findings.

Given the environmental context, it is worth pondering the notion that humans do not instinctively fear high horizontal speeds in the same way they fear the risks associated with vertical gravitational forces, despite the fact that both can, under certain circumstances, result in fatal outcomes. Figure 1.2, showing a painting by the Swedish artist Carl Jilg, provides an impression of this dilemma.

In conclusion, infrastructure can influence drivers' behaviour in such a way that risks are underestimated because of the impressions created by road

Figure 1.2 Horizontal versus vertical (Carl Jilg).

design, for example, a smooth and clean road surface, straight and wide roads, and the absence of trees along the roadside. "A road that looks like a motorway will be used like a motorway." Infrastructural measures to promote safe behaviour may include elements that support communication (Falk et al., 2003), measures that reduce speed, especially at locations where interaction between road users is required, and arrangements that serve as reminders of traffic rules (e.g. speed humps). When considering infrastructure, crucial questions arise: What kind of support does the infrastructure provide for various modes of transport? How does it influence the relationships between different categories of road users, and what communication processes does it facilitate? What power relations does it reflect? Both perceptual impressions and physical conditions influence behaviour. Let us end this chapter with a psychological term: Receiving a jolt from underneath the car after passing over a speed hump too quickly functions as contingent reinforcement with negative valence (or a contingent punishment), which is highly effective.

The vehicle or mode

Concerning the effects of the vehicle or the chosen mode, two different perspectives are psychologically relevant. One of them refers to the question how the vehicle we drive or ride affects our behaviour viz. how the choice of this vehicle is related to our own character. Possession and use of a strong and fast car or motorcycle will of course be related to what we do and how we act as road users. Psychologists need to help understand this relationship and integrate it in education and campaign measures. The other perspective concerns the work of planners. They need to consider how infrastructure and traffic management impact the needs and influence the behaviour of different types of road users – car drivers, motorcyclists, bicyclists, and walkers – and they need help from psychologists in order to carry out this work appropriately.

Conclusion

Traffic psychology is a scientific discipline dedicated to understanding and influencing human behaviour within the complex traffic system, with the ultimate goal of achieving a sustainable traffic system and enhancing road safety. This involves supporting individual road users and shaping the physical and social environment, while respecting social, economic, and ecological criteria. The field's broad scope is comprehensively illustrated by the Diamond Model, which posits that road user behaviour is influenced by a dynamic interplay of individual characteristics, interaction with others (communication), cultural and societal preconditions, infrastructure features, and the chosen transport mode or vehicle. At the individual level, behaviour in traffic is shaped by a wide array of personal factors such as attitudes, habits, interests, personality, emotions, and socio-demographic characteristics. A central concept in this regard is Quality of Life (QoL), understood as a combination of objective conditions and the subjective assessment of well-being, encompassing emotional, social, and physical aspects. Traffic can negatively impact QoL, and promoting sustainable mobility is linked to improving it. Psychology can help encourage a shift from habitual car use to more rational behaviour, emphasising the fitness and health benefits of walking, cycling, and public transport, which can contribute significantly to daily recommended physical activity. This requires understanding road users' motives, wishes, and interests.

References

Chaloupka-Risser, C., Risser, R., & Zuzan, W. D. (2011). *Verkehrspsychologie: Grundlagen und Anwendungen*. Literas Universitätsverlag.

Deane, F. P., Taylor, J., & Podd, J. (2000). Determining the focus of driving fears. *Journal of Anxiety Disorders*, *14*(5), 453–470.

Falk, A., Fehr, E., & Fischbacher, U. (2003). On the nature of fair behavior. *Economic Inquiry*, *41*(1), 20–26.

Forward, S. (1994). *Theoretical models of attitudes and the prediction of drivers' behaviour* (TRID database No. 434). TRID.

Hakamies-Blomquist, L., & Jutila, U. (1997). *"General problems of pedestrians and cyclists": Report from WALCYNG – WP 3*. University of Helsinki.

Hydén, C., Nilsson, A., & Risser, R. (1998). *WALCYNG: How to enhance walking and cycling instead of shorter car trips and to make these modes safer. Final report* (Bulletin 165). Lunds Tekniska Högskola, Institutionen för Trafikteknik.

Kim, J. K., Kim, S., Ulfarsson, G. F., & Porrello, L. A. (2007). Bicyclist injury severities in bicycle–motor vehicle accidents. *Accident Analysis & Prevention*, *39*(2), 238–251.

Kuiken, M., van der Horst, R., & Theeuwes, J. (2012). *Designing safe road systems: A human factors perspective*. Ashgate Publishing, Ltd.

Le Vine, S. E., Jones, P. M., & Polak, J. W. (2009). Has the historical growth in car ownership come to an end in Great Britain. In *Proceedings of European Transport Conference* (Vol. 10). Association for European Transport (AET).

Reyes-Muñoz, A., & Guerrero-Ibáñez, J. (2022). Vulnerable road users and connected autonomous vehicles interaction: A survey. *Sensors (Basel)*, *22*(12), 4614.

Risser, R. (2006). Der Diamant: Systematisches Auffinden von Interventionsansätzen. In C. Chaloupka-Risser & R. Risser (Eds.), *Mehr sicheres Verhalten im Straßenverkehr: Psychologische, juristische und technische Aspekte* (in motion, 3). INFAR.

Risser, R., Chaloupka, C., Gluske, H., Reetz, A., Riedel, R., Tscherteu, G., & Tuma, F. (1990). *Strassenverkehr und Lebensqualität*. Literas Universitätsverlag.

Risser, R., Haindl, G., & Ståhl, A. (2010). Barriers to senior citizens' outdoor mobility in Europe. *European Journal of Ageing*, *7*(2), 69–80.

Risser, R., Schmeidler, K., Steg, L., Forward, S., & Martincigh, L. (2006). Assessment of the quality of life in cities: Environmental conditions and mobility. *Urbani Izziv*, *17*(1–2), 187–193.

Rosenbloom, T. (2009). Crossing at a red light: Behaviour of individuals and groups. *Transportation Research Part F: Traffic Psychology and Behaviour*, *12*(5), 389–394.

Snoek, F. J. (2000). Quality of life: A closer look at measuring patients' well-being. *Diabetes Spectrum*, *13*(1), 24.

Taylor, J. E., & Paki, D. (2008). Wanna drive? Driving anxiety and fear in a New Zealand community sample. *New Zealand Journal of Psychology*, *37*, 31–37.

Theeuwes, J. (2021). Self-explaining roads: What does visual cognition tell us about designing safer roads? *Cognitive Research: Principles and Implications*, 6(1), 1–15.
Theeuwes, J., & Van Der Horst, R. (2017). *Designing safe road systems: A human factors perspective*. CRC Press.
Wong, C. K., Fung, C. S., Siu, S. C., Wong, K. W., Lee, K. F., Lo, Y. Y., Fong, D. Y., & Lam, C. L. (2012). The impact of work nature, lifestyle, and obesity on health-related quality of life in Chinese professional drivers. *Journal of Occupational and Environmental Medicine*, 54(8), 989–994.

2 Aim and scope of traffic psychology

Matúš Šucha and Ralf Risser

Introduction

Traffic psychology is a sub-field of psychology that studies the behaviour of road users and the psychological processes underlying that behaviour. Traffic psychology aims to identify the determinants of road user behaviour in order to develop effective measures for changing human behaviour in the context of transportation and travel (Rothengatter, 1997). This includes scientific research into the background and origins of traffic behaviour, such as skills and knowledge, motives, beliefs, attitudes, values, and personality traits of road users; as well as research into how this behaviour is influenced by the physical environment (infrastructure), the mode of transport (vehicle), the social environment (interaction with other road users), and social or cultural factors (Risser, 2004).

In this way, traffic psychology seeks to understand, describe, predict, and influence the behaviour of road users. In the context of this chapter, the focus of traffic psychology is primarily on promoting safe behaviour in traffic and thereby reducing road casualties.

Another major challenge for traffic psychology is the impact of the transport system on the environment in terms of air pollution, traffic congestion and land use, mobility, quality of life, public health, combined with the general goal to steer transport mode choice and travel behaviour in a wished-for direction.

Traffic and transport system

Participation in traffic involves more than just travelling from point A to point B. It is a dynamic and constantly evolving process of interaction between road users, with each using a particular mode of transport within a physical environment. By constructing a predictable and recognisable infrastructure, governments aim to guide this dynamic behaviour in a safe and smooth manner. The *engineering* field continuously seeks to develop safer vehicles through the introduction of both passive and active safety systems. *Education* – in all its forms, including road safety education, public campaigns, and driver

DOI: 10.4324/9781003608394-2

training – aims to provide road users with the necessary knowledge and competences, as well as to foster a safety-oriented attitude. A positive attitude towards safety is arguably the most important precondition for responsible and safe behaviour in traffic. Rules and laws are essential for managing traffic and regulating vehicles. These need to be supported by effective *enforcement*, which helps compensate for shortcomings in road design and addresses deviant behaviour. In the literature, these three approaches – Engineering, Enforcement, and Education – are commonly referred to as "the three E's". Together, they are intended to promote appropriate behaviour and/or to eliminate, or to mitigate, unsafe, aggressive, or dangerous driving.

A road crash may be seen as the consequence of a chain of events involving the human factor, vehicle, and infrastructure (see Figure 2.1). Various factors, such as the condition of the vehicle, the road, and the road user, play an important role in traffic accidents. However, as shown in Figure 2.1, the human factor may be considered as the most critical one. In fact, road user behaviour along with human errors (whether intentional or unintentional) is recognised worldwide as the main "culprit" in connection with traffic crashes. Sabey and Taylor (1980) estimate that faulty road user behaviour is a contributing factor in approximately 94% of accidents and the sole factor in about 65% of all accidents, thus arriving at values similar to those of Rumar (Figure 2.1).

Human errors relate to skills (not possessing the technical ability to control the vehicle, but also higher-order skills such as risk perception and anticipation), knowledge (not knowing the traffic rules or the risks of driving under the influence of alcohol, drugs, or medication), motivation (disobeying traffic rules, sensation-seeking, risk-taking, etc.), emotion (being angry, depressed, etc.), temporary dysfunction (inattention, drowsiness, and being under the influence of psychoactive substances), or permanent dysfunction (e.g. limited visual perception, attention disorders, diabetes, epilepsy, or similar conditions).

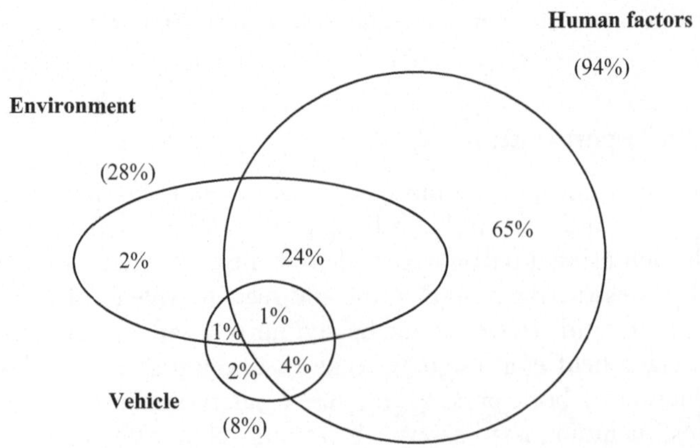

Figure 2.1 Causes of traffic accidents (Rumar, 1985).

Integrative, competence-based approach

Preparing people to become safe and responsible road users involves more than teaching rules and learning the skills required to behave safely. An early analysis of the driver's task and of accidents had already shown that adequate psychomotor skills, knowledge, and physiological functioning are not sufficient for competent and safe performance as a driver (Näätänen & Summala, 1974). Hatakka et al. (1999) were the first to integrate a wider range of aspects in their GDE (Goals for Driver Education) matrix, which was developed for car drivers.

Following the competence-based view of driving introduced by the GDE matrix, there is a growing emphasis on the need to train and assess higher-order aspects of driving: risk tolerance, reflection of one's own behaviour in traffic, and perceiving and recognising hazards. This, of course, applies to road users in general. Within the framework of the Dutch "lifelong learning approach", GDE matrices with specific learning objectives have also been developed for other important groups of road users, such as schoolchildren, experienced drivers, and elderly road users (Vissers et al., 2007).

One main ingredient of traffic psychology's efforts to reduce the risk of causing or being involved in an accident consists of promoting self-awareness as a road user. Such efforts include helping road users reflect their own attitudes, motives, habits, and personality traits in order to be able to act and react appropriately in road traffic.

Improving traffic safety involves multiple aspects. Clearly, there is an interaction between traffic infrastructure, the vehicle, and the road user (see Figure 2.2). Improvements are necessary in all three pillars of road safety. Road infrastructure needs to be enhanced (function, classification and layout,

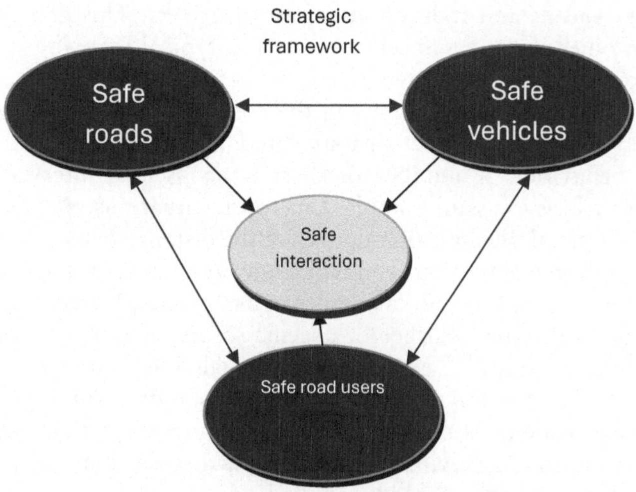

Figure 2.2 The three pillars of road safety (OECD/ITF, 2008).

surface quality, visibility, etc.), vehicles must be made safer (speed-controlling measures, seat belts, airbags, safety cell construction, driver assistance systems, etc.), and the capabilities of road users must be improved. Any effective safety improvement programme is hardly possible without the concerted adoption of various measures across all relevant areas – from infrastructure and vehicle design to driving skills and driver behaviour. An *integrative approach* to road safety is required where all three pillars should use traffic psychological know-how.

Research domains

Traffic psychology plays an important role in all three components of the traffic system. Research into how these components can be approached in order to improve the behaviour of different road users is a major area of interest. Some examples:

- *Safe roads*
  A road design that prevents accidents must ensure that road users recognise and respond to relevant cues, an approach referred to as the "self-explaining road". It also needs to ensure that these cues prompt road users to consistently engage in behaviour corresponding to the road design, a strategy which is known as "behavioural engineering". While the concept of "Sustainable Safety" implies that the cue is the standardisation of road design. An issue that still requires elaboration is how this standardisation can be modelled in such a way as to standardise roads to a degree that meets drivers' expectations (Wegman et al., 2006).

  For instance, the image a road conveys is, in some cases, more important than its actual structural layout. If a road gives the impression that it is easy to drive on, therewith making drivers feel too safe, this can lead to higher speeds and reduced attention of drivers. This delicate balance between objective and subjective safety is a typical issue that psychology needs to deal with.

  In several domains, especially in the commercial world of advertising, psychology has developed various "nudging" techniques to influence human behaviour. A number of these subconscious intervention techniques can be successfully applied in traffic situations. A few examples include * optical illusions (manipulating the distance between road markings – e.g. dash length of broken white line) as a way of influencing vehicle speed; * subliminal stimuli (also known as "priming"), such as children's drawings on the road surface or specific scents, as a way of moderating driver behaviour; and * extra positive signals (also known as "prompting") as a way of enforcing compliance with standard rules.

  All these techniques need to be analysed in terms of their effectiveness in improving road safety, and it needs to be discussed in which situations and in which way they could be applied.

- *Safe vehicles*
Developments in car technology are advancing rapidly. An increasing number of systems are being developed to support the driver in making the right decisions. The idea of autonomous vehicles has become highly realistic. In comparison to other transport sectors, however, full automation is likely to remain a utopia. Even in aviation and most railway systems, where a high level of automation is technically feasible, human involvement is still necessary in certain situations. As well, the introduction of fully automated vehicles would be preceded by transitional phases with vehicles where the driver will continue to play an essential role in ensuring safety. Moreover, it is difficult to imagine a traffic system where there are no pedestrians, cyclists, or micromobility users, none of them conceivable as part of an autonomous mobility network. The development of a theory of "meaningful human control" over automated driving systems (ADS) and its translation into design and driver guidelines pose a challenge for engineers and even more so for traffic psychologists. Important questions arise regarding driver education, attention span, perception of safety, and related issues.
- *Safe road users*
Attention to the human factor in both road and vehicle design is of paramount importance. While traffic psychology should address all the components of the traffic system, the road user remains the focal point of its interest. Historically, this professional tradition pertained to diagnosis, testing, and driver rehabilitation. However, since the 1980s, traffic psychology has increasingly broadened its scope to include research on human behaviour in traffic from a more general perspective. Phenomena such as attention and risk perception have been studied with growing intensity. Behavioural models to explain traffic behaviour have been developed to offer deeper insights into the interactions between the three core components of the traffic system. The first attempt to classify all the existing models was undertaken by Michon in 1985 (Michon, 1985). Probably, the most widely explored model is that of Ronis et al. (1989), which was based on the theory of planned behaviour formulated by Fishbein and Ajzen (1975). Fylan (2017) expanded this model by incorporating more specific factors (see Figure 2.3).

A further model, Fuller's task-capability interface (TCI) model (Fuller et al., 2005) describes the dynamic interaction between the determinants of task demands and driver capabilities, offering a valuable understanding of traffic behaviour. The model is based on the assumption that drivers adopt a preferred level of driving difficulty, which results from the relationship between perceived task demands and their perceived personal capacities. The difficulty of the task is shaped by external factors such as traffic density and weather conditions, but it can also be influenced by the driver's own behaviour, particularly in relation to speed choice, concentration, and safety

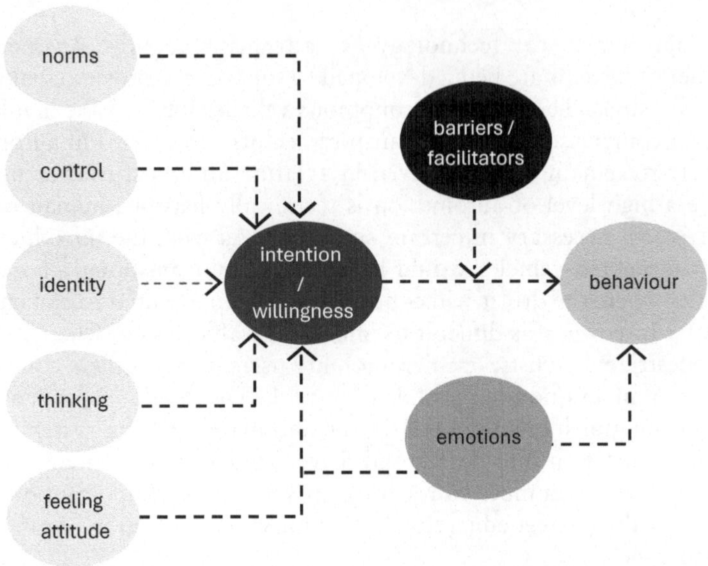

Figure 2.3 Integrated driver model (Fylan, 2017).

margins. The process of adjusting the balance between task demands and one's own capacities is also known as "calibration" (Kuiken & Twisk, 2001).

Evidence-based approach

In general, traffic psychology adopts an evidence-based approach. Sackett et al. (1997) define this as the integration of individual (clinical) expertise with the best available external evidence from systematic research. Briner (1998) describes this approach as comprising three steps: first, one must gain a clear understanding of all the aspects involved in the creation and development of the problem, and the information then needs to be translated into answerable questions. The next step involves a search for the best possible solutions based on empirical data and evidence. Finally, one must critically assess the chosen intervention, evaluate the results, and contribute to the growing body of knowledge.

In the Netherlands, an intervention checklist was developed in 2011 (Vissers, 2013). Inspired by the "Intervention Mapping" methodology (Bartholomew et al., 2011), this ten-item checklist provides a step-by-step plan for both the design and evaluation of road safety interventions. It is intended to encourage designers and funding organisations to develop evidence-based programmes and place greater emphasis on quality improvement and evaluation.

Intervention–evaluation cycle

As traffic psychology is a relatively new field in many countries, new evidence is constantly being generated. This emerging scientific knowledge will hopefully influence practice. While the range of academic research continues to grow, it is not always possible for practitioners to access all the results. However, anyone following evidence-based approaches in their practice can learn through action. Monitoring and evaluation must be integrated into every phase of an intervention, from planning to implementation and beyond. For monitoring and evaluation to be effective, strategic planning and the development of monitoring and evaluation strategies should go hand in hand. This approach is illustrated in the intervention–evaluation cycle (see Figure 2.4).

As traffic psychology is still an emerging discipline in many countries, a continuous stream of new scientific findings is being produced, offering the potential to significantly inform and improve practice. To fully harness this potential, it is essential to integrate monitoring and evaluation systematically from the earliest stages of intervention planning, ensuring these processes

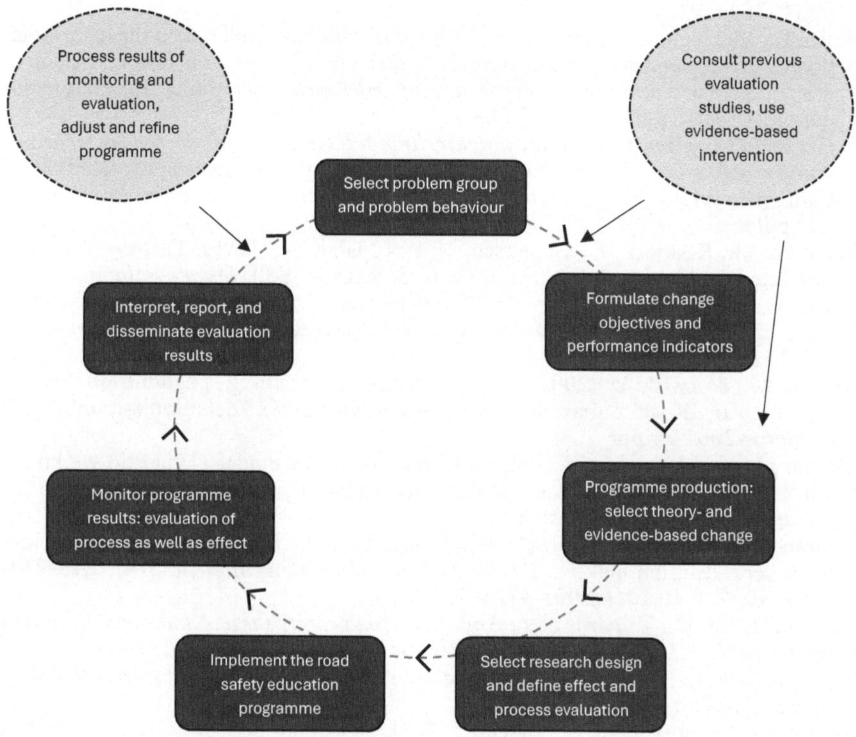

Figure 2.4 Intervention–evaluation cycle (European Commission, Mobility and Transport Directorate-General, n.d.).

are aligned with the evolving goals and context. The intervention–evaluation cycle (see Figure 2.4) illustrates how structured feedback mechanisms can facilitate learning through action and support adaptive, evidence-based decision-making.

References

Bartholomew, L. K., Parcel, G. S., Kok, G., Gottlieb, N. H., & Fernández, M. E. (2011). *Planning health promotion programs. An intervention mapping approach* (3rd ed.). Jossey-Bass.

Briner, R. B. (1998). *What is an evidence-based approach to practice and why do we need one in occupational psychology?* Paper presented at the BPS Occupational Psychology Conference. www.researchgate.net/publication/242424550_What_is_an_evidence-based_approach_to_practice_and_why_do_we_need_one_in_occupational_psychology

European Commission, Mobility and Transport Directorate-General. (n.d.). *Monitoring and evaluation. In European road safety observatory: Road safety management.* Retrieved March 7, 2025, from https://road-safety.transport.ec.europa.eu/document/download/6ea433bb-4e2e-4d6c-bc02-8af523c67949_en?filename=road_safety_management.pdf

Fishbein, M., & Ajzen, I. (1975). *Belief, attitude, intention, and behavior: An introduction to theory and research.* Addison-Wesley. https://people.umass.edu/aizen/f&a1975.html

Fuller, R., McHugh, C., & Pender, S. (2005). Task difficulty and risk in the determination of driver behavior: La difficulté de la tâche et le risque dans le comportement des conducteurs. *European Review of Applied Psychology, 58*(1), 13–21. https://doi.org/10.1016/j.erap.2005.07.004

Fylan, F. (2017). *Using behaviour change techniques: Guidance for the road safety community.* RAC Foundation. www.racfoundation.org/wp-content/uploads/2017/11/Using_behaviour_change_techniques_Guidance_for_the_road_safety_community.pdf

Hatakka, M., Keskinen, E., Gregersen, N. P., & Glad, A. (1999). Theories and aims of educational and training measures. In S. Siegrist (Ed.), *Driver training, testing and licensing – toward theory-based management of young drivers' injury risk in road traffic: Results of EU project GADGET* (Working Package 3, BFU Report 40). BFU.

Kuiken, M., & Twisk, D. (2001). *Safe driving and the training of calibration.* SWOV Institute for Road Safety Research. www.swov.nl/sites/default/files/publicaties/rapport/r-2001-29.pdf

Michon, J. A. (1985). A critical view of driver behavior models: What do we know, what should we do? In L. Evans & R. C. Schwing (Eds.), *Human behavior and traffic safety* (pp. 485–524). Springer.

Näätänen, R., & Summala, H. (1974). A model for the role of motivational factors in drivers' decision-making. *Accident Analysis & Prevention, 6*(3–4), 243–261. https://doi.org/10.1016/0001-4575(74)90003-7

OECD/ITF. (2008). *Towards zero: Ambitious road safety targets and the safe system approach.* OECD Publishing. https://doi.org/10.1787/9789282101964-en

Risser, R. (2004). Philosophy of traffic calming. *The Asian Journal: Journal of Transport and Infrastructure, 11*(1), 1–9.

Ronis, D. L., Yates, J. F., & Kirscht, J. P. (1989). Attitudes, decisions, and habits as determinants of repeated behavior. In A. R. Pratkanis, S. J. Breckler, & A. G. Greenwald (Eds.), *Attitude structure and function: The third Ohio State University volume on attitudes and persuasion* (pp. 213–239). Lawrence Erlbaum Associates.

Rothengatter, J. A. (1997). Psychological aspects of road user behaviour. *Applied Psychology: An International Review*, 46(3), 223–234. https://doi.org/10.1111/j.1464-0597.1997.tb01227.x
Rumar, K. (1985). *The role of human error in accidents* (Report No. 277). Swedish Road and Traffic Research Institute (VTI).
Sabey, B. E., & Taylor, H. (1980). The known risks we run: The highway. In R. C. Schwing & W. A. Albers (Eds.), *Societal risk assessment* (pp. 43–70). Springer.
Sackett, D. L., Richardson, W. S., Rosenberg, W., & Haynes, R. B. (1997). *Evidence-based medicine: How to practice and teach EBM*. Churchill Livingstone.
Vissers, J. A. M. M. (2013). *Checklist verkeerseducatie: Werkwijze en procedure* [Checklist road safety education: Method and procedure]. Royal Haskoning DHV.
Vissers, J. A. M. M., Van Betuw, A. M., Nägele, R. C., Kooistra, A., & Harteveld, M. (2007). *Learning goals for lifelong road safety education*. Royal Haskoning DHV.
Wegman, F., Dijkstra, A., Schermers, G., & van Vliet, P. (2006). *Sustainable safety in The Netherlands: the vision, the implementation, and the safety effects* (R-2005-5). Paper presented at the 3rd International Symposium on Highway Geometric Design, 26 June to 2 July 2005, Chicago. www.swov.nl/sites/default/files/publicaties/rapport/r-2005-05.pdf

3 Traffic psychology in the world

Mikuláš Toman

Introduction

When referring to the field of traffic psychology, we could hardly find a subject more relevant than the human factor in transport. After all, the study of human behaviour is the very essence of psychology as a scientific discipline. If psychologists are to apply their expertise and research efforts to the field of transport, the human factor is likely to be the first topic that comes to mind – in terms of both its influence on traffic situations and, conversely, the effect of traffic situations on individuals.

These will be the key topics addressed in this chapter. It should be noted from the outset that, for obvious reasons, the issue of the human factor in transport is predominantly considered in the context of traffic safety. This perspective will also be adopted in the sections that follow.

A brief historical overview

Let us start from the beginning: when do we first find the issue of the human factor in transport as a topic in professional publications? If we search using the popular Scopus database, the first article we find dates back to 1927. Focusing primarily on urban planning, this paper by Strong also mentions the human factor as one of the elements that should determine regional planning. It is argued that the human factor always follows the path of least resistance (including in the context of the assessment of optimal routes) (Strong, 1927). While not always explicit, this natural human tendency to seek the path of least resistance remains a major theme in traffic psychology, for example, in relation to advanced driver assistance systems. It is therefore fitting that this tendency was pointed out in one of the earliest articles on the human factor in transport.

The very oldest article dealing with the topic of our interest was found using the equally popular Google Scholar. Dodge (1923) very aptly highlighted several aspects that future research should address, noting that, despite their clear impact on road safety, they had not been the subjects of any scientific interest. He argued in favour of greater efforts being put into

DOI: 10.4324/9781003608394-3

exploring the issue of attention on the part of both drivers and pedestrians and their interactions and pointed out the need to take into account and systematically analyse the variability and individuality of road users, as well as suggesting modifications to traffic police uniforms to make them more visible in poor light and darkness. Finally, Dodge also noted the need to subject drivers to standardised tests and educate the general public about road safety, which is an area where psychologists could play a particular role.

However, Dodge's article (1923) was really just one of those "swallows" that "does not make a summer". Indeed, according to the Scopus database, only a few dozen papers per year featured the human factor in traffic among their keywords until the 1970s. It was not until the turn of the second millennium that hundreds of papers on the subject of our interest began to appear annually, and from around 2008 onwards, more than a thousand papers per year have been published on this topic (Scopus, 2022). However, this is hardly a groundbreaking revelation, considering that while automobiles have existed since the late 19th and early 20th centuries (with research on road traffic accounting for a significant portion of publications), road safety – and, by extension, the human factor in transport – only gained public attention much later. A suitable reference point in this case is the introduction of the seat belt, a pioneering safety feature capable of reducing the severity of road accidents by up to 55% (National Highway Traffic Safety Administration, 1984). Attributed mainly to engineers at Volvo, the widespread use of seat belts did not occur until the early 1960s (Andréasson et al., 2000). However, in some cases, the legal requirement to wear seat belts was introduced several decades later, depending on the legislation of each country.

As has been mentioned several times, when examining the human factor in transport from a psychological perspective, we are largely studying the human factor in road traffic. Should one wonder why this is the case, the explanation is straightforward: road traffic is the most common mode of transporting people and, consequently, the most statistically hazardous. Therefore, it is most logical to focus on the human factor in road transport.

With respect to what has been stated earlier, we must also bear in mind other areas of transport where the human factor is investigated. Beyond road traffic, air transport (e.g. Wise et al., 2016) and rail transport (e.g. Kecklund et al., 2011) are also areas of research interest, with a key unifying element across all these domains being the focus on the driver, pilot, or train operator. While each study takes place in a different setting and field, psychologists primarily concentrate on the same core subject: the human being as an individual interacting with a machine and its surroundings and the possible ways of describing, explaining, or shaping this interaction.

Most commonly studied human factors

The following sections address the individual human factors that are most frequently encountered in psychological research. The order of the sections is intentionally random, as it is neither possible nor desirable to establish a definitive ranking of these factors in terms of their impact on road safety. If different factors eventually lead to similar risks, such an ordering would not be particularly meaningful. Finally, the decision not to hierarchise the factors is further supported by the fact that they often complement, stimulate, or reinforce one another.

Stress

As in any other aspect of life, stress is undoubtedly a major human factor in transport and road safety. Since we intend to cover stress as the specific topic of one of the sections, however, we will forgo a broader theoretical introduction to this area and focus solely on its relevance to transport and road safety.

A brief definition of stress is provided for the sake of clarity: it is a *state of physiological, psychological, or emotional strain experienced by an individual in which actual or perceived demands lead to an increased mobilisation of resources in an effort to cope with those demands* (Lazarus & Folkman, 1984).

Stress in the context of transport

We should remember to define stress in its full scope. According to Selye's (1976) identification, the effects of stress on the body can be divided into *eustress* and *distress*, that is, positive and negative stress, respectively.

Eustress occurs almost immediately after an individual has experienced even the slightest strain in any form. Eustress turns into distress when the perceived pressure becomes subjectively overwhelming. According to Selye's (1976) theory, when experiencing the maximum amount of eustress without the strain subjectively crossing over into distress, an individual achieves peak efficiency in any activity they are engaging in, whether it be working, studying, or driving a car.

This division, or rather the spectrum of stress, is crucial in the context of road safety because it directly and unequivocally links stress with attention, boredom, fatigue, and aggression behind the wheel. We have just connected the present section with the following ones, and that is no coincidence. In the following pages, we will also encounter similar references linking various human factors to one another. If we were to search for a single "original human factor", previous studies would not provide a satisfying answer. Because of the relatively complex nature of the individual factors, existing research is simply unable to determine whether the proverbial egg or the chicken came first.

Let us return to stress. In the context of road safety, Healey and Picard (2005) present a clearer notion of the eustress-distress spectrum by extending it to include accompanying phenomena that belong at the assumed extremes of the continuum. If an individual is at the very beginning of experiencing eustress, or if no eustress occurs at all, we may observe a loss of attention and fatigue or the onset of boredom. Conversely, at the other extreme, where distress is at its maximum, we observe, (not only) in the transport context, the presence of fear, nervousness, exhaustion, anxiety, impatience, and anger. When it comes to stress, we are dealing with a really broad spectrum of different factors that influence a range of traffic situations. However, we will continue by discussing those arising from distress, that is, from an abnormally elevated level of stress.

Traffic-related stressors

So, what causes stress for drivers? According to Chung et al. (2019), these factors can be divided into four main groups: the driver's physical and mental state, road and traffic conditions, the condition of the vehicle, and external disturbances.

With regard to the driver's physical and mental state, a wide range of individual stress-inducing factors can be identified by the very nature of the term. Physical factors include a lack of sleep, leading to driver drowsiness, which will be discussed in more detail later. An example of psychological factors is the subjectively perceived level of driving experience, which can contribute to stress (Kompier & di Martino, 1995).

Traffic congestion is hardly a surprising stressor as far as the road and traffic conditions are concerned. Wiesenthal et al. (2000) identified a correlation between increasing road traffic density and elevated levels of experienced stress. Another factor within this category is road width. Schießl (2007) suggests that a narrowing road affects the level of stress experienced by the driver; in other words, the need to drive on a narrow, winding road imposes greater psychological strain on the driver compared to wider or straighter roads.

It is also no surprise that the condition of the vehicle contributes to the level of stress experienced by the driver. The stressors related to this include various unusual sounds that the vehicle may begin to make, or technical issues such as engine malfunctions (Wiesenthal et al., 2000).

The final group consists of external disturbances, a phrase used to refer to a relatively wide range of potential stressors, both inside and outside the car, many of which depend solely on the specific situation. Engström (2008) mentions the presence of a passenger in the car as potentially stressful, with "potentiality" playing a key role, as under certain circumstances, the passenger can have a positive effect on the driver (e.g. in situations where the driver is tired or bored). However, if the passenger demands the driver's attention in the middle of rush hour in a large city, they become an obvious source of stress. A similar effect can be observed with driver assistance systems. Reimer et al. (2010) note that the mere presence of these systems, especially when they are fully operational, reduces driver stress. However, if the system fails to cooperate, it becomes a stressor. Similarly, it is perceived as a source of stress if the system works but the driver does not fully understand its functionality. Weather is a significant external disturbance. Hill and Boyle (2007) state that rainy, foggy, and generally adverse weather conditions alone also increase the level of stress experienced by the driver.

Looking into stress susceptibility in terms of driver characteristics, Hill and Boyle (2007) found that women typically reported higher levels of stress while driving in comparison to men, older drivers were more likely to experience stress-related tension than their younger counterparts, and individuals with a history of crashes experienced greater stress levels than those without previous involvement in road accidents.

Measurement of stress

The methods for measuring stress while driving can be divided (similarly to the methods of measuring aggression, fatigue, attention, or boredom) into three main areas: *psychological assessment*, *psychophysiological measurement*, and *observation of driver behaviour*.

Since stress is certainly nothing new in psychological research, it is not difficult to find several specific-purpose questionnaires aimed at both measuring stress itself and assessing stress in the context of driving. As noted by Chung et al. (2019), across traffic psychology studies, we find simple specific-purpose scales used to measure current subjectively perceived levels of stress (usually in terms of assessing the perceived level of stress verbally on a scale of 1 to 10, where 1 is no stress at all and 10 is maximum stress; e.g. Reimer et al., 2010), as well as full-fledged standardised questionnaires, such as the Driving Stress Inventory (DSI), Driving Behaviour Inventory (DBI), and Stress Arousal Checklist (SACL). There are also questionnaires, including the Dundee Stress State Questionnaire (DSSQ) and NASA Task Load Index (NASA-TLX), which do not focus directly on stress, but measure the levels of cognitive load which are closely related to stress.

When focusing on physiological indicators of stress, studies most commonly involve electrocardiography or electroencephalography, as well as galvanic skin response and photoplethysmography. This type of scientific exploration of stress appears to be gaining increasing popularity, both because psychophysiology provides "hard data" and because it inherently represents (to a greater or lesser degree, depending on the specific study) an interdisciplinary approach to research. Let us now have a closer look at several recent studies in this field. Electroencephalography was employed in a research project by Affanni et al. (2022), who developed and tested an EEG headband designed to detect driver stress (primarily on the basis of beta wave activity), using a combination of fuzzy logic and case-based reasoning approaches from cognitive science. This method can thus be used in a wide range of follow-up studies. Electrocardiography in relation to stress was explored by Cardone et al. (2020), discussing the stress index (Baevsky & Chernikova, 2017), a specific indicator of heart rate variability reflecting the stress level of an individual. Galvanic skin response (GSR) was investigated by Kim et al. (2020). They developed a statistical model capable of analysing driver stress purely through GSR on the basis of binary logistic regression. Finally, photoplethysmography is typically used as a complementary physiological indicator rather than a stand-alone method (e.g. Rastgoo et al., 2018; Singh et al., 2011). In any case, physiological stress detection does not generally rely on a single simple indicator such as "more stress = higher heart rate", but rather on a thorough analysis of multiple indicators together (such as the "stress index" earlier).

The remaining approach to measuring driving-related stress involves the observation of driver behaviour, which can take place at different levels. Drivers can be observed visually, either through camera recordings or *in vivo*, or by means of various indicators, typically integrated into OBD-II (an onboard diagnostic unit that tracks processes such as steering wheel position or pedal movements). Such systems are often developed by researchers for their studies (e.g. Lee et al., 2016). Specific indicators of increasing driver stress explored in research typically include changes in steering wheel angle, adjustments to driving speed, and variations in reaction time (Abou Elassad et al., 2020).

Aggression

Aggressive driving has been a central topic for traffic psychologists for decades. Some even argue that it is possibly the most researched driving style (Sagberg et al., 2015), as aggression and the aggressive driving style it leads to are among the major contributing factors for countless traffic accidents worldwide. Berdoulat et al. (2013) note that aggressive driving is perceived by the general population as one of the key road safety problems. Moreover, such behaviour is becoming increasingly common.

Aggression in traffic psychology

In the context of traffic psychology, aggression generally implies *aggressive driving* or *driving aggression*. In line with the definition proposed by the AAA Foundation for Traffic Safety (2009), aggressive driving involves *any dangerous driving behaviour, performed deliberately and with ill-intention or disregard for safety*.

This is a fairly general definition that can "shelter" a wide range of specific driving behaviours. Therefore, we propose further dividing aggressive driving into *hostile* aggression and *instrumental* aggression, as per Shinar (2017).

Hostile aggression behind the wheel is understood as involving any antagonistic behaviour directed at other road users that is purposeless, that is, it ultimately has no effect on the mobility of the individual concerned. Hostile aggression includes what is commonly known as road rage (Shinar, 2017).

Instrumental aggression, on the other hand, is a collective term for all aggressive behaviours aimed at reaching the destination more quickly. It includes overtaking, speeding, weaving through traffic, tailgating, and running red lights (Shinar, 2017).

Sagberg et al. (2015) provide additional input by poignantly observing that some specific driving behaviours that fall under aggressive driving can be either hostile or instrumental, depending on the situation. For example, when a driver honks, they may do so because they are simply frustrated with the current traffic situation and need to let off steam, or because they want to prompt the driver ahead of them to take action and stop preventing them from reaching their destination quickly.

In order to define aggression behind the wheel at least in basic negative terms, we can compare aggressive driving with risky driving, which is a major topic (not only) in traffic psychology and closely related to aggressive driving. So what is the difference? Du et al. (2018) propose that drivers often engage in risky driving because they are seeking excitement (additional stimulation in fact) which might lead to speeding, for example. In contrast, the main motivation behind aggressive driving is the release of frustration and aggression. At this point, we can observe a potential overlap in meaning between risky driving and instrumental aggression. However, it is important to bear in mind that while instrumental aggression is driven by the desire to reach a destination sooner, it still involves an aggressive driving style. It therefore carries a negative emotional charge, unlike risky driving, which simply strives for as many new and exciting experiences as possible.

Who is an aggressive driver?

It will hardly come as a surprise that aggressive driving is closely linked to the driver's personality, specifically their level of aggression. Research findings consistently suggest that if an individual generally exhibits heightened levels of aggression, this tendency is likely to manifest itself in their driving style (Berdoulat et al., 2013). According to Lajunen and Parker (2001), aggressive driving can be considered more or less a trait, that is, a relatively stable way of expressing this aggression.

However, it is also important to consider the distinction between different types of aggressive driving, that is, hostile and instrumental aggression. Drivers exhibiting hostile aggression mainly want to harm other road users. This is a form of reactive aggression: the driver tends to take revenge in the first place. In comparison, instrumental aggression is proactive. It involves behaviour featuring a certain emotional charge and is driven by the desire to reach the destination more quickly (Pouliot et al., 2007).

When one is attempting to identify an aggressive driver, personality traits are undoubtedly a viable and frequently chosen key. As noted by Skippon et al. (2012), driving styles can be directly compared to specific personality profiles within the "Big Five" model. For example, when a young man wishes to show off his youthfulness, spontaneity, and masculinity to women, he may do so through an aggressive driving style. Conversely, if an older man wants to present himself as mature and suitable for a long-term relationship, he is more likely to drive cautiously. A cautious driving style thus correlates with high scores for conscientiousness and agreeableness, whereas aggressive driving is associated with low scores for these traits. Additionally, Lajunen and Summala (1995) identified a relationship between aggressive driving and neuroticism, Type A personality, and, to some extent, low self-esteem and a low sense of coherence (a salutogenic concept of personal resilience; Antonovsky, 1993).

If we focus more on demographic indicators, we find that younger drivers tend to be generally more aggressive in their approach than older drivers (Sagberg et al., 2015) and that men are typically more aggressive than women (Berdoulat et al., 2013).

Measuring driver aggression

As implied by the previous section, driver aggression is predominantly examined using psychodiagnostic tools, specifically purpose-built and personality questionnaires. Although aggression can be relatively easily related to one of the basic emotions, namely anger, it is rarely measured by means of physiological indicators in traffic psychology research (e.g. Mesken et al., 2007). This may be due to the fact that driver aggression is often measured in order to achieve the most accurate possible identification of individuals who may pose a potential risk to other road users. The remaining part of this paragraph draws on a review by Sagberg et al. (2015). Tools specifically designed to assess driving aggression or aggressive driving include the Multidimensional Driving Style Inventory (MDSI), Driving Vengeance Questionnaire, Driving Anger Scale (DAS), Driving Anger Expression Inventory (DAI), Driving Aggression Indicators Scale (DAIS), and Propensity Towards Angry Driving (PAD). Examples of other instruments measuring aggression in the driving context, although not exclusively focused on aggression, are the Driving Behaviour Questionnaire (DBQ), the Dula Dangerous Driving Inventory (DDDI), the Driver Behaviour Inventory (DBI), and the Multidimensional Driving Style Inventory (MDSI).

As in the case of stress, another approach to measuring aggression is the observation of driving behaviour. In this context, aggressive driving will again involve specific driving behaviours mentioned earlier, such as speeding, frequent use of the fast lane, flashing other drivers, or the "brake-accelerate" driving style (Sagberg et al., 2015). On the basis of their research involving observation of driving behaviour, MacAdam et al. (1998) even introduced a "driving aggressivity index" expressing the ratio of occurrences of the various observed behaviours.

Fatigue

Fatigue is another essential topic in traffic psychology. While the term "fatigue" conveys a similar meaning to most people, it may be useful to clarify its precise characteristics. Fatigue is a neurobiological process that regulates the circadian rhythm and the human need for sleep. It represents a transitional state between wakefulness and sleep. Its most notable effect is physical lassitude. It manifests itself as a *state of diminished readiness to respond, impairing both cognitive and psychomotor functions*. As a result, fatigue leads to an overall decrease in individual performance and prevents tasks from being performed efficiently (Duta et al., 2004; Lal & Craig, 2001). This is also the reason why fatigue can be defined as the inability to maintain a certain level of performance over an extended period (Larue, 2010).

Fatigue in transport

In the field of transport safety, particularly, fatigue remains a persistent focus of research. Similar to aggression, fatigue results in a significant number of road accidents annually, which makes it one of the most critical threats to road safety. In Europe and the United States, fatigue is responsible for approximately 21% of all fatal accidents (Zheng et al., 2019).

For the purposes of specific research, fatigue can be further divided into *active*, *passive*, and *sleep-related*, according to May and Baldwin (2009). Active fatigue refers to any activity that requires the active involvement of the individual, such as high-density road traffic, driving in poor visibility, or the need to perform any secondary tasks while driving. Passive fatigue is affiliated with monotonous situations which do not generate enough stimulation for the driver, for example, in traffic jams or when driving on an empty straight highway. Finally, sleep-related fatigue occurs as a result of natural physiological processes and the circadian rhythm. It is natural for each of us to feel more alert at certain times of the day than at others. The periods between 2:00 and 4:00 AM and between 1:00 and 3:00 PM are perceived by adults as the time intervals with the most significant energy drop and, accordingly, the highest levels of fatigue (Sikander & Anwar, 2018).

How can driver fatigue be identified? It manifests itself in various ways, such as yawning, slowed reaction times, drooping eyelids, gaze diversion, increased blinking, head tilting, weakened grip on the steering wheel, inattention, or reduced sensory functions. However, it is important to bear in mind that different symptoms and levels of fatigue can be observed. Therefore, it is not advisable to rely on a single "absolute" indicator but, instead, multiple indicators should be considered for greater measurement accuracy (Li et al., 2017; Sikander & Anwar, 2018).

It is also important to remember that fatigue and its physiological signs do not arise independently, but are closely related to the driver's motivation, boredom, and monotony. As has been mentioned several times, human factors clearly support and complement one another. We must therefore take this fact into account when studying them (Larue, 2010).

Fatigued driver

When exploring fatigue, we are primarily talking about the state of the individual rather than a personality trait. This makes it difficult to identify a clear-cut demographic group or personality profile that would be "sleepier" than others when driving. On the other hand, we could hardly claim that there are no studies seeking answers to such questions.

In their work focused primarily on professional drivers, Tàpia-Caballero et al. (2022) suggest that personality characteristics do play a certain role in the tendency to become fatigued more quickly. Specifically, they refer to openness to experience, noting that the higher a driver scores on this scale, the more likely they are to succumb to fatigue. They go on to explain that individuals who are more open to experience find it much harder to tolerate monotony, a key factor in passive fatigue at least. Larue (2010) adds two additional traits: (i) extraversion, supported by the argument that extroverts are more prone to boredom, a frequent precedent of fatigue; and (ii) sensation-seeking, which is explained in terms similar to the openness to experience earlier.

The only relevant demographic indicator is the driver's age, which, as evidence shows, correlates negatively with susceptibility to fatigue. Younger drivers are more likely to become overwhelmed by fatigue while driving in comparison to older drivers. Researchers attribute this to older drivers being more aware of the risks and, therefore, more motivated to engage fully in the task of driving. As a result, they are less prone to drowsiness (Tàpia-Caballero et al., 2022).

Measuring driver fatigue

Studies involving physiological measurements clearly predominate among research efforts aimed at investigating driver fatigue. It therefore seems pertinent to address physiological indicators of fatigue first. As indicated earlier, fatigue is a complex phenomenon which can hardly be effectively measured using a single indicator (Xiao & Abas, 2021). The most common methods for measuring physiological indicators include electrocardiography (ECG), electroencephalography (EEG), electrooculography (EOG), and electromyography (EMG). In ECG measurements, heart rate variability (HRV) serves as a key metric. The growing availability of smartwatches has also made it easier to obtain such data, often providing relatively sophisticated cardiac diagnostics. EEG-based fatigue detection relies on specific brain wave patterns: delta and theta waves become more prevalent, while alpha and beta waves decrease in fatigued individuals. Belakhdar et al. (2016) developed a single-channel automated driver drowsiness detection method based on artificial neural networks, identifying drowsiness from one EEG channel using a Fast Fourier Transform. In the case of EOG, Ma et al. (2016) designed a wearable drowsiness detection system that transmits oculomotor muscle activity data via Bluetooth to a smartphone. Finally, EMG-based measurements focus on frequency changes, with a sharp drop in frequency signalling driver boredom.

Another common method for measuring fatigue is observing driver behaviour. As with stress, fatigue can be assessed both by directly monitoring the driver's behaviour and by using onboard diagnostic devices (Xiao & Abas, 2021).

When detecting driver behaviour, researchers rely on previously mentioned indicators of fatigue. Changes in eye and mouth movements, head position, and facial expressions are examined. Specifically, eye-related fatigue is characterised by a reduced frequency of blinking, a shift in the open-to-closed eye ratio, and overall narrower eye openings. These signs of fatigue are most commonly assessed by means of percentile eyelid closure (PERCLOS), blink rate (BR), and maximum closure duration (MCD) (Xiao & Abas, 2021). When looking for mouth-related measures of fatigue, yawning is the most obvious indicator. Other signs that may be used for more sophisticated identification methods include mouth shape, mouth corner position, and the extent to which the mouth is opened (Shi et al., 2017). In comparison to the aforementioned methods, head movement detection provides a relatively straightforward means of identifying fatigued driving by measuring head nodding.

If researchers seek to identify fatigue using onboard diagnostic systems, they focus on lane position tracking, steering wheel angle (SWA), acceleration, and braking patterns. Lane-keeping detection systems assess the driver's ability to maintain their lane position. SWA analysis offers a more nuanced approach by tracking irregularities in steering behaviour. This is typically done using steering wheel angle data, which makes it possible to establish a baseline for a given driver and subsequently look for any deviations from this standard (Li et al., 2017). Finally, acceleration and braking patterns are monitored for any relevant variations (Wang et al., 2016).

Psychodiagnostic tools should also be brought up at this point. As mentioned earlier, fatigue is primarily not associated with personality profiles in research. Consequently, assessment tools in this area are relatively scarce and largely limited to self-report scales. The most commonly used measure is the Karolinska Sleepiness Scale (KSS) (Gillberg et al., 1994), with the Stanford Sleepiness Scale (SSS) (Shahid et al., 2012) also being frequently employed. Personality questionnaires are rarely used in this context, which is understandable given the distinction that fatigue is a state, not a trait.

Attention and distraction

Attention, inattention, and distraction – or, more commonly, distracted driving – are often discussed together in the academic literature as "two sides of the same coin". Regardless of the terminology, the topic remains central to researchers concerned with traffic psychology. As noted in the introduction to this chapter, human factors in transport do not exist in isolation but rather influence one another to varying degrees. If there is one aspect relevant to all the human factors discussed, it is attention. Its significance is further underscored by the fact that driver inattention and distraction are the focus of a biennial international conference, which consistently attracts strong participation (8th International Conference on Driver Distraction and Inattention, 2022).

All three terms mentioned earlier, including their interrelationship, can be explained in more detail by the following definition: distracted driving is often described as a specific form of driver inattention, which involves a diversion of attention from activities critical for safe driving towards a competing activity (Regan et al., 2008). We will now primarily focus on distraction, even though it is regarded as a phenomenon secondary to attention. While the vast majority of studies in this field use all three terms – attention, inattention, and distraction – their content ultimately addresses distraction, with (in)attention mainly serving as a logical contextual reference. This is because attention, by its nature, is a broad concept that covers a wide range of research on various subtopics.

Distraction and attention in transport

However, the definition provided by Regan et al. (2008) is just one of many, and it should be noted that other definitions may differ to varying degrees. In this regard, Edwards and Wundersitz (2019) aptly point out that such differences in definitions can lead to distortions in statistical data about the impact of driver inattention and distraction on road safety. In this respect, they refer to the research undertaken by Beanland et al. (2013), where inattention was operationalised as giving insufficient attention to a driving task, including driving under the influence and fatigue. On the other hand, distraction was defined as a diversion of attention away from driving, such as engaging in conversation with a passenger. While the differences between individual studies may seem relatively minor, they can nevertheless lead to different interpretations of the same statistical data.

Inattention (or distraction) can be classified into visual, auditory, biomechanical, and cognitive types. Visual distraction includes any activity in which the driver averts their eyes from the road and focuses on something unrelated to driving for an extended period (typically, research works with durations exceeding two seconds). Auditory distraction involves any activity that diverts the driver's auditory attention away from driving (e.g. talking on the phone). Biomechanical distraction occurs when the driver physically manipulates something other than the vehicle's primary controls (such as the steering wheel, gear shift, and pedals in this context); on the contrary, other vehicle controls may often be sources of distraction themselves. Finally, cognitive distraction encompasses anything unrelated to driving that absorbs the driver's mind (Young & Regan, 2007).

With a basic definition of distracted and inattentive driving established, we can now focus on specific behaviours that operationalise these concepts. In their review study, Edwards and Wundersitz (2019) compiled a list of activities identified in the scholarly literature as manifestations of driver inattention and distraction. These include eating and drinking, talking to passengers, adjusting AC or radio controls, checking social media, reading/writing, watching videos, checking the Internet for emails or other web content, operating GPS navigation, consulting a paper map, looking through a bag/backpack, gazing at the scenery/glancing away from the road ahead/looking at roadside objects, signs, or people, reaching for objects in the car, attending to children/pets in the car, being distracted by animals/insects, applying make-up or engaging in other self-grooming, smoking, and cognitive distractions (such as daydreaming, becoming lost in thought, or experiencing emotional distress).

A typical study on this topic seldom covers the entire spectrum of these behaviours. However, if we were to identify the "number one behaviour" in typical research on driver inattention, it would undoubtedly be phone use. Its significance in this field is evident from the fact that all the studies referenced in this section mention it to some extent (e.g. Qi et al., 2020; Ranney, 2008). Moreover, phone use is particularly risky in that it constitutes multiple types of distraction: biomechanical (handling the phone), auditory (conversation), visual (reading a message), and cognitive (thinking about the content of a call or message) (Kashevnik et al., 2021).

Last but not least, distracted and inattentive driving often occurs simply because the driver needs additional stimulation at a given moment, as the task of driving alone may not provide sufficient cognitive engagement. The impact of non-driving-related activities (e.g. telephone conversations) on overall driving performance varies significantly, depending on the driving environment: such behaviour poses a much lower risk during a leisurely countryside drive than in heavy urban traffic (Strayer et al., 2003). Distracted driving often becomes risky only when external conditions change dramatically. This observation is supported by Hungund et al. (2021), who studied the tendencies towards distracted driving related to the increased use of advanced driver assistance systems (ADAS), which reduce the cognitive demands of driving. They reached a clear conclusion that the more assistance systems are used, the greater the general tendency to distracted driving behaviour is.

Inattentive drivers

Research findings have been largely consistent in this regard. Younger and less experienced drivers are generally more prone to distracted and inattentive driving, with most studies attributing this to a lack of experience and awareness of the potential consequences. For example, Ranney (2008) reports that young drivers are more than twice as likely to engage in secondary activities while driving. This aligns with the conclusions of Sullman et al. (2015), who found that middle-aged and older drivers (>30 years) were significantly less likely to engage in distracting activities compared to younger drivers.

In terms of personality characteristics, individuals who perceive driving as being under their control – those with excessive self-confidence – are generally more prone to distraction behind the wheel.

Measuring driver inattention and distraction

While inattention and distraction are studied in the same terms as the previously discussed human factors, certain specific aspects are unique to the present topic.

Driver behaviour observation is arguably the most common method used to measure distraction and inattention, but it is examined in variations that have not been extensively covered in this chapter. A typical approach involves naturalistic driving studies, where driver behaviour is video-recorded during regular vehicle use and later analysed to identify activities conducive to distraction. Alternatively, statistical analyses of crash data serve to process all available official records on the causes of accidents. Another frequently used method is direct roadside observation of a large number of passing drivers. Finally, vehicle diagnostics data, such as steering wheel angle, acceleration and braking patterns, and lane-keeping behaviour, are also employed for observational studies (Dingus et al., 2016; Foss & Goodwin, 2014; Kashevnik et al., 2021; Ranney, 2008; Sullman et al., 2015).

Psychophysiological tools used to measure driver inattention and distraction include EEG, ECG, and EMG, as well as electrodermal activity (EDA) and a pulse oximeter. As with the previously discussed human factors, all of these are applied for research purposes in various combinations rather than as stand-alone indicators. Physiological correlates and measurements of inattention and distraction are explored in greater detail by Kashevnik et al. (2021).

Psychodiagnostic tools are rarely employed to study distracted driving. However, self-report questionnaires are used for such purposes. Their unique nature makes them particularly valuable, in that, unlike the other measurement methods mentioned, they help uncover the underlying motives behind a given behaviour (Edwards & Wundersitz, 2019).

Boredom

Both in traffic psychology research and in psychological research generally, boredom as a human factor tends to be overlooked. We believe this is because boredom is closely related to other human factors that have been studied in more depth and over a longer period, making them easier to operationalise and measure. In this context, boredom is most often associated with driver fatigue, inattention, and monotony.

How do we define boredom? Surprisingly, there are several theories that define boredom, and these can be categorised into four main groups: psychodynamic theories, existential theories, attention theories, and arousal theories (Fahlman et al., 2013). Without delving into the different nuances of each, we will adopt a definition that synthesises all of these approaches, as proposed by Fahlman et al. (2013). They conceive of boredom as an *aversive experience of unfulfilled desire to be engaged in a satisfying activity*.

Although it still cannot be determined with certainty whether boredom precedes fatigue or is merely a concomitant phenomenon accompanying it, boredom is undoubtedly a sufficiently distinct and specific phenomenon to warrant greater research attention.

Driver boredom

For a better understanding of the entire issue, let us now take a closer look at what constitutes boredom. Steinberger et al. (2016) identify four basic building blocks: *stimulation*, *attention*, *engagement*, and *arousal*. Arousal is linked to the physiological domain, as it refers to excitation in the body – when a person is experiencing boredom, their level of arousal is low. Stimulation builds upon arousal, since the purpose of stimulation is to increase arousal and, consequently, overall excitation. Both arousal and stimulation are insufficient during boredom. Attention has been covered in greater detail earlier, but let us just note that in boredom situations, attention is not fully focused on driving. Finally, engagement is closely related to attention. Steinberger et al. define it as the extent to which an individual is involved in an activity (i.e. the driving task). Like stimulation, engagement is low during boredom.

This more specific breakdown of boredom is presented to illustrate how closely boredom is interconnected with related human factors. However, what we wish to emphasise above all is how boredom may in fact explain why these factors arise in the first place. If boredom consists, among other things, of fatigue, then the observed fatigue may very well be merely a result of experienced boredom. If boredom includes inattention, then the observed inattention may similarly result from boredom. In short, we believe that boredom behind the wheel could be the right key to understanding the causes of a significant amount of undesirable and risky driving behaviour. Indeed, boredom may hold the answer to "why", not just "what".

No less than 79.5% of all road accidents on Czech roads in 2020 were caused by driver error (Straka & Pelešková, 2021). As noted by Steinberger et al. (2017), young people account for a significant proportion of those at fault. A considerable percentage of the accidents they cause result from their use of smartphones and social media while driving. These drivers often resort to such distractions because the current traffic situation does not offer them sufficient stimulation or engage them enough.

As suggested by Csikszentmihalyi's work (2000) and his concept of "flow", a person performs best during those moments when they are optimally stimulated, facing a challenge that is sufficient but not excessive.

Before discussing the manifestations of boredom, we should clarify two terms that anybody interested in international research on boredom will inevitably encounter: *state boredom* and *trait boredom*. The former is something everyone experiences at times – it refers to any situation in which an individual feels bored. The latter concerns how prone a person is to boredom, that is, how likely they are to become bored in a given situation sooner than others. In this context, the term *boredom proneness* is used. Both state boredom and trait boredom are relevant concepts in traffic psychology (Hunter & Eastwood, 2018; Ng et al., 2015).

What are the actual manifestations of driver boredom? Mapping various strategies adopted by young drivers to cope with boredom while driving, Toman (2020) identified the following behaviours: singing, listening to music, talking to a passenger or playing games with them, engaging in thinking-related activities such as mindfulness, prayer, or daydreaming, speeding, engaging in dynamic/aggressive driving, driving with a focus on maximising mileage, eating or drinking, smoking, looking around, observing passing vehicles, monitoring distance travelled or the traffic situation, listening to audiobooks/podcasts/sermons, engaging in telephone conversations, adjusting vehicle controls, such as changing the radio station, opening windows, and repositioning the seat, engaging in sexual acts, such as masturbation or oral sex, and watching videos. Evidently, the effects of boredom on drivers are numerous and it is fair to say that none of these behaviours are inherently positive or supportive of safe driving.

Bored driver

As noted earlier in this section, the research focusing specifically on driver boredom remains limited. Therefore, studies that deal with boredom outside the driving context will also be referenced.

To start with demographic data, Sundberg et al. (1991) found that adult men generally experience boredom more frequently than women do. This is attributed to the assumption that men are typically expected to be more proactive in their pursuits, which may ultimately lead to them finding more situations boring in comparison to women.

As for personality factors, evidence suggests that emotional stability and the ability to entertain oneself predispose individuals to a lower propensity for boredom. Conversely, extraversion and emotional negativity appear to be linked to a heightened susceptibility to boredom (Barnett & Klitzing, 2006).

Measuring boredom

Boredom may be more challenging to measure than other human factors. It depends on whether we want to measure boredom as a state or as a trait. In the former case, tools for physiological measurement and observation rather than psychodiagnostic tools will be used.

The only exception is specific-purpose scales, occasionally used to supplement information obtained through observation or physiological measurements, and a very specific method developed by Fahlman et al. (2013), the Multidimensional State Boredom Scale (MSBS), which features 29 items designed to objectify the current experience of state boredom.

Physiological indicators of state boredom include EEG, ECG, EMG, electrodermal activity (EDA), and cortisol levels. As regards specific correlates of boredom within these indicators, state boredom is associated with increased heart rate and a decrease in electrodermal activity (both occurring simultaneously), a rise in cortisol levels, and a decrease in the activity of alpha, beta, and theta waves collected in EEG (Klarkowski, 2017; Merrifield & Danckert, 2014).

On the other hand, it comes as no surprise that trait boredom can be measured using only psychodiagnostic tools. To assess driver boredom, tools such as the Boredom Susceptibility Scale (BSS) and the Boredom Proneness Scale (BPS) are applied.

The individual in interaction with other road users

Up to this point, we have discussed ways of defining, operationalising, and distinguishing human factors from one another. However, as is often the case with reality, every further reduction may provide more specific and structured information, but it also strips away nuances that ultimately make reality what it is. For this reason, this section will focus on what has not been considered in previous ones: the individual in the context of others.

According to Aristotle, human beings are inherently social creatures (Johnson, 2018), and just as in any other field of psychological research, society and culture play an essential and inseparable role in traffic psychology. Human behaviour does not occur in a vacuum but within a social environment. Likewise, human factors in transport do not manifest themselves in isolation but always within a societal context. Therefore, it is impossible to address road safety without at least briefly considering the influence of culture and society as a whole.

The influence of culture on driving style

Culture is undoubtedly a powerful driving force in society, and transport is no exception. For example, it is widely recognised that driving behaviour varies between Northern and Southern Europe, with cultural differences being more to blame for this variation than differing climate zones.

When walking around Rome, for instance, one is unlikely to come across a car without at least some visible signs of previous minor collisions with other parked vehicles, something that is far from the norm in countries such as the Czech Republic. Differences in driving culture can also be observed at the regional level. Drivers navigating small towns and villages are likely to drive differently from those primarily using cars in large urban areas. This notion is supported by Ward (2007), who examined differences in accident rates between rural and urban areas in the United States. His study found that while accidents are more common in cities, those that occur outside urban areas tend to have more severe, often fatal, consequences. Despite the obvious reasons, such as generally higher speed limits and narrow shoulders with ditches in rural areas, Ward also points to distinct driving behaviours typical of rural drivers as a significant contributing factor.

Traffic culture is therefore a crucial factor in road safety. But how do we identify traffic culture? Benzaman et al. (2022) propose three possible methods for measuring traffic culture.

One method is to correlate specific socio-economic factors with traffic safety. This approach was applied by Melinder (2007), who found that traffic safety may correlate with religious affiliation and economic wealth.

The second approach involves developing purpose-built self-assessment questionnaires aimed at capturing respondents' specific values and beliefs. An example is a study by Gaygısız (2010) which investigated the perceived quality of public governance across 51 different countries to find that this perceived quality correlated with accident statistics in the countries under scrutiny.

The third option is to derive culture from "hard" statistical data, or from the difference between estimates and actual performance measures. Page (2001) suggests that the quality of a national traffic culture is implied by the difference between the number of road fatalities estimated on the basis of objective statistical indicators (e.g. population size, percentage of young drivers, alcohol consumption per capita, percentage of the population living in urban areas) and the actual numbers measured. The resulting difference between these indicators can then serve as the basis for identifying positive and negative traffic cultures.

Specific figures from Page (2001) show that from 1980 to 1994, Greece, Belgium, the United States, and Spain were clearly representatives of a negative traffic culture, while countries with a clearly positive traffic culture included the Netherlands, Sweden, Switzerland, Norway, and the United Kingdom.

Parental influence on driving style

The influence of society on an individual can hardly be discussed without considering the family. Given the significant impact that parents have on shaping an individual's behaviour and experiences in general, they inevitably play a crucial role in determining how a person will drive.

The impact of family climate and parenting styles can be most directly observed in young and novice drivers. Following the example of Taubman-Ben-Ari (2014), the next section will mainly focus on research related to this population.

Evidence suggests that parents' history of crash involvement or even just traffic violations predicts similar behaviours among their offspring (Brookland et al., 2014; Wilson et al., 2006). A resemblance between the driving styles of parents and their adolescent children was confirmed by Prato et al. (2009), who monitored driver behaviour using in-vehicle recording devices. Similarly, Schmidt et al. (2014) found that when parents tend to drive aggressively, under the influence of alcohol or drugs, inattentively, or in violation of traffic laws in the presence of their children, the latter are highly likely to adopt the same behaviours in their own driving practices.

On the other hand, children raised in a supportive environment are less likely to engage in risky behaviours (Parker & Benson, 2004). The same holds true for adolescents from generally well-functioning families (Taubman-Ben-Ari & Katz-Ben-Ami, 2013). Likewise, open parent–children communication that is supportive and informative in relation to safe driving styles tends to foster a positive attitude towards safe driving among adolescents (Yang et al., 2013).

As the previous two paragraphs imply, the correlation between parents' and their children's driving behaviour, as well as the overall parental influence on adolescents' driving styles, has been well demonstrated by decades of research. While thorough descriptions of what happens are available, a full understanding of the mechanisms generating these effects remains elusive. In other words, we know the "what", but not the "why". On the one hand, the above phenomena can be attributed to the general tendency of children to model their behaviour, including in driving, after their parents. On the other hand, this proposition appears to contradict adolescents' tendency to distinguish themselves from their parents and pursue their own individuality (Taubman-Ben-Ari, 2014). A satisfying answer to the "why" thus remains to be seen.

What we already have, however, is a tool to measure the influence of family climate on an individual's anticipated driving style. Developed by Taubman-Ben-Ari and Katz-Ben-Ami (2013) and later refined and validated by Burns et al. (2020), the Family Climate for Road Safety Scale (FCRSS) is capable of predicting driving style from information about family climate on the basis of five factors.

Risky drivers and vulnerable road users

In traffic psychology research, there are essentially two primary approaches to studying the human factor in transport: either individual human factors are explored as separate phenomena, potentially related to personality types or demographic characteristics, or the human factor in traffic is examined as a whole, typically by studying risky drivers and vulnerable road users. Since this chapter primarily adopts the former approach to defining the human factor, the next section will aim to complement the overall picture drawn from the previous sections, specifically in the context of risky drivers and vulnerable road users.

Risky drivers

So who exactly is considered a "risky driver"? In their systematic review, Bowen et al. (2020) explored this issue through five driving-related areas (emotions, personality, nature of work, commuting, and mental health). They discovered that emotions, particularly anger, have a significant impact on risky driving practices. Negative personality traits were likewise associated with unsafe driving behaviours. On the other hand, more positive personality traits and altruistic approaches have a protective effect. It was also shown that one's perception of one's work may influence dangerous driving practices, as a negative view of work is related to undesirable driving styles. However, evidence suggests that when a person faces challenges at work, they may be more likely to drive safely. In terms of mental health, anxiety and stress have a negative impact on driving behaviour.

Looking at the previous paragraph, it becomes evident that a deeper exploration of personality traits would inevitably lead us back to the earlier sections, where individual human factors are addressed.

However, the most specific and evidence-based indicator of risky driving behaviour should be sought in demographics. The expert audience unanimously agrees that the most at-risk population consists of young men under the age of 25 (e.g. Brown et al., 2017; Craig & Levulis, 2017). Young male drivers are particularly susceptible to peer pressure, generally more easily influenced, and tend to "show off" in the presence of their peers, with risky driving being an obvious way of doing so. They also tend to overestimate the benefits of such behaviour while underestimating its drawbacks and are even more likely to make riskier decisions around peers than when alone (Engström, 2003; Gardner & Steinberg, 2005; Huang & Winston, 2011).

If we were to ask why young men, in particular, exhibit such behaviour, the answer lies in neuropsychology. Evidence provided by Porter (2011) shows that significant changes in brain activity and the development of certain brain regions continue until around the age of 25 (and in some cases, even up to 29). It is therefore no exaggeration to state that while many jurisdictions recognise adulthood at 18 years of age, the brain of an average 18-year-old male is not yet fully mature in a developmental sense. Porter (2011) further provides a detailed list of these neurological changes and their consequences, including the development of the cortical and limbic systems, the cerebellum, hippocampus, amygdala, corpus callosum, and prefrontal cortex. Clearly, these are significant changes that help shed light on their behavioural implications.

Those interested in exploring this topic in greater depth are encouraged to use the keywords *risky drivers* and *risky driving* when searching for international studies. The identification of risky driving styles in research most commonly involves the use of the previously mentioned Driver Behaviour Questionnaire (DBQ) (Bowen et al., 2020).

Vulnerable road users

Another group which traffic psychology research frequently focuses on comprises vulnerable road users. But who exactly falls into this category?

The answer to this question can be found in the literature. Existing knowledge on vulnerable road users was synthesised by Holländer et al. (2021), who reviewed research conducted between 2000 and 2020 in an effort to create a comprehensive taxonomy. They conclude that pedestrians, cyclists, and motorcyclists are those most commonly classified as vulnerable road users. Nevertheless, some definitions in the literature consider any road user without external protection to be vulnerable.

It is further noted that a road user's vulnerability is determined not only by their choice of travel mode but also by specific conditions or needs. Consequently, Holländer et al. also include children (individuals under 18), the elderly (aged 60 and above), and people with visual, cognitive, hearing, and mobility impairments in the vulnerable road user category.

However, this broader classification reflects the results of a single very recent study. The majority of research on vulnerable road users typically adopts definitions that include "only" pedestrians, cyclists, and motorcyclists.

The psychological impact of climate change on individual mobility

We first addressed the human factor in the context of the individual, then in relation to others. It is therefore only fitting to conclude this chapter by considering the human factor in the broader context of the environment. Indeed, as the topic of climate change and its impact on various aspects of life continues to gain prominence (Hussey, 2023), it has also become an area of focus in research on the human factor in transport, particularly in relation to travel mode choice, specifically a preference for public transport over private car use.

We will start by explaining the principles on which psychological factors influence transport mode choice and then discuss the extent to which the impact of environmental anxiety is reflected in statistics.

Attitudes versus behaviour

Pro-environmental thinking can be regarded as a latent construct, that is, something that cannot be directly observed in an individual. However, it can be inferred from a combination of a person's attitudes and behaviours. These two aspects and their interaction are key to our discussion. While closely related, these concepts remain distinct: behaviour is an observable action (e.g. disposing of a plastic cup in a recycling bin), whereas attitude refers to the subjectively perceived importance an individual attributes to a given issue (e.g. the degree to which one believes that active engagement in waste separation has a tangible impact on the environment). In most cases, specific attitudes ("waste separation is meaningful") lead to corresponding behaviours

("I dispose of plastic in the plastic bin"). However, the phrase "in most cases" is more important in this respect than we might expect. Research consistently shows that people occasionally act contrary to their beliefs, particularly in situations where pro-environmental behaviour is financially costly, less convenient, or when they feel that the majority does not behave in the same way (Roberts et al., 2018; Vieira et al., 2023).

A typical example in this regard might be shorter car journeys: if our workplace is not within walking distance, there is no nearby public transport stop, or the timetables do not align with our working hours, but we have access to a car, we may well end up choosing that car for our commute. And we will do so despite our sincere belief that using public transport helps save the environment. Such an experience then creates cognitive dissonance, the uncomfortable feeling resulting from the conflict between attitude and behaviour, leading to one of two shifts: either we change our behaviour and start using public transport or we adjust our attitude and come to believe that using a car just for commuting to work will not have as significant an impact on the environment as we initially thought (De Vos & Singleton, 2020).

The impact of pro-environmental attitudes on transport mode choice in numbers

If attitudes do not necessarily translate into behaviour, and if we tend not to follow our attitudes when it is inconvenient for us, to what extent does pro-environmental thinking really influence transport mode choice?

Roberts et al. (2018) analysed data obtained from approximately 40,000 households with the help of the UK Household Longitudinal Study. They found that individuals with pro-environmental attitudes are generally more likely to use public transport rather than private cars for daily commuting. A closer examination of the data reveals that attained higher education correlates with both stronger ecological attitudes and behaviours. Moreover, a similar effect is observed with better availability of public transport and with a higher probability of traffic congestion during peak hours.

In addition, the data indicates that while higher household income is associated with more frequent ecological attitudes, it correlates with less frequent pro-environmental behaviour. This finding aligns with the previously mentioned notion that behaviour is more dependent on external circumstances and convenience than attitudes are, and that attitudes may therefore contradict behaviour. This is further supported by the observation that, despite holding similarly pro-environmental attitudes, men use public transport more often than women, simply because women are more likely to drive children to and from school, go shopping on their way to and from work, and so on (Roberts et al., 2018).

In their study of 937 Portuguese respondents, Vieira et al. (2023) focused on the conflict between environmental attitudes and behaviour. They found that, paradoxically, pro-environmental behaviours that have a smaller carbon footprint (e.g. reducing the consumption of water, electricity, or plastic) were more frequently aligned with pro-environmental attitudes than behaviours with a higher carbon impact (e.g. flying or car use). In fact, car use was the only behaviour under study that was not significantly influenced by pro-environmental attitudes. Possible explanations include the social status associated with car ownership, convenience, and the (un)availability of public transport.

Bouscasse et al. (2018) examined the influence of environmental concern on transport mode choice among 1,061 French respondents. In line with the previous studies, they found that individuals with a high level of environmental concern perceive public transport use as easier, more useful, and more pleasurable than those without such pro-environmental attitudes. Conversely, a low level of environmental concern generates affective and symbolic motivations for car use, although this relationship does not appear to be particularly robust.

Coates et al. (2024) focused specifically on climate anxiety and its impact on the choice between "green" and less eco-friendly modes of transport. Having approached climate anxiety and general anxiety from the same perspective, they hypothesised that the highest levels of pro-environmental behaviour would be observed among individuals exhibiting moderate climate anxiety, while a decrease in pro-environmental choices would be found among those with low and high levels of climate anxiety. They justified this hypothesis by drawing parallels with the effects of anxiety in performance-oriented tasks such as professional sports or high-pressure work where a certain degree of anxiety enhances activation and efficiency, but beyond a certain threshold, it diminishes both performance and its quality. Their research ultimately upheld this hypothesis by showing that the increase in pro-environmental behaviour follows an inverted U-shaped curve in relation to climate anxiety.

In conclusion, pro-environmental thinking and attitudes have a clear impact on the extent to which individuals engage in environmentally friendly behaviour. However, this relationship is not straightforward; rather, it is shaped by additional factors such as cultural and social influences, the convenience and availability of transport options, and habitual behaviours. Moreover, as noted earlier in this section, the influence is bidirectional: not only do attitudes shape behaviour but behaviour also shapes attitudes through efforts aimed at reducing cognitive dissonance. Consequently, while an individual with pro-environmental attitudes is more likely to use public transport than someone who does not prioritise ecological concerns, this does not mean they will always or consistently make that choice as often as one might expect.

References

8th International Conference on Driver Distraction and Inattention. (2022). *About DDI*. https://ddi2022.org/ddi2018/about-ddi/

AAA Foundation for Traffic Safety. (2009). *Aggressive driving research update*. AAA Foundation for Traffic Safety.

Abou Elassad, Z. E., Mousannif, H., Al Moatassime, H., & Karkouch, A. (2020). The application of machine learning techniques for driving behavior analysis: A conceptual framework and a systematic literature review. *Engineering Applications of Artificial Intelligence*, *87*, 103312.

Affanni, A., Aminosharieh Najafi, T., & Guerci, S. (2022). Development of an EEG headband for stress measurement on driving simulators. *Sensors*, *22*(5), 1785.

Andréasson, R., Bäckström, C.-G., & Vattenfall, A. B. (2000). *The seat belt: Swedish research and development for global automotive safety*. Cultural Preservation Committee, Vattenfall.

Antonovsky, A. (1993). The structure and properties of the sense of coherence scale. *Social Science & Medicine*, *36*(6), 725–733.

Baevsky, R. M., & Chernikova, A. G. (2017). Heart rate variability analysis: Physiological foundations and main methods. *Cardiometry*, *10*, 66–76.

Barnett, L. A., & Klitzing, S. W. (2006). Boredom in free time: Relationships with personality, affect, and motivation for different gender, racial and ethnic student groups. *Leisure Sciences*, *28*(3), 223–244.

Beanland, V., Fitzharris, M., Young, K. L., & Lenné, M. G. (2013). Driver inattention and driver distraction in serious casualty crashes: Data from the Australian national crash in-depth study. *Accident Analysis and Prevention*, *54*, 99–107.

Belakhdar, I., Kaaniche, W., Djmel, R., & Ouni, B. (2016). Detecting driver drowsiness based on single electroencephalography channel. In *2016 13th international multi-conference on systems, signals & devices (SSD)* (pp. 16–21). IEEE.

Benzaman, B., Ward, N. J., & Schell, W. J. (2022). The influence of inferred traffic safety culture on traffic safety performance in US States (1994–2014). *Journal of Safety Research*, *80*, 311–319.

Berdoulat, E., Vavassori, D., & Sastre, M. T. M. (2013). Driving anger, emotional and instrumental aggressiveness, and impulsiveness in the prediction of aggressive and transgressive driving. *Accident Analysis & Prevention*, *50*, 758–767.

Bouscasse, H., Joly, I., & Bonnel, P. (2018). How does environmental concern influence mode choice habits? A mediation analysis. *Transportation Research Part D: Transport and Environment*, *59*, 205–222.

Bowen, L., Budden, S. L., & Smith, A. P. (2020). Factors underpinning unsafe driving: A systematic literature review of car drivers. *Transportation Research Part F: Traffic Psychology and Behaviour*, *72*, 184–210.

Brookland, R., Begg, D., Langley, J., & Ameratunga, S. (2014). Parental influence on adolescent compliance with graduated driver licensing conditions and crashes as a restricted licensed driver: New Zealand Drivers Study. *Accident Analysis & Prevention*, *69*, 30–39.

Brown, T. G., Ouimet, M. C., Eldeb, M., Tremblay, J., Vingilis, E., Nadeau, L., Pruessner, J., & Bechara, A. (2017). The effect of age on the personality and cognitive characteristics of three distinct risky driving offender groups. *Personality and Individual Differences*, *113*, 48–56.

Burns, A. B., Garner, A. A., Chang, A., Becker, S. P., Kofler, M. J., Jarrett, M. A., Luebbe, A. M., & Burns, G. L. (2020). Factor structure of the family climate for road safety scale in emerging adults in the United States. *Accident Analysis & Prevention*, *142*, 105563.

Cardone, D., Perpetuini, D., Filippini, C., Spadolini, E., Mancini, L., Chiarelli, A. M., & Merla, A. (2020). Driver stress state evaluation by means of thermal imaging: A supervised machine learning approach based on ECG signal. *Applied Sciences*, *10*(16), 5673.

Chung, W. Y., Chong, T. W., & Lee, B. G. (2019). Methods to detect and reduce driver stress: A review. *International Journal of Automotive Technology*, *20*(5), 1051–1063.

Coates, Z., Kelly, M., & Brown, S. (2024). The relationship between climate anxiety and pro-environment behaviours. *Sustainability*, *16*(12), 5211.
Craig, C. M., & Levulis, S. J. (2017, September). The relationship between global and information processing factors and self-perceived risky driving among older adults. In *Proceedings of the human factors and ergonomics society annual meeting* (Vol. 61, No. 1, pp. 1447–1451). Sage.
Csikszentmihalyi, M. (2000). The contribution of flow to positive psychology. In J. E. Gillham (Ed.), *The science of optimism and hope: Research essays in honor of Martin E. P. Seligman* (pp. 387–395). Templeton Foundation Press.
De Vos, J., & Singleton, P. A. (2020). Travel and cognitive dissonance. *Transportation Research Part A: Policy and Practice*, *138*, 525–536.
Dingus, T. A., Guo, F., Lee, S., Antin, J. F., Perez, M., Buchanan-King, M., & Hankey, J. (2016). Driver crash risk factors and prevalence evaluation using naturalistic driving data. *Proceedings of the National Academy of Sciences*, *113*(10), 2636–2641.
Dodge, R. (1923). The human factor in highway regulation and safety. *Highway Research Board Proceedings*, *2*(32), 73–78.
Du, X., Shen, Y., Chang, R., & Ma, J. (2018). The exceptionists of Chinese roads: The effect of road situations and ethical positions on driver aggression. *Transportation Research Part F: Traffic Psychology and Behaviour*, *58*, 719–729.
Duta, M., Alford, C., Wilson, S., & Tarassenko, L. (2004). Neural network analysis of the mastoid EEG for the assessment of vigilance. *International Journal of Human-Computer Interaction*, *17*(2), 171–195.
Edwards, S., & Wundersitz, L. (2019). Distracted driving: Prevalence and motivations. *Accident Analysis and Prevention*, *54*, 99–107.
Engström, I. (2003). Passenger influence on young drivers. In H. J. Chapman & G. Underwood (Eds.), *Driver behaviour and training* (Vol. 2, pp. 191–200). Ashgate.
Engström, I. (2008). *Young drivers and their passengers: Crash risk and group processes* [Doctoral dissertation, Linköping University Electronic Press].
Fahlman, S. A., Mercer-Lynn, K. B., Flora, D. B., & Eastwood, J. D. (2013). Development and validation of the multidimensional state boredom scale. *Assessment*, *20*(1), 68–85.
Foss, R. D., & Goodwin, A. H. (2014). Distracted driver behaviors and distracting conditions among adolescent drivers: Findings from a naturalistic driving study. *Journal of Adolescent Health*, *54*(5), S50–S60.
Gardner, M., & Steinberg, L. (2005). Peer influence on risk taking, risk preference, and risky decision making in adolescence and adulthood: An experimental study. *Developmental Psychology*, *41*(4), 625.
Gaygısız, E. (2010). Cultural values and governance quality as correlates of road traffic fatalities: A nation level analysis. *Accident Analysis & Prevention*, *42*(6), 1894–1901.
Gillberg, M., Kecklund, G., & Åkerstedt, T. (1994). Relations between performance and subjective ratings of sleepiness during a night awake. *Sleep*, *17*(3), 236–241.
Healey, J. A., & Picard, R. W. (2005). Detecting stress during real-world driving tasks using physiological sensors. *IEEE Transactions on Intelligent Transportation Systems*, *6*(2), 156–166.
Hill, J. D., & Boyle, L. N. (2007). Driver stress as influenced by driving maneuvers and roadway conditions. *Transportation Research Part F: Traffic Psychology and Behaviour*, *10*(3), 177–186.
Holländer, K., Colley, M., Rukzio, E., & Butz, A. (2021). A taxonomy of vulnerable road users for HCI based on a systematic literature review. In *Proceedings of the 2021 CHI conference on human factors in computing systems* (pp. 1–13). ACM.
Huang, P., & Winston, F. K. (2011). Young drivers. In B. E. Porter (Ed.), *Handbook of traffic psychology* (pp. 315–338). Academic Press.

Hungund, A. P., Pai, G., & Pradhan, A. K. (2021). Systematic review of research on driver distraction in the context of advanced driver assistance systems. *Transportation Research Record, 2675*(9), 756–765.

Hunter, A., & Eastwood, J. D. (2018). Does state boredom cause failures of attention? Examining the relations between trait boredom, state boredom, and sustained attention. *Experimental Brain Research, 236*(9), 2483–2492.

Hussey, D. (2023, September). Exploring the impact of actionable guidance & feedback on pro-environmental behaviors, climate anxiety, and environmental efficacy. In *Proceedings of the human factors and ergonomics society annual meeting* (Vol. 67, No. 1, pp. 1735–1740). Sage.

Johnson, C. (2018). Aristotle politics: A new translation. With introduction and notes. *Ancient Philosophy, 38*(1), 218–222.

Kashevnik, A., Shchedrin, R., Kaiser, C., & Stocker, A. (2021). Driver distraction detection methods: A literature review and framework. *IEEE Access, 9*, 60063–60076.

Kecklund, L., Mowitz, A., & Dimgard, M. (2011). Human factors engineering in train cab design – prospects and problems. *Human Modelling in Assisted Transportation*, 327–333.

Kim, J., Park, J., & Park, J. (2020). Development of a statistical model to classify driving stress levels using galvanic skin responses. *Human Factors and Ergonomics in Manufacturing & Service Industries, 30*(5), 321–328.

Klarkowski, M. W. (2017). *The psychophysiological evaluation of the player experience* [Doctoral dissertation, Queensland University of Technology].

Kompier, M. A., & Di Martino, V. (1995). Review of bus drivers' occupational stress and stress prevention. *Stress Medicine, 11*(1), 253–262.

Lajunen, T., & Parker, D. (2001). Are aggressive people aggressive drivers? A study of the relationship between self-reported general aggressiveness, driver anger and aggressive driving. *Accident Analysis and Prevention, 33*(2), 243–255. https://doi.org/10.1016/s0001-4575(00)00039-7

Lajunen, T., & Summala, H. (1995). Driving experience, personality, and skill and safety-motive dimensions in drivers' self-assessments. *Personality and Individual Differences, 19*(3), 307–318.

Lal, S. K. L., & Craig, A. (2001). A critical review of the psychophysiology of driver fatigue. *Biological Psychology, 55*(3), 173–194.

Larue, G. S. (2010). *Predicting effects of monotony on driver's vigilance* [Doctoral dissertation, Queensland University of Technology].

Lazarus, R. S., & Folkman, S. (1984). *Stress, appraisal, and coping*. Springer Publishing Company.

Lee, D. S., Chong, T. W., & Lee, B. G. (2016). Stress events detection of driver by wearable glove system. *IEEE Sensors Journal, 17*(1), 194–204.

Li, Z., Li, S. E., Li, R., Cheng, B., & Shi, J. (2017). Online detection of driver fatigue using steering wheel angles for real driving conditions. *Sensors, 17*(3), 495.

Ma, Z., Li, B. C., & Yan, Z. (2016). Wearable driver drowsiness detection using electrooculography signal. In *2016 IEEE topical conference on wireless sensors and sensor networks (WiSNet)* (pp. 41–43). IEEE.

MacAdam, C., Bareket, Z., Fancher, P., & Ervin, R. (1998). Using neural networks to identify driving style and headway control behavior of drivers. *Vehicle System Dynamics, 29*(S1), 143–160.

May, J. F., & Baldwin, C. L. (2009). Driver fatigue: The importance of identifying causal factors of fatigue when considering detection and countermeasure technologies. *Transportation Research Part F: Traffic Psychology and Behaviour, 12*(3), 218–224.

Melinder, K. (2007). Socio-cultural characteristics of high versus low risk societies regarding road traffic safety. *Safety Science, 45*(3), 397–414.

Merrifield, C., & Danckert, J. (2014). Characterizing the psychophysiological signature of boredom. *Experimental Brain Research*, *232*(2), 481–491.
Mesken, J., Hagenzieker, M. P., Rothengatter, T., & De Waard, D. (2007). Frequency, determinants, and consequences of different drivers' emotions: An on-the-road study using self-reports, (observed) behaviour, and physiology. *Transportation Research Part F: Traffic Psychology and Behaviour*, *10*(6), 458–475.
National Highway Traffic Safety Administration (NHTSA), Plans and Programs, Office of Planning and Analysis. (1984, July 11). *Passenger car front seat occupant protection*. NHTSA.
Ng, A. H., Liu, Y., Chen, J. Z., & Eastwood, J. D. (2015). Culture and state boredom: A comparison between European Canadians and Chinese. *Personality and Individual Differences*, *75*, 13–18.
Page, Y. (2001). A statistical model to compare road mortality in OECD countries. *Accident Analysis & Prevention*, *33*(3), 371–385.
Parker, J. S., & Benson, M. J. (2004). Parent-adolescent relations and adolescent functioning: Self-esteem, substance abuse, and delinquency. *Adolescence*, *39*(155), 519–530.
Porter, B. E. (Ed.). (2011). *Handbook of traffic psychology*. Academic Press.
Pouliot, S., Vallieres, E., Bergeron, J., & Vallerand, R. (2007). Processus psychologiques impliqués dans l'agressivité au volant de type proactif ou réactif: Une revue de littérature des modèles théoriques et des connaissances empiriques [Psychological processes underlying proactive and reactive aggressive driving: A literature review of theoretical models and empirical knowledge]. *Recherche, transports, sécurité*, *97*, 319–338.
Prato, C. G., Lotan, T., & Toledo, T. (2009). Intrafamilial transmission of driving behavior: Evidence from in-vehicle data recorders. *Transportation Research Record*, *2138*(1), 54–65.
Qi, Y., Vennu, R., & Pokhrel, R. (2020). *Distracted driving: A literature review*. Illinois Center for Transportation.
Ranney, T. A. (2008). *Driver distraction: A review of the current state-of-knowledge*. National Highway Traffic Safety Administration.
Rastgoo, M. N., Nakisa, B., Rakotonirainy, A., Chandran, V., & Tjondronegoro, D. (2018). A critical review of proactive detection of driver stress levels based on multimodal measurements. *ACM Computing Surveys (CSUR)*, *51*(5), 1–35.
Regan, M. A., Lee, J. D., & Young, K. (2008). *Driver distraction: Theory, effects, and mitigation*. CRC Press.
Reimer, B., Mehler, B., & Coughlin, J. F. (2010). *An evaluation of driver reactions to new vehicle parking assist technologies developed to reduce driver stress*. New England University Transportation Center, Massachusetts Institute of Technology.
Roberts, J., Popli, G., & Harris, R. J. (2018). Do environmental concerns affect commuting choices? Hybrid choice modelling with household survey data. *Journal of the Royal Statistical Society Series A: Statistics in Society*, *181*(1), 299–320.
Sagberg, F., Selpi, Bianchi Piccinini, G. F., & Engström, J. (2015). A review of research on driving styles and road safety. *Human Factors*, *57*(7), 1248–1275.
Schießl, C. (2007, May). Stress and strain while driving. In *Proceedings of the young researchers seminar-European conference of transport research institutes, Brno, Czech Republic* (pp. 27–30). DLR.
Schmidt, S., Morrongiello, B. A., & Colwell, S. R. (2014). Evaluating a model linking assessed parent factors to four domains of youth risky driving. *Accident Analysis & Prevention*, *69*, 40–50.
Scopus. (2022, August 24). *Scopus – document search results*. Scopus. www.scopus.com/results/results.uri?sort=plf-f&src=s&st1=human+AND+factors+in+traffic
Selye, H. (1976). Stress without distress. In *Psychopathology of human adaptation* (pp. 137–146). Springer.
Shahid, A., Wilkinson, K., Marcu, S., & Shapiro, C. M. (Eds.). (2012). *STOP, THAT and one hundred other sleep scales*. Springer Science & Business Media.

Shi, S. Y., Tang, W. Z., & Wang, Y. Y. (2017). A review on fatigue driving detection. In *ITM web of conferences* (Vol. 12, p. 01019). EDP Sciences.

Shinar, D. (Ed.). (2017). *Traffic safety and human behavior*. Emerald Group Publishing.

Sikander, G., & Anwar, S. (2018). Driver fatigue detection systems: A review. *IEEE Transactions on Intelligent Transportation Systems*, *20*(6), 2339–2352.

Singh, R. R., Conjeti, S., & Banerjee, R. (2011, October). An approach for real-time stress-trend detection using physiological signals in wearable computing systems for automotive drivers. In *2011 14th international IEEE conference on intelligent transportation systems (ITSC)* (pp. 1477–1482). IEEE.

Skippon, S., Diels, C., & Reed, N. (2012). *Driving style as a fitness indicator*. Proceedings of the 5th International Conference on Driver Behaviour and Training, Paris, France.

Steinberger, F., Moeller, A., & Schroeter, R. (2016). The antecedents, experience, and coping strategies of driver boredom in young adult males. *Journal of Safety Research*, *59*, 69–82.

Steinberger, F., Schroeter, R., & Watling, C. N. (2017). From road distraction to safe driving: Evaluating the effects of boredom and gamification on driving behaviour, physiological arousal, and subjective experience. *Computers in Human Behavior*, *75*, 714–726.

Straka, J., & Pelešková, J. (2021). *Ročenka nehodovosti na pozemních komunikacích za rok 2020* [Annual report of road accidents for 2020]. The Directorate of the Traffic Police Service of the Police Praesidium of the Czech Republic.

Strayer, D. L., Drews, F. A., & Johnston, W. A. (2003). Cell phone-induced failures of visual attention during simulated driving. *Journal of Experimental Psychology: Applied*, *9*(1), 23–32.

Strong, H. (1927). Regional planning and its relation to the traffic problem. *The Annals of the American Academy of Political and Social Science*, *133*(1), 215–221.

Sullman, M. J., Prat, F., & Tasci, D. K. (2015). A roadside study of observable driver distractions. *Traffic Injury Prevention*, *16*(6), 552–557.

Sundberg, N. D., Latkin, C. A., Farmer, R. F., & Saoud, J. (1991). Boredom in young adults: Gender and cultural comparisons. *Journal of Cross-Cultural Psychology*, *22*(2), 209–223.

Tàpia-Caballero, P., Serrano-Fernández, M. J., Boada-Cuerva, M., Sora, B., & Boada-Grau, J. (2022). Influence that job characteristics, personality and burnout have on fatigue in professional drivers. *International Journal of Occupational Safety and Ergonomics*, *28*(3), 1331–1341.

Taubman-Ben-Ari, O. (2014). The parental factor in adolescent reckless driving: The road ahead. *Accident Analysis & Prevention*, *69*, 59–66.

Taubman-Ben-Ari, O., & Katz-Ben-Ami, L. (2013). Family climate for road safety: A new concept and measure. *Accident Analysis & Prevention*, *54*, 1–14.

Toman, M. (2020). *Vyrovnávací strategie užívané řidiči pro potlačení nudy za volantem* [Strategies used by drivers to cope with driver boredom] [Unpublished diploma thesis, Palacky University Olomouc].

Vieira, J., Castro, S. L., & Souza, A. S. (2023). Psychological barriers moderate the attitude-behavior gap for climate change. *PLOS One*, *18*(7), e0287404.

Wang, M. S., Jeong, N. T., Kim, K. S., Choi, S. B., Yang, S. M., You, S. H., Lee, J. H., & Suh, M. W. (2016). Drowsy behavior detection based on driving information. *International Journal of Automotive Technology*, *17*(1), 165–173.

Ward, N. J. (2007). The culture of traffic safety in rural America. *Improving Traffic Safety Culture in the United States*, *241*.

Wiesenthal, D. L., Hennessy, D. A., & Totten, B. (2000). The influence of music on driver stress 1. *Journal of Applied Social Psychology*, *30*(8), 1709–1719.

Wilson, R. J., Meckle, W., Wiggins, S., & Cooper, P. J. (2006). Young driver risk in relation to parents' retrospective driving record. *Journal of Safety Research*, *37*(4), 325–332.

Wise, J. A., Hopkin, V. D., & Garland, D. J. (Eds.). (2016). *Handbook of aviation human factors*. CRC Press.
Xiao, Y., & Abas, A. B. (2021). A review on fatigue driving detection. *ASP Transactions on Internet of Things*, *1*(3), 1–14.
Yang, J., Campo, S., Ramirez, M., Krapfl, J. R., Cheng, G., & Peek-Asa, C. (2013). Family communication patterns and teen drivers' attitudes toward driving safety. *Journal of Pediatric Health Care*, *27*(5), 334–341.
Young, K., & Regan, M. (2007). Driver distraction: A review of the literature. In I. J. Faulks, M. Regan, M. Stevenson, J. Brown, A. Porter, & J. D. Irwin (Eds.), *Distracted driving* (pp. 379–405). Australasian College of Road Safety.
Zheng, W. L., Gao, K., Li, G., Liu, W., Liu, C., Liu, J. Q., Wang, G., & Lu, B. L. (2019). Vigilance estimation using a wearable EOG device in real driving environment. *IEEE Transactions on Intelligent Transportation Systems*, *21*(1), 170–184.

4 Human factors in traffic

Ralf Risser

Introduction

This chapter describes which factors steer human behaviour, why knowing is not enough to change behaviour, and that it takes more than just receiving or delivering the correct information. It discusses the (perceived) difficulties which prevent people from "doing the right thing", such as choosing sustainable mobility modes or behaving safely. The difficulties associated with doing the right thing are explained and suggestions for planners and decision-makers as to how to deal with these issues are provided.

Road user behaviour is the (visible) product of multiple factors, which can be divided into those that affect the road user from the inside and those that exert an effect from the outside. To achieve changes by targeting the factors that influence behaviour from the inside involves medium- and long-term interventions, such as campaigning, education, and training. However, this will not suffice. The factors that influence behaviour from outside the person need to be dealt with as well.

What influences human behaviour: some theories

The *cognitive dissonance theory* by Festinger (1957) is most relevant in this respect. It refers to cognitions, that is, sensations, thoughts, beliefs, opinions, attitudes, and how they are related. These relations can be consonant or dissonant. Dissonant relations – for example, I want to be safe but I need to speed because I am in a hurry – provide a feeling of mental discomfort. The intensity of the dissonance among different cognitions is expressed by the index:

$$\text{Dissonance intensity} = \frac{n\ (\text{dissonant relations})}{N\ (\text{dissonant + consonant relations})}$$

The strategies to reduce dissonance are (1) to change the cognitive elements so that previously dissonant relations become consonant and (2) to introduce new cognitive elements (e.g. arguments). Frequently, we apply a combination

DOI: 10.4324/9781003608394-4

of both alternatives. The changes made often represent not reality but rationalisations, or justifications of one's own beliefs and behaviours.

Dissonance produces pressure. If we do things that are in conflict with our beliefs – for example, using the car because it is convenient, knowing that this is bad for the environment – we may strive for consonance by adding, eliminating, or changing our beliefs, or by changing our behaviour. To support a mode shift in the population, authorities may choose both ways; changing the "concert of arguments", for instance, by campaigning, or providing incentives for a change of behaviour. It is important to show, though, that the "new" behaviour will provide cognitive consonance: for example, using sustainable modes is good for the environment, but it also produces personal benefits. People who never tried out such behaviour usually do not know of those benefits. Incentives are needed that make them try this "new" behaviour and discover its advantages for themselves.

Another theory, the *theory of reactance*, was introduced by Jack W. Brehm in 1966. It assumes that restrictions on one's own freedom of choice are unpleasant, causing reactance. We want to have multiple options and to choose from these following our own decisions. Otherwise, resistance is provoked. In a state of resistance, we might even increase our perception of the value of the behaviour that we are not allowed to choose. Inhibiting car use may make it more desirable. Being forced to respect speed limits may be felt to be a restriction of one's autonomy, resulting in intentional non-compliance with the law. Communication is needed in order to make citizens understand the reasons for law enforcement with respect to speeding. In the end, people should avoid speeding because they have understood the necessity of doing so, and not because they are forced to do so.

Attribution theory (Meyer, 2003) states that our reaction to the behaviour of others – or more generally to things happening around us – depends on what reasons we attribute to those behaviours or happenings. For instance, if somebody steps on our toes, we could attribute this to an unfriendly attitude or to an unintentional slip. The former assumption would produce negative feelings, while the latter would cause fewer problems, especially when followed by an excuse. If no excuse follows, we cannot be sure what the reason for their stepping on our toes was. If we attribute the wrong reason, the communication process with the other person might develop in a wrong way. In theory, assuming that there are usually positive attitudes underlying the actions of others will support positive attitudes towards other people, according to Meyer (2003). In the same way, if positive characteristics are attributed, for example, to safe behaviour as a road user, attitudes towards abiding by the law may change in a positive direction. Usually, campaigns try to achieve this.

Something that is mentioned very frequently is the *theory of planned behaviour*, formulated by, among others, Ajzen (1991). There, the concept of *attitude* refers to the preparedness of individuals to react to persons, objects, or situations. Attitudes may represent more or less reflected judgements, including, of course, prejudices. The concept is that attitudes can be "made visible" by asking people questions concerning their assumptions, beliefs, interpretations, attributions, etc., resulting in the understanding of behavioural intentions. If this concept works well, we skip to investigating real behaviour, which is usually a laborious process. But independently of the question whether individuals who are being interviewed tell the truth concerning their own attitudes and intentions, the theory touches on further aspects that are very important for understanding human behaviour. Especially, it proposes that expectations of how "significant others" look at things influence our own attitudes and behaviour, and it points out that we need the feeling of having control over a certain behaviour in order to consider choosing it.

In sum, the theory lets us assume that we can expect a change of behaviour in a wished-for direction if attitudes towards this behaviour are, or become, positive, if it can be shown that significant others adhere to this behaviour, and that it will be easy to adopt this behaviour.

The *situational model of behaviour* by Rosenstiel (2003) proposes that both personal and situational variables affect our behaviour. The questions concerning whether we want to adopt a certain behaviour and if we are able to do so refer to the personal domain, and the ones concerning whether we are obliged or allowed to do so under certain circumstances refer to the situational domain.

According to the theory, an individual system of values develops according to the attitudes and values which are societally tolerated or even desirable. Behaviour that is in accordance with this system of values will be accepted and, in the end, internalised. However, this will only be possible if preconditions permit adherence to this behaviour. For instance, for people to respect speed limits, it will be necessary to provide infrastructure that credibly reflects the necessity of doing so. A road that looks like a motorway will be used like a motorway and low speed limits will not be respected.

Situational variables will contribute to an individual's desire to behave in certain ways in certain situations. However, when thinking of the public space in which traffic and mobility take place, it is necessary to consider that physical and mental conditions, experience, and learned abilities and skills are not equally distributed among citizens. Certain situations might be too difficult to cope with for part of the population while posing no problem at all for others. As the public space needs to be accessible and usable by all citizens, we necessarily end up at the conclusion that this public space needs to be adjusted for easy use by the weakest groups of the population. We might even conclude that such adjustment may result in advantages for the whole population.

The *theory of social comparison processes* (Festinger, 1950, 1954) proposes that we are driven to evaluate and compare what we think and what we do by looking at others. For instance, to answer the question how sustainable, or how safe, our own behaviour is, we may resort to comparing what we do with what others do in similar situations. The closer to us the persons we compare ourselves to are, the more important this comparison becomes. According to Festinger, social comparison processes constitute one main reason why we seek the company of others. Especially in new situations, the behaviour of others delivers a benchmark of how to behave. Behaving differently from others may cause discomfort or even conflicts. The "behaviour of others" can also be demonstrated in the media, such as television, newspapers, or social media, thus increasing the proneness of the public to behaving in this way.

The *model of ecological behaviour* developed by Fietkau and Kessel (1981) explores ways to enhance ecological behaviour. It proposes that five major factors need to be considered to achieve this: (1) environmental *knowledge*, (2) pro-environmental *attitudes and values*, (3) *opportunities to act* pro-environmentally, (4) *incentives* for pro-environmental behaviour, and (5) the *perceived consequences* of behaviour.

Knowledge is needed in order to adopt the desired behaviour. Pro-environmental attitudes and values need to be enhanced, for example, by education and campaigns. Opportunities to act pro-environmentally refer to appropriate preconditions. Adequate incentives support making use of those opportunities, whereby it has to be ensured that making use of opportunities leads to positive feedback.

To consider these theories in connection with all steps taken should lead to more sustainable behaviour of citizens in connection with traffic and mobility.

Šucha (2019) adds to the above thoughts the assumption that human behaviour is a result of internal factors – attitudes, etc. – and of situational external factors, such as the quality of the infrastructure, aesthetics, and accessibility. These thoughts draw on the *Diamond Model* by Ralf Risser. According to this model, people's behaviour in traffic is conditioned by a range of factors. A comprehensive approach is needed to influence road user behaviour. Among other things, one may expect that a voluntary change of behaviour needs to be connected to an experienced improvement of the individual's quality of life. There is no universally accepted definition of quality of life, but most definitions refer to well-being, satisfaction, and happiness. Quality of life may be affected by economic, social, and environmental conditions. Policies that strive for a change of behaviour in connection with transport and mobility need to see to it that such policies increase an individual's quality of life in connection with the use of public spaces (Steg & Gifford, 2005).

Risser's model (2004) proposes analysing both the physical and social aspects of the public space and the individual's own position. This involves five areas: (1) individual characteristics; (2) communication with other people or road users; (3) socialisation agents, culture, social establishment, and media (summarised as social aspects); (4) the infrastructure of the public

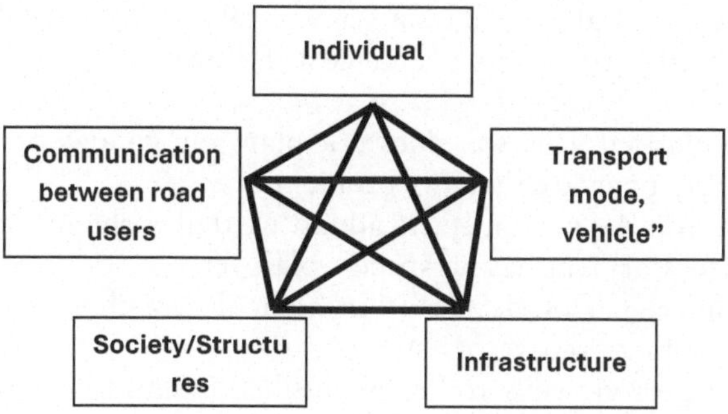

Figure 4.1 Traffic as a socio-technical system – the Diamond Model (Risser, 2004).

space; and (5) vehicle or mode characteristics. These areas (Figure 4.1) interact.

The model suggests that over time, individuals develop attitudes, values, considerations, motives, and specific points of view which have been, and continuously are, influenced by the other levels of the diamond. The interaction between road users has a direct and immediate impact on individual behaviour, but also produces a communication history constituted by experiences in the past. Infrastructure provides the physical conditions for all activities in the public space. The modes of transport are connected with different advantages and disadvantages, not least in the framework of the interaction between road users.

Finally, society/structures comprise media and the related public discourse, as well as the private interaction with family, friends, social environment, and peer groups and how the role of different forms of traffic participation is viewed and discussed there, and how all this is mirrored in laws and regulations and in informal norms.

How to approach different groups of citizens

The *transtheoretical model* (TTM) proposes that people find themselves in different states of awareness concerning specific areas of their behaviour (Prochaska & DiClemente, 1982). This involves, for instance, the willingness to do something about possibly problematic behaviour. Problematic behaviour could be behaviour that is harmful for oneself, for society, or for both. For example, a lack of physical activity may cause damage to our health. However, when a physician suggests using the car less and walking more in order to provide some physical activity, he or she may encounter a range of reactions, from complete denial to full agreement with the suggestions.

The TTM claims that strategies aimed at producing a change in a person's behaviour must take account of the attitudinal state of the person being addressed. The TTM also includes a stage of relapse and seeks to prevent it. The model describes the stages of change as follows:

- Pre-contemplation: a person does not plan any change of behaviour or may be totally unaware that any problem exists;
- Contemplation: the person starts admitting that a change in a behaviour might be necessary but has no specific plan yet;
- Preparation: the person decides to try to achieve a change in a behaviour and takes the first preparatory steps;
- Action: the person seeks to realise his/her intention and engages in a desired behaviour, or avoids an undesired behaviour;
- Maintenance: the person seeks to sustain the change in behaviour; precautions must be taken to prevent relapse;
- Change adopted: the desired behaviour has developed into a habit; no urge to resume the previous behaviour is experienced.

The model suggests that a change in behaviour involves a learning process which may take place in several rounds before a sustained change is accomplished. This process may be conceived as a spiral: a certain starting point is left in an effort to change. When one round of change is successfully completed, a person finds him- or herself on a higher level of the spiral.

External measures to support the change must then be adjusted to the given level of awareness, or insight. There, the most important point is to detect the stage of awareness (e.g. interviews with target groups or surveys) and to apply interventions accordingly. For example, if our target group is in the pre-contemplation stage, it means that they do not perceive that there is a problem or do not realise that there are alternatives to their behaviour. For this target group, an appropriate intervention would be to cause a dissonance (see also dissonance theory). For example, facts can be presented that show how beneficial walking is, both for society and, especially, for themselves. This will cause a dissonance for people who usually try to avoid walking. One way to dissolve this dissonance is to walk more. The first step – from pre-contemplation to contemplation – may be that this person starts to seek more information concerning the advantages of walking. Once attention is directed there, appropriate information can be presented. Information about the advantages of walking – and discussion of how to cope with the disadvantages – will be followed by more practical advice: city plans for walkers, times to destinations at different speeds, traffic rules, safety problems and how to cope with them, etc. Then we need to make sure that we do not lose "new" walkers, for example, because of negative experiences with bad infrastructure, that is, we need to make sure that the interventions do not exclude those who already adhere to wished-for behaviour.

Institutional framework for practitioners

At a closer look, Schneider (2013) actually puts these theories together in order to form one strategy for enhancing sustainable transport modes. His

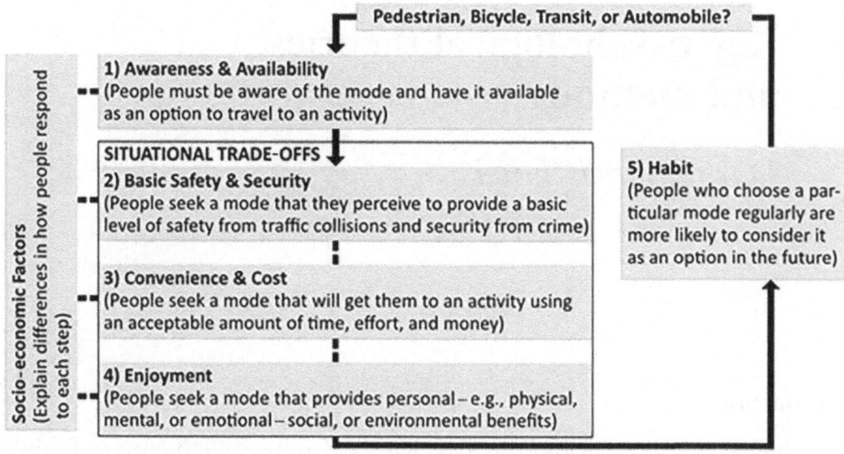

Figure 4.2 Proposed theory of routine mode choice decisions (Schneider, 2013).

theory of routine mode choice decisions is displayed as a model in Figure 4.2. It actually tells practitioners what they should try to do in order to achieve their goals.

References

Ajzen, I. (1991). The theory of planned behavior. *Organizational Behavior and Human Decision Processes*, *50*, 179–211.

Brehm, J. W. (1966). *A theory of psychological reactance*. Academic Press.

Festinger, L. (1950). Informal social communication. *Psychological Review*, *57*(5), 271–282.

Festinger, L. (1954). A theory of social comparison processes. *Human Relations*, *7*, 117–140. https://doi.org/10.1177/001872675400700202

Festinger, L. (1957). *Theory of cognitive dissonance*. Stanford University Press.

Fietkau, H. J., & Kessel, H. (1981). *Umweltlernen: Veränderungsmöglichkeiten des Umweltbewußtseins*. Hain.

Meyer, W. U. (2003). *Einige grundlegende Annahmen und Konzepte der Attributionstheorie*. Universität Bielefeld.

Prochaska, J. O., & DiClemente, C. C. (1982). Transtheoretical therapy: Toward a more integrative model of change. *Psychotherapy: Theory, Research & Practice*, *19*(3), 276–288.

Risser, R. (2004). Philosophy of traffic calming. *The Asian Journal: Journal of Transport and Infrastructure*, *11*(1), 1–9.

Rosenstiel, L. (2003). *Grundlagen der Organisationspsychologie. Basiswissen und Anwendungshinweise* (5th ed.). C. E. Poeschel Verlag.

Schneider, R. J. (2013). Theory of routine mode choice decisions: An operational framework to increase sustainable transportation. *Transport Policy*, *25*, 128–137.

Steg, L., & Gifford, R. (2005). Sustainable transportation and quality of life. *Journal of Transport Geography*, *13*(1), 59–69.

Šucha, M. (2019). *The human factor in traffic – possible ways of influencing people's behaviour* [Habilitation thesis, Charles University in Prague].

5 Key psychological theories and methods

Darina Havlíčková

Introduction

This chapter presents various theories that have influenced the ways of thinking about human functioning in the traffic context. These theories have been gradually formed and developed since the 1970s. They are often general psychological theories (including the transtheoretical model of behaviour change; Prochaska & DiClemente, 1983) adapted to the context of transport. Some theories, however, originate from other disciplines, such as environmental sciences (e.g. the value-belief-norm theory; Stern, 2000). Many of these theories have become the subject of extensive debate and even controversy. Nevertheless, such criticism has often led to them being further refined and, in some cases, finally superseded.

Despite the existence of numerous thematically related sources, it remains very difficult to find a general theory that explains human behaviour in the traffic context. The main reason is the complexity of the traffic situation. On the human side, numerous factors come into play: personality traits (such as dominance, anxiety, and impulsivity), intelligence, education, and experience. These factors influence driver behaviour in multiple ways. For example, they equip drivers with strategies for acquiring information and the ability to adapt to a changing traffic environment. A driver's capacities can be affected in real time by many other variables, such as emotions, mental or physical illness, or fatigue. At the same time, the individual in traffic is interacting with other road users and their variable inputs, as well as with traffic infrastructure.

The multifaceted nature of traffic psychology makes it difficult to articulate a comprehensive theory that could provide a thorough explanation of traffic behaviour. Existing theories most often address traffic safety (defined as the reduction of crash rates) and health in transport (defined as the choice of healthier and more environmentally friendly modes of transport). However, there is no universally used or clearly established classification of these theories. For this reason, an overview of different theories is presented next, organised according to common areas of application. The acceptance of

DOI: 10.4324/9781003608394-5

multifactoriality and variability in research on traffic behaviour can also be considered a theoretical framework *per se*.

Categorisation of theories in traffic psychology

Despite the large number of sources on the subject, it is very difficult to find a comprehensive classification of theories. There are several reasons for this. Firstly, many authors tend to focus more on specific factors than on the overall condition of an individual. Another reason is the strong interconnection between various factors, which influence and complement each other. Theories relevant to traffic psychology can be categorised into the following groups according to common areas of application:

Theories of behavioural change in traffic

These theories share an emphasis on actively promoting desirable behavioural change in traffic – most commonly in the direction of reducing road crash rates or encouraging the choice of more environmentally friendly and healthier modes of transport.

- The theory of planned behaviour (Ajzen, 1991)
- The transtheoretical model of behaviour change (Prochaska & DiClemente, 1983)
- The model of goal-directed behaviour (Carrus et al., 2008)
- The norm activation model (Schwartz, 1977)
- Value-belief-norm theory (Stern, 2000)
- The comprehensive action determination model (Klöckner & Blöbaum, 2010; Klöckner & Friedrichsmeier, 2011)

Theories of risk acceptance

These theories address risk perception and the resulting safety or riskiness of traffic behaviour. The aim is generally to identify the factors involved in the perception and evaluation of traffic-related risk.

- Risk homeostasis theory (Wilde, 1988, 1994)
- The danger compensation model (O'Neill, 1977)
- Subjective versus objective safety theory (Sørensen & Mosslemi, 2009)
- The hierarchical risk model (Van der Molen & Bötticher, 1988)

Education theories

These theories focus on the process of driver education and training. They emphasise the breadth of areas that must be addressed during driver instruction. This is not just about mastering the technical operation of a vehicle but also about broader contexts, such as lifestyle in general. The following theories fall under this category:

- The GDE matrix (Goals for Driver Education; Hatakka et al., 2002)
- The graduated driver licensing system (Vaa et al., 2015)

Theories of processing demands in the traffic environment

These theories deal with factors that influence the ability to cope with demands in challenging traffic situations. This category includes:

- The generic transactional model of driving stress and fatigue (Matthews et al., 2005)

Theory of planned behaviour

The *theory of planned behaviour* (TPB; Ajzen, 1991) is a successor to the *theory of reasoned action* (e.g. Ajzen & Fishbein, 1980). TPB assumes three core predictors of behaviour: (1) attitudes towards engaging in a behaviour (e.g. an individual views a particular mode of transport positively); (2) subjective norms, reflecting perceived social pressure (e.g. an individual feels pressured to use a particular mode of transport); and (3) perceived behavioural control (e.g. an individual believes they possess the necessary skills and opportunities to use a given mode of transport). All three of these elements form the intention, which directly precedes actual behaviour. In this framework, the choice of traffic behaviour (e.g. the choice of travel mode) is understood as reasoned action (see Figure 5.1). The theory has been widely used to explain behaviour guided by perceived self-control (Ajzen, 1991).

A key component of the model is the behavioural intention, which is influenced by the individual's belief in the likelihood that the behaviour will produce the expected outcome, and by the subjective evaluation of the risks and benefits of that outcome. TPB has been successfully applied to predict and explain a wide range of behaviours and intentions, including smoking, drinking, and proneness to risky driving. According to TPB, the achievement of a particular behaviour depends on both motivation (intention) and ability (perceived behavioural control). The theory distinguishes between three types of beliefs: behavioural, normative, and control beliefs. If an individual believes that change is within their power, has a strong sense of self-efficacy, perceives the behaviour as socially approved, and holds a positive attitude towards it, then the likelihood that the behaviour will actually occur increases.

Nevertheless, the correlations between the levels of these three variables and actual behaviour are not particularly strong; they can explain only around 25% of the variance in real-world behaviour (Åberg et al., 2008).

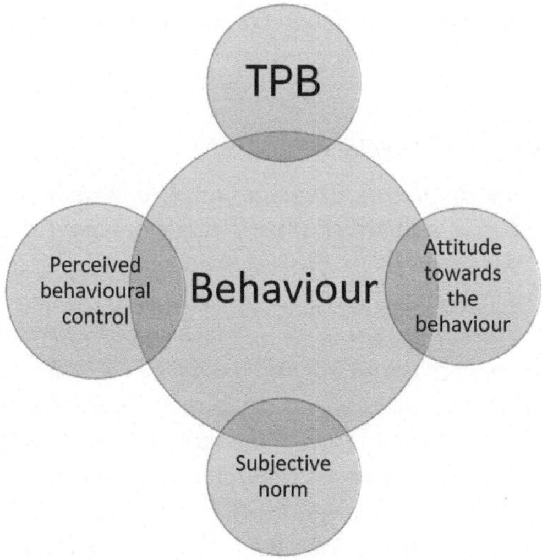

Figure 5.1 Theory of planned behaviour.

For instance, in their evaluation of TPB, Paris and Van den Broucke (2008) concluded that actual driver behaviour (e.g. speeding) can only be partially predicted from the TPB constructs, and that the cognitive determinants of safe driving identified by the model need to be supplemented by additional factors, such as less "conscious" cognitive characteristics. These include individual driving habits, personality variables, and external factors such as the texture of the road surface. Moreover, there are other variables that influence intention, behaviour, and motivation: fear, perceived threat, the driver's mood, and experience. Verplanken et al. (1997) demonstrated that intentions are significantly associated with behaviour only when the habits relevant to mode choice are weak. Similarly, Bamberg and Schmidt (2003) found that including a measure of habit increases the accuracy of predicting both intentions and actual behaviour. While TPB accounts for normative influences, it still fails to consider environmental or economic factors that may also affect and help to predict human behaviour. The effectiveness of an extended TPB model incorporating *environmental concern*, *moral norms*, and *descriptive norms* was demonstrated by Donald et al. (2014) in relation to transport mode choice.

Thus, TPB appears to be a partially useful theory for explaining personal attitudes and perceived social norms, which may distort drivers' judgement. These may, in turn, foster social identification with more adaptive modes of transport behaviour and *promote the development of self-control in high-risk drivers through the enhancement of perceived self-efficacy in traffic contexts*. Although Fishbein and Ajzen provided a solid theoretical foundation, as the influence of social psychology began to wane, the theory came under increasing criticism. The observed correlations between attitudes and behaviour proved to have limited predictive value.

Specific applications of TPB in empirical research

Using Ajzen's theory of planned behaviour (1985) as a framework, Elliott et al. (2007) found that subjective norms (i.e. perceived pressure or acceptance of a given behaviour from others) were associated with an increased intention to exceed the speed limit (as measured by a self-report questionnaire) and more frequent speeding in a simulator. It may be assumed that a person's normative beliefs about social consensus on risky driving behaviour (e.g. "I can exceed the speed limit on an empty road because everyone does it and it's normal") can serve to justify the enactment of risky behaviour. The application of the theory of planned behaviour to motorcycling in a survey of motorcyclists in the United Kingdom (Chorlton et al., 2012) supports this assumption. The respondents were asked about their intention to engage in seven instances of risky riding behaviour. It was ascertained that attitudes towards risky riding were strongly associated with the intention to carry out such behaviour. To assess subjective norms, the participants were asked to evaluate how four specific groups – the police, other road users, the family, and fellow-motorcyclists – would feel about them engaging in risky road behaviour. Those who reported no intention to behave riskily perceived more negative pressure from each of these groups. These findings support the TPB assumption that perceived normative pressure may facilitate behaviour in traffic contexts. Elliott and Thomson (2010) further examined motorcyclists' intentions to speed on roads with both low and high speed limits. Overall, their findings were consistent with TPB: attitudes, normative and subjective social pressure, and perceived behavioural control were all found to be important predictors of riding above the speed limit.

As regards the perceived control component within the TPB model, the task-capability interface (TCI) theory provides a relevant framework. The TCI model proposes a homeostatic hypothesis of control: drivers continuously make real-time decisions to keep the perceived difficulty of the driving task within acceptable bounds – including, but not limited to, by adjusting their speed. In other words, an increase in the perceived difficulty of the driving task, such as when driving in snow or darkness, leads to a corresponding decrease in speed. A large proportion of drivers report driving more slowly than usual when task difficulty increases: for instance, in fog (98%), heavy rain (96%), or on unfamiliar roads (88%) (Campbell & Stradling, 2003). Conversely, when perceived task difficulty decreases, such as when roads are empty at night, driving speed tends to increase (Broughton, 2005).

The *transtheoretical model* of behavioural change (TTM) developed by Prochaska and DiClemente (1983) conceptualises behavioural change as a stage-based process, in contrast to the theory of planned behaviour, which views such change as a continuous process. According to TTM, behavioural change occurs through a series of successive steps. The model comprises *six stages of change* (see Figure 5.2):

1. Pre-contemplation – the individual has no intention of changing their behaviour.
2. Contemplation – the individual begins to recognise the existence of a problem and its consequences.
3. Preparation – the individual starts to prepare for change.
4. Action – behavioural change is already taking place, although a high risk of relapse into previous patterns remains.
5. Maintenance – it is only at this stage that the new behaviour becomes habitual.
6. Termination – the new behaviour is so firmly established that a return to former behavioural patterns is unlikely.

TTM offers an explanation as to why, despite its successful implementation among the target group, a new behaviour may ultimately fail to manifest itself. It also provides the opportunity for tailoring interventions to the varying needs of individuals along the different stages of change. The model emphasises the role of habit and behavioural automation, highlighting the

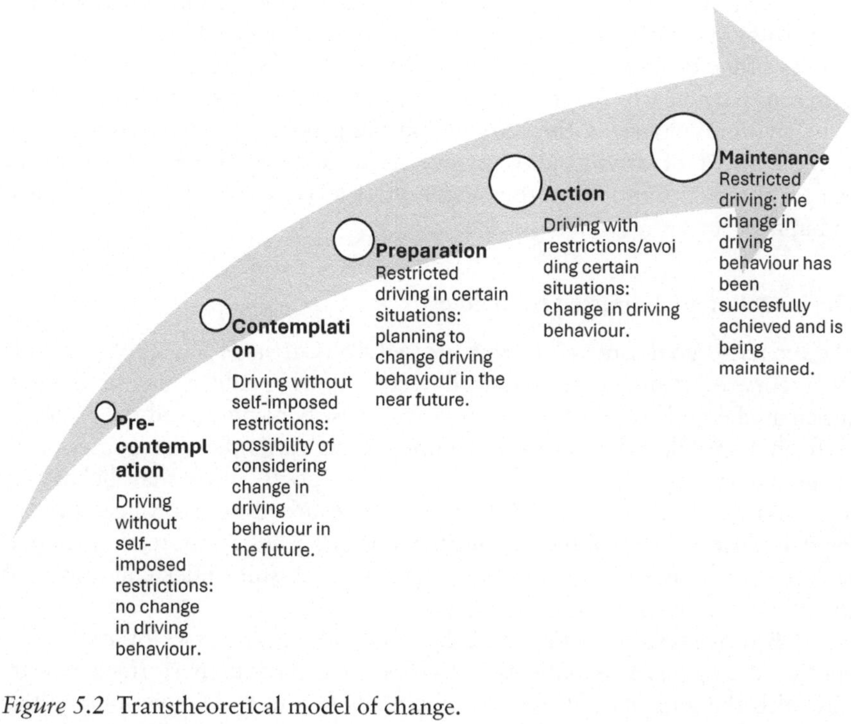

Figure 5.2 Transtheoretical model of change.

importance of the maintenance stage, which follows the desired behavioural change as a path towards the establishment of a new habit. For example, Forward (2014) empirically demonstrates that respondents' behaviour can be considered automatic in both the first and, subsequently, the final stage. Despite the benefits brought about by the change, it remains essential to account for the potential negative consequences of that change and allow sufficient time and space for the consolidation of the new behaviour, if it is to become permanent.

TTM focuses particularly on the development of strategies aimed at *increasing* an individual's *awareness*, with the ultimate goal of motivating them to change their behaviour in a specific direction. If an individual fails to recognise the importance of safety measures (e.g. unaware of the risks associated with drink-driving or texting while driving), they are unlikely to feel any need to modify their behaviour. This model thus acknowledges that, in most cases, behavioural change involves a lengthy trajectory, with individual decision-making processes playing a pivotal role.

However, TTM is not the only *stage-based model*. Another example is the *precaution adoption process model* (PAPM) by Weinstein (1988). PAPM includes both an *unaware* stage, in which individuals do not recognise their problem, and an *unengaged* stage, in which individuals may be fully aware of the problem but lack motivation to alter their behaviour (Weinstein, 1988). A frequent cause of this lack of engagement is the individual's low perception of personal vulnerability. A combination of multiple factors causes a large proportion of individuals to linger in the PAPM unengaged stage.

Like other behavioural theories, TTM has been subject to considerable *criticism*, particularly for its alleged artificial construction of distinct stages of behavioural change. Critics argue that the process of behavioural change in the context of driving may involve more discrete phases. Nonetheless, from a practical standpoint, the model offers a concept that is easy to grasp and apply.

The model of goal-directed behaviour

The *model of goal-directed behaviour* (MGB) (Carrus et al., 2008) extends the theory of planned behaviour by incorporating additional predictors: anticipated emotions (positive or negative), past behaviour, and the desire to perform a specific behaviour (see Figure 5.3). It views the core components of the theory of planned behaviour in terms of goals rather than behaviour (Chatzisarantis & Hagger, 2009). Desire is regarded as a stronger predictor than attitudes and subjective norms (Sutton, 1998). Suppressing routine (habitual) responses is difficult and often unsuccessful under conditions of cognitive load (Passafaro et al., 2014).

MGB thus seeks to understand decision-making processes by *emphasising the role of habit*. A habit is a decision or behaviour performed repeatedly with the aim of achieving specific goals (Aarts & Dijksterhuis, 2000; Verplanken et al., 1997). Through repeated enactment, the association

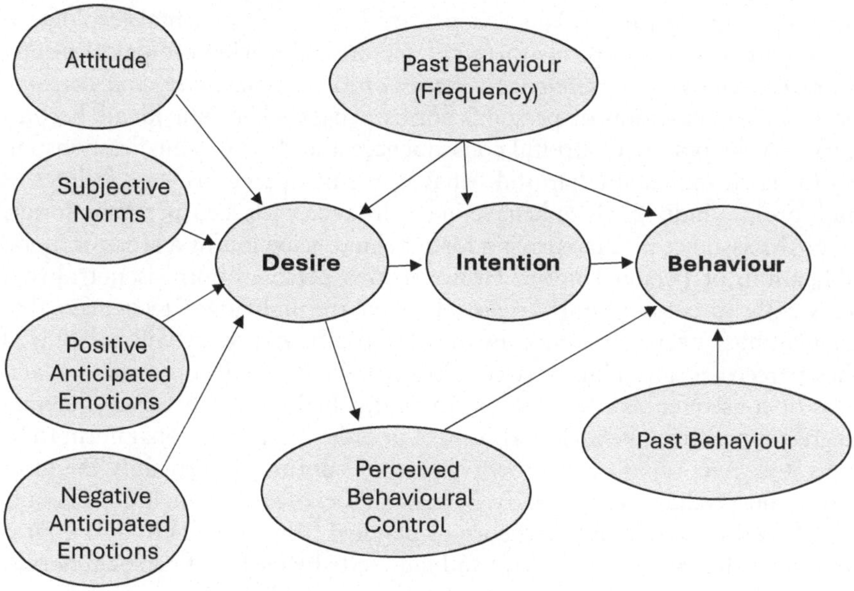

Figure 5.3 The model of goal-directed behaviour.

between the goal and the decision (or behaviour) becomes stronger (Wood et al., 2002). As habitual behaviour tends to require less time and cognitive effort, it may be assumed that individuals are less likely to monitor their decisions in goal-directed action consciously. Research has suggested that some long-term travel-related decisions may often be governed by unconscious habit rather than short-term daily motivations (Aarts & Dijksterhuis, 2000; Fuji et al., 2002; Verplanken et al., 1997; Murtagh et al., 2012; Friedrichsmeier et al., 2013). Concerning travel mode choice, this was confirmed by Aarts and Dijksterhuis (2000), who used the self-reported habit strength index in their study. The results indicate that travel purpose is closely linked to travel mode choice. Specific purposes or goals automatically triggered specific mode choices, and these habit-induced choices proved difficult for their participants to override. Their study thus demonstrated the automatic nature of travel mode choice, showing that such decisions are under the direct control of automatic associations with habitual behaviour and therefore involve less conscious deliberation. Since such decisions have become habitual over time, commuters, for example, are therefore less likely to change their preferred travel choices (Aarts et al., 1998; Verplanken et al., 1997).

Norm activation model

The *norm activation model* (*NAM*) (Schwartz, 1977) focuses on the processes through which moral norms are activated and transformed into action. These processes are used to explain altruistic and pro-environmental (ecological) behaviour. The model posits three key factors influencing ecological behaviour: *awareness of consequences, ascription of responsibility, and personal norms*. The activation of personal norms occurs when individuals become aware of the potential harmful consequences and their personal responsibility for environmentally harmful behaviour. Anticipated pride or feelings of guilt prompt individuals to behave in accordance with their personal norms.

NAM assumes that the driving force behind behaviour is a sense of moral obligation, or personal norm. However, this personal norm is not always active; the process is only triggered when the individual recognises that change must occur (becomes aware of a need). A person will act only if they perceive a causal link between their behaviour and its outcomes (awareness of consequences). In addition, the individual must experience a certain degree of perceived behavioural control in order for the personal norm to be activated. According to this theory, personal norms are probably the most important predictor of normatively oriented behaviour, though not the only one. Social norms (as in the theory of planned behaviour, TPB) also appear to shape behaviour, both directly and indirectly by means of personal norms. NAM (like TPB) assumes that perceived behavioural control is important for the enactment of behaviour, and that social norms are significant predictors of behaviour (Barradell, 2013).

Hunecke et al. (2001) developed a modified version of NAM, in which the personal norm is considered the central integrative variable in the process of pro-environmental behaviour (see Figure 5.4). It refers to the specific motivational basis for action, expressed through a sense of moral obligation. According to Schwartz, personal norms, unlike social norms, can be described as internal moral beliefs that are upheld regardless of others' expectations. For this reason, the behavioural relevance of personal norms is limited to behaviour that includes a moral dimension. Given the moral implications often associated with pro-environmental behaviour, reconstructing

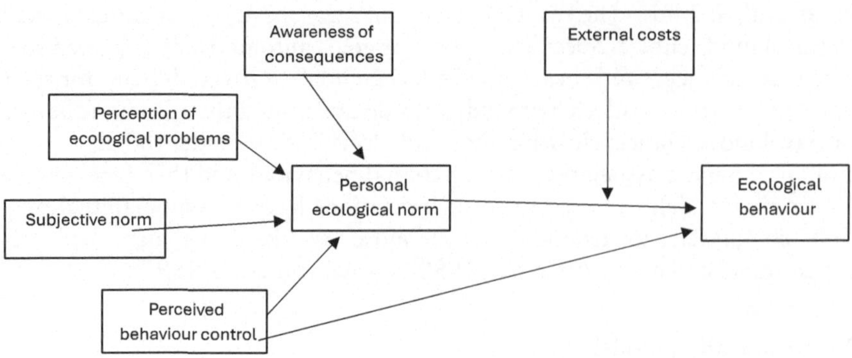

Figure 5.4 Modified norm activation model.

such behaviour through the lens of NAM enables a deeper insight into ecological action. Awareness of consequences is postulated as a predictor of personal ecological norms and represents a central variable derived from the original NAM. It is awareness of the consequences of one's behaviour that serves as a predictor of pro-environmental behaviour. Hunecke et al. (2001) confirmed that social norms play a crucial role in shaping personal moral obligations and travel mode choice. Finally, financial considerations and perception of environmental problems must also be taken into account (Blöbaum et al., 1997).

The *theory of reasoned action* (Fishbein & Ajzen, 1975), the *theory of planned behaviour* (Ajzen, 1991), and Schwartz's *norm activation model* (1977) in its later revision (Stern & Dietz, 1994) are often proposed as theoretical foundations in traffic psychology research. While these theories have much to offer in terms of goal setting and self-regulation, they do not capture the change processes *per se*.

Value-belief-norm theory

The *value-belief-norm (VBN)* theory (Stern et al., 1999; Stern, 2000) represents a synthesis of existing theoretical concepts of environmentally significant behaviour. According to VBN, pro-environmental behaviour is determined by three types of value orientations: egoistic, altruistic, and biospheric (Stern, 2000; Oreg & Katz-Gerro, 2006). People with strong *egoistic values* will carefully consider and prioritise the consequences of behaviour that affects their personal resources. In contrast, those with strong *altruistic values* will focus more on the consequences of their actions for others, while individuals with strong *biospheric values* will be greatly concerned about the impacts on nature and the environment. The VBN theory examines the relationship between these three value orientations and an ecological world view, that is, people's general beliefs about the relationship between humans and the environment. Recent studies suggest that it may be beneficial to expand the model to include a hedonic value orientation, whereby people focus on personal pleasure or satisfaction (Tolppanen & Kang, 2021). This leads to the definition of four types of values: hedonic, egoistic, altruistic, and biospheric (Tolppanen & Kang, 2021). Hedonic values may be particularly relevant in understanding motivations to reduce car use, which is often associated with hedonic benefits such as enjoyment of driving, comfort, and speed.

In general, values tend to predict behaviour indirectly by means of specific beliefs and personal norms (De Groot & Steg, 2008a; Nordlund & Garvill, 2003; Poortinga et al., 2004; Steg et al., 2005). The theory of environmental values, beliefs, and norms thus assumes that values influence behaviour through a norm activation process. More specifically, it posits that different types of behaviour (including car use and the acceptability of transport policies as a form of non-activist behaviour in the private sphere) are the result of personal norms, that is, feelings of moral obligation to engage in specific actions (e.g. Black et al., 1985; De Groot & Steg, 2009, 2008a; Harland et al., 1999; Schwartz, 1977; Steg & De Groot, 2010; Stern, 2000).

Comprehensive action determination model

The *comprehensive action determination model* (*CADM*) (Klöckner & Blöbaum, 2010; Klöckner & Friedrichsmeier, 2011) represents an attempt to integrate various theoretical approaches. This model of ecological behaviour synthesises notions laid down in the *theory of planned behaviour* (TPB), the *norm activation model* (NAM), the *theoretical concept of habit*, and the *ipsative theory of behaviour* (derived from Frey, 1988). Within this framework, individual behaviour is directly determined by three sources: intentional processes, situational influences, and habitual processes. Indirectly, it is shaped by personal and social norms, which exert their effects through intentional and habitual processes. All four sources interact with one another. Habitual and situational processes interfere with intentional processes, thereby attenuating the influence of intentions on behaviour. Perceived behavioural control activates normative and intentional processes, which in turn affect situational influences. Klöckner and Blöbaum (2010) compared this model with the theory of planned behaviour, the norm activation model, and a combination of the two. CADM has been shown to provide the most comprehensive explanatory power. In the field of transport, it is primarily applied to ecological behaviour, particularly the use of environmentally friendly modes of transport.

Specific applications of CADM in empirical research

CADM was tested using structural equation modelling on a sample of 389 university students in the context of travel mode choice. Its performance was compared with established less comprehensive models (specifically TPB, NAM, and a combination of the two). The results showed that all the models demonstrated good to excellent model fit and explained a substantial proportion of the variation in travel mode choice. However, CADM accounted for the highest percentage of variation (65%) (Klöckner & Blöbaum, 2010). Within the CADM framework, both subjective and objective situational constraints were responsible for the majority of variation in travel mode choice. Nevertheless, intentions and habits also played a significant role. The influence of social and personal norms was mediated by habits and intentions, while habits were found to moderate the relationship between intention and behaviour. This implies that, from an intervention perspective, altering situational conditions appears to be a promising strategy for modifying behaviour. In addition, normative, intentional, and habitual processes have the potential to counteract this modification and therefore need to be taken into consideration. An analysis of overall effects in the study revealed that the primary determinants of the students' travel mode choices were situational constraints (such as car availability), intentions, and habits.

(Klöckner & Blöbaum, 2010)

TPB assumes that travel mode choice is directly influenced by the intention to use a car (or not) and by perceived behavioural control (Bamberg & Schmidt, 2003). It further proposes that intention is generated by perceived behavioural control, social norms, and attitudes. Extensions of TPB that incorporate the concept of habit (e.g. Verplanken et al., 1997) show that car use habits directly influence travel mode choice and moderate the relationship between intention and behaviour. Bamberg et al. (2007) suggest that personal norms are an additional predictor of intentions, but not a direct predictor of behaviour. With reference to NAM, personal norms are assumed to be influenced by awareness of the need for change, awareness of consequences, and perceived behavioural control, as these are necessary to activate a sense of moral obligation (Hunecke et al., 2001). Moreover, social norms are expected to influence personal norms through their internalisation and adaptation to an individual's personal value system. CADM does not assume a direct effect of social norms on behaviour, as previous research into travel mode choice has not identified them as significant predictors (Bamberg et al., 2007; Bamberg & Möser, 2007; Klöckner & Matthies, 2009).

Risk Homeostasis theory

The *homeostasis risk theory* (*HRT*) (Wilde, 1988, 1994) was developed in response to empirical evidence indicating that safety measures are not as effective as had been anticipated on the basis of prior theoretical calculations (Mayou et al., 1991; Rumar et al., 1976; Geller et al., 1987; Klen, 1997); see Figure 5.5.

Empirical research insights

A number of studies have shown that the reduction in fatalities following the introduction of mandatory seat belt or motorcycle helmet laws was

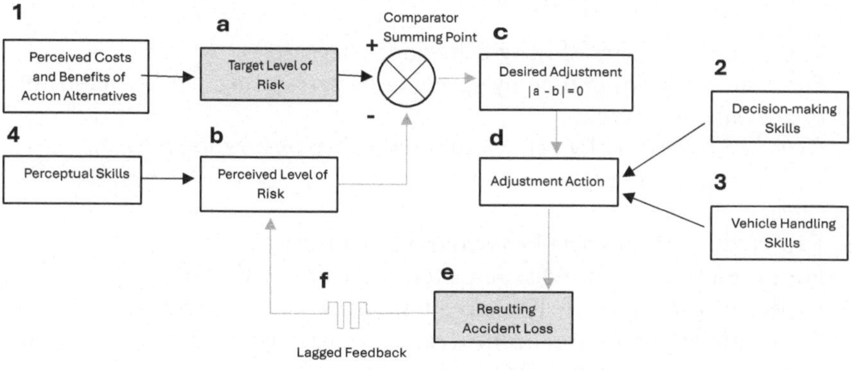

Figure 5.5 Risk homeostasis theory.

> generally smaller than anticipated on the basis of increased usage rates (see, for example, Jamieson & Kelly, 1973; Janssen, 1994; Robinson, 1996). The less-than-expected effect of seat belts was explained by the selective non-use of belts among certain segments of the driver population (Janssen, 1994). It was found, for instance, that the better or more advanced the equipment mountaineers use is, the greater the risks they tend to take (Drasdo, 1969). Mohr and Clemmer (1989) found that while an automatic tool changer in drilling machines reduced the number of work-related injuries, it also led to an increase in other types of accidents. According to Rumar et al. (1976), vehicles equipped with studded tyres were driven faster, yet drivers' safety margins were greater than when driving without studs, suggesting that the risk compensation was only partial. Taxis equipped with ABS in Munich were driven at higher speeds and were more frequently involved in crashes than other vehicles (Biehl et al., 1987). Streff et al. (1987) reported that drivers who had previously neglected to wear seat belts drove faster once they began to use them. An experimental study (Janssen, 1994) found that beginner seat-belt wearers showed signs of continuing behavioural adaptation, such as increased speed or a tendency to follow other vehicles more closely. It therefore appears that, paradoxically, even after the introduction of protective measures, the level of risk tends to remain relatively constant. This is why such theories are referred to as homeostatic.

According to this theory, every individual has their own personal risk threshold. Therefore, if the risks involved in the transport system are reduced, drivers will respond by engaging in riskier behaviour, or will take on additional risks, thereby maintaining their individual level of risk. The degree of risk that different people are willing to take during any given time period and during different time periods is determined by a variety of factors (Wilde, 2014).

If the expected benefits of risky behaviour are high and the anticipated costs are perceived as relatively low, the target level of risk will be high. The term target is used here synonymously with the "preferred, desired, accepted, tolerated, and subjectively optimal" level of risk that individuals are willing to take when driving. The target level of risk is determined by four categories of motivational (i.e. subjective utility) factors:

1. Expected *advantages* of risky behaviour alternatives:
   For example, gaining time by speeding or performing a risky manoeuvre to alleviate boredom.
2. Expected *costs* of risky behaviour alternatives *on a rational level*:
   For example, car repair expenses or increased insurance premiums for causing an accident.
3. Expected *benefits* of safe behaviour alternatives:
   For example, insurance discounts for accident-free driving.
4. Expected *costs* of safe behaviour alternatives *on a personal level*:
   For example, discomfort from wearing a seat belt or peer pressure that may label one a "wimp" for buckling up.

The higher the values in categories 1 and 4, the higher the target level of risk. Conversely, the target level of risk decreases as values in categories 2 and 3 increase. Some of the motivational factors are economic in nature, while others are cultural, social, or psychological. Typically, these factors are so internalised that individuals are unaware of them when choosing a particular behaviour.

It is also assumed that the mechanism by which a driver achieves equilibrium includes three *distinct behavioural choices* (Wilde, 1988). When there is a change in the level of environmental risk, the individual exposed to the risk may first respond by modifying their behaviour within the given environment. In road traffic, this may include driving faster or more slowly, overtaking less frequently, changing overtaking speed, and increasing attention.

The second way of achieving homeostasis can be referred to as a "shift in mobility mode", or a transition from one travel mode to another. For example, in adverse weather conditions, a motorcyclist may decide to travel by train to avoid the risk of a crash. Finally, if the target (i.e. acceptable) level of risk and the actual "objective" level of risk are in balance (e.g. during a windstorm), the individual may choose to stay home and not to undertake any journey at all.

The model is based on a *mathematical formula* that theoretically determines the degree of behavioural adaptation a driver exhibits in response to improved safety in their environment. Behavioural adaptation, or risk compensation, refers to the anticipated changes in individual behaviour following the introduction of safety measures, while it is assumed that these behavioural changes reduce the potential safety benefits of the interventions (Janssen, 1994; Wagenaar, 1992; O'Neill, 1977; Wilde, 1982; Wilde & Murdoch, 1982).

However, HRT has been criticised for various reasons, particularly the vague definition of the risk construct and the contradiction between "perceived" (subjective) risk and "actual" (objective) risk (Michon, 1989). Moreover, Ajzen (1991) argued that Wilde's theory does not fundamentally differ from the principles of cognitive self-regulation, which Ajzen formulated in the context of behavioural prediction. Indeed, the theory of planned behaviour also assumes that people weigh the costs and benefits of their actions and the utility of their outcomes. Other criticisms concern the vague nature of the theory itself; it is unclear whether risk homeostasis occurs at the individual or community level. Similarly, Wilde's theory is ambiguous in the time frame it assumes to be required for perceived risk to stabilise. While it is beyond doubt that perceived risk is just one of the factors underlying behavioural change, the empirical evidence upon which HRT is based remains valid. The shortcomings of HRT lie more in the fact that some of its aspects are insufficiently elaborated, rather than in the theory itself being fundamentally flawed.

The theoretical model of danger compensation

The *theoretical model of danger compensation* (*TMDC*) (O'Neill, 1977) incorporates elements from several models: the *risk homeostasis theory*, the *theory of travel utility maximisation*, and *behavioural feedback models*. The model of danger compensation assumes that the driver behaves as a rational person, that is, someone with relatively *stable goals* who seeks to *maximise the overall expected value of their actions*. As with HRT, the model is based on a mathematical formula which can be used to determine theoretically the degree of compensation a driver exhibits in response to an improvement in the safety of their environment. Presuming that all drivers act as rational and informed human beings, the theory predicts that all drivers should either increase or reduce crash rates in response to enhanced safety. However, it can be difficult to support such a broadly formulated assumption with tenable arguments.

Subjective versus objective safety theory in transport

Subjective versus objective safety theory (SOST) (Sørensen & Mosslemi, 2009) relates to individuals who *do not feel safe in traffic*, or in more general terms, to their concerns that they may pose a danger to themselves or other road users. A subjective sense of unsafety in traffic can cause road users to *limit their mobility and social activities*. However, research has shown that the link between objective and subjective traffic safety is very weak. Some studies (Menkehorst et al., 1987; Summala & Näätänen, 1988; Fuller, 2008) even suggest that a certain level of perceived danger may have a desirable effect in the form of more alert and therefore safer behaviour. In such cases, locations may become objectively safer precisely because people do not feel safe. It is also possible that people selectively avoid subjectively unsafe situations – for example, by choosing a different route, travelling at a different time, or opting for another travel mode. In extreme cases, this may even lead to people giving up driving altogether. In such instances, subjective safety affects freedom of movement, which is generally considered undesirable. However, it may benefit objective safety in those locations, as lower traffic volumes lead to fewer crashes – the positive correlation between subjective and objective safety thus decreases.

Empirical evidence

Dutch research shows that feelings of danger typically arise when there are frequent speeding violations in 30 km/h zones and where there is heavy traffic or complex and unclear traffic situations develop in such sections (Sørensen & Mosslemi, 2009; Van Haaf, 2002; Vlakveld, 2008). Findings from other fields exploring subjective and objective safety (e.g. criminology) also support the weak correlation between the two. This implies that situations perceived by people as dangerous are not necessarily objectively dangerous, i.e. not the ones in which traffic crashes actually occur. Earlier research indicated that only half of objectively dangerous situations were also perceived as such (Menkehorst et al., 1990). The role of subjective safety varies greatly among different road users. Some experience little or no fear in traffic, while others find the very thought of participating in traffic so intimidating that they choose to give up driving.

Sørensen and Mosslemi (2009) identified a number of factors that influence people's sense of subjective safety in Norway. These include traffic volume (greater traffic density reduces perceived safety), speed (higher speed reduces perceived safety), and the presence of heavy vehicles (increased presence lowers the sense of subjective safety). For the complete list, see Sørensen and Mosslemi (2009). They also identified 13 traffic safety measures which they referred to as ambivalent in that they may either increase or reduce a sense of subjective and/or objective safety or their effects are unclear. These were further categorised into four groups: infrastructure for vulnerable road users (e.g. cycle lanes), improvements to road cross sections (e.g. lane width), cyclist equipment (e.g. helmets and reflective materials), and regulation of heavy vehicles (e.g. speed limits and weight restrictions).

The hierarchical risk model

The *hierarchical risk model* (*HRM*) (Van der Molen & Bötticher, 1988) for traffic participants was developed in response to inconsistent empirical findings from some of the previously described risk theories. The key issue addressed by this approach is that drivers adjust their behaviour across different strategic layers on the basis of environmental factors and psychological processes. The attempt to explain driver behaviour in terms beyond mere risk reduction led to the development of a body of "hierarchical" theories. Van der Molen and Bötticher's (1988) hierarchical model of risk for traffic participants explicitly characterises the hierarchy of driver behaviour in terms of three traffic task levels: strategic, tactical, and operational. The strategic level covers activities such as route planning, the choice of transport mode, and the desired travel speed. The tactical level involves more specific manoeuvres, such as overtaking. Actions selected at the tactical level are completed at the operational level, through physical driving inputs such as braking and emergency reactions (see Figure 5.6).

This theory makes it possible to *calculate behavioural alternatives* and the subjective aspects of their probability and utility. It also describes the perceptual, judgemental, and decision-making processes involved in the driving task, taking into account subjective correlates of outcome probabilities and values. Importantly, it explicitly distinguishes risk judgement from other types of judgement. Perception of the physical environment is continuously influenced by an internal representation of similar situations. It also contains knowledge about one's own limitations and capabilities, as well as interactions with the environment. Internal representation influences evaluative processes through

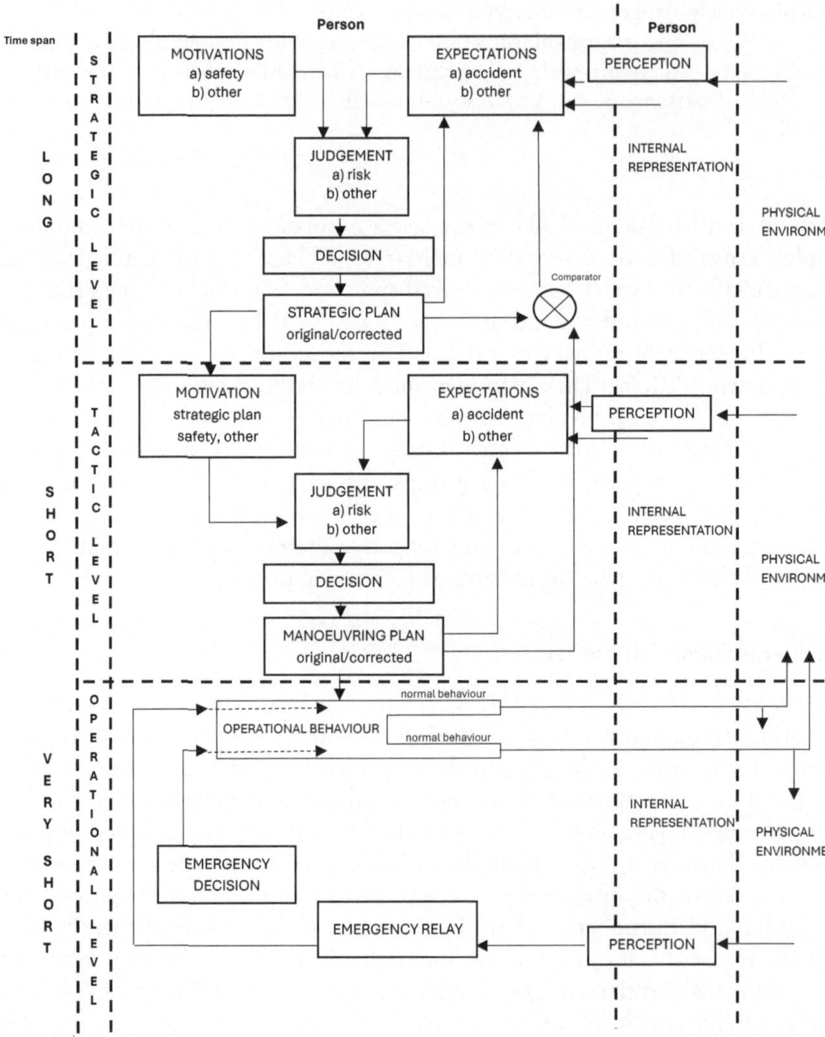

Figure 5.6 Hierarchical risk model.

expectations. A driver's motivations determine the subjective importance of possible outcomes associated with behavioural alternatives.

However, the model is only a *"meta-model" that provides no explanation for the processes occurring within the structure as described.* It appears to assume that in equally attractive situations at the three different levels a driver will not choose a riskier alternative if a lower-risk option is available. Finally, its authors define interrelationships and outline processes while acknowledging that their model is unsubstantial, as they do not specify how the aforementioned processes actually work.

The GDE matrix (guidelines/goals for driver education)

The *GDE* (Hatakka et al., 2002) represents the first comprehensive and holistic approach defining the core focus areas for traffic psychology and driver training. It outlines key directions for the future development of driver education and training. As a body of knowledge titled *Goals for Driver Education*, the GDE summarises the essential components and levels of driver behaviour in order to make driver education more effective (in reducing the crash risk among novice drivers, for example). The GDE introduces a four-level descriptive model in which *driver behaviour is conceptualised as a hierarchy*, with the *driver's goals and motives playing a central role*. The most recent constructivist ideas from mainstream pedagogy and the psychology of learning are combined with a hierarchical approach to driver behaviour. Two key conclusions can be drawn from this comprehensive framework for the goals and content of driver education. First, a conceptual analysis highlights the need to place greater emphasis on motivational aspects in driver education – more so than is currently the case. Second, pedagogical methods need to be re-evaluated in order to achieve the intended goals. For instance, active learning methods and the use of self-reflection should be encouraged in driver instruction.

The GDE framework proposes *four levels of driving competence*: (1) Vehicle control – managing the speed, direction, and positioning of the vehicle; (2) Mastery of traffic situations (driving in traffic) – adapting to the demands of specific traffic contexts; (3). Goals and context of driving – trip purpose, driving environment, social and societal context; (4). Goals for life and skills for living – self-regulation, the meaning of driving and the car in one's personal development. Vehicle control and managing traffic situations constitute the fundamental skills for safe participation in traffic and should be thoroughly taught during driver training. At each level of the hierarchy, there are *potential risks* that the driver must be able to identify – for example, risks associated with road types, personal motives, or behavioural tendencies. However, the nature of these risks varies depending on the level of the hierarchy (see Table 5.1).

The core idea of the matrix structure is that driver training should aim to progress *from the bottom left-hand corner to the top right-hand corner*. The matrix identifies all the factors essential for safe driving. Domains such as introspective competencies, personal motives, and self-critical thinking can be better understood and acquired in practice if a psychologist facilitates the learning process by highlighting their importance and implications for road safety (Hatakka et al., 2002).

It should be noted that the GDE is not a static model but one characterised by a high degree of dynamism, which must be taken into account when developing and evaluating driver education programmes (Hatakka et al., 2002). Research suggests that the GDE model is currently not being utilised as effectively as it could be in the design and implementation of professional education and training for young drivers (Rodwell et al., 2018; Watson-Brown

Table 5.1 GDE matrix

	Knowledge and skills	*Risk-increasing aspects*	*Self-assessment*
IV. Goals for life and skills for living	"I" Lifestyle, age, group, culture, social situation, etc. related to driving behaviour	"K" Sensation-seeking Risk acceptance Group norms Peer pressure	"L" Introspective competence Personal preconditions Impulse control
III. Goals and context of driving	"G" Choice of travel mode Choice of travel time Role of motives Route planning	"H" Alcohol, fatigue Low friction Rush hours Young passengers	"I" Personal motives Self-critical thinking
II. Driving in traffic (identification and management of traffic situations)	"D" Traffic rules Cooperation Hazard perception Automatised reflexes	"E" Non-compliance Close-following Low friction Vulnerable road user	"F" Assessment of driving skills Personal driving style
Vehicle control	"A" Vehicle functioning Protection systems Vehicle handling Physical laws	"B" Non-use of seat belts Failure of vehicle subsystems Worn tyres	"C" Assessment of car control skills

et al., 2018; Mirman & Kay, 2012). Finally, the GDE also offers at least a basic logical framework for guiding and educating both parents of learner drivers and professional driving instructors.

Graduated driver licensing

Graduated driver licensing (GDL) (Vaa et al., 2015) represents a practical application of the GDE model. Its aim is to establish a system in which young drivers, immediately after becoming licensed, are subject to certain restrictions for a defined period (typically two years). These restrictions apply to driving activities considered particularly high-risk, such as night-time driving or carrying passengers. The approach is supported by findings that drivers aged 16–19 have a nine to ten times higher crash risk compared to drivers aged 55–64. Proposals for graduated driver education systems began to emerge in the 1980s. Within the European Union, the system was implemented during the 1990s.

The generic transactional model of driving stress and fatigue

The *generic transactional model of driving stress and fatigue* (Matthews et al., 2005) defines the role of personality in processing demands arising from the traffic environment. In this context, personality reflects stable self-knowledge stored in long-term memory, encompassing beliefs, attitudes, and explanatory tendencies used by the driver when interpreting the traffic environment. Individual differences in this self-knowledge are mediated by a constellation of *five personality traits relevant to the driving context: aggression, dislike of driving, hazard monitoring, fatigue proneness, and thrill-seeking* (Matthews, 2002).

Dislike of driving is associated with driving-related anxiety and an absence of enjoyment of driving. In extreme cases, it may take the form of driving phobia. Aggression is related to feelings of anger and frustration and a willingness to engage in risky behaviours (such as tailgating or overtaking). In its extreme form, it may manifest itself as *road rage*. Hazard monitoring refers to searching actively for danger and threats, even when driving has become routine. Fatigue proneness denotes a vulnerability to symptoms of physical and mental fatigue. Thrill-seeking involves deriving pleasure from risk and danger (Matthews, 2001).

Being linked with both negative emotional traits and a lack of confidence in one's driving competence, dislike of driving appears to be the most significant predictor of stress vulnerability (Dorn & Matthews, 1995). Drivers with high levels of dislike of driving tend to engage in emotional coping, self-criticism, and wishful thinking (Matthews et al., 2005). They also display a form of *depressive realism* (Alloy & Abramson, 1979), meaning they are able to estimate their own driving abilities more realistically, unlike the majority of drivers who tend to overestimate themselves and underestimate others (Dorn & Matthews, 1995). In addition to dislike of driving, aggression and fatigue proneness are key factors in stress vulnerability. This has been confirmed by both field and simulator-based studies (Matthews et al., 1998; Matthews, 2001, 2002).

Each trait predicts specific patterns of driving behaviour. For instance, aggression and thrill-seeking are associated with higher levels of accident involvement, traffic violations, and self-reported rule-breaking and driving errors. Dislike of driving correlates with self-reported driving errors but not with self-reported speeding, nor with an elevated accident rate. Dislike of driving, aggression, and fatigue proneness have a negative effect on cognitive performance and the execution of driving tasks. To assess these defined traits, the Driver Stress Inventory (DSI) was developed (Matthews et al., 1997; Matthews, 2002), which meets standard psychometric criteria (Matthews et al., 1997). In general, the transactional model suggests that interventions may be effective in targeting environmental factors such as traffic congestion, mitigating stress reactions, and enhancing drivers' abilities to cope with episodic stressors and challenges (Matthews et al., 2005).

Empirical findings

Drivers generally tend to overestimate their own judgement and skills and to underestimate the risk of a crash, especially in comparison to their evaluations of other drivers. However, drivers who score high on the dislike of driving scale, i.e. those prone to emotional difficulties, report similar evaluations of themselves and their peers. This suggests that stress-prone drivers display a form of "depressive driving stress realism" (Alloy & Abramson, 1979). While they may have a more accurate perception of their personal safety and driving competence, the cost of this realism may be increased vulnerability to emotional stress. Drivers with a high level of dislike of driving also tend to engage in emotion-focused coping, i.e. they use strategies such as self-criticism. Stress also exerts indirect effects: worry and prolonged use of emotion-focused coping are likely to overload attention and disrupt driving performance (see Matthews & Wells, 1999; Wells & Matthews, 1996). The behaviour of aggressive drivers may also provoke hostile reactions from other road users. This can in turn reinforce the negative social perception of aggressive drivers.

(Deffenbacher et al., 2003; Emo et al., 2004)

Many drivers show latent tendencies towards maladaptive coping styles in traffic situations. However, these inclinations only manifest themselves as dangerous behaviour when triggered by specific environmental conditions. For example, an aggressive personality becomes hazardous mainly when the driver is frustrated or obstructed by other drivers. Dislike of driving leads to cognitive interference especially when the driver's sense of vehicle control is threatened by poor driving conditions. Fatigue proneness appears problematic only during longer, more tiring journeys. Aggression and thrill-seeking are the strongest predictors of crash involvement, speeding, and rule violations, while dislike of driving shows the strongest association with driving errors. Similar associations have been observed using objective behavioural measures in driving simulator studies (Matthews, 2001; Matthews et al., 2005). Stress-related factors have also been linked to maladaptive driving behaviour in bus, taxi, and lorry drivers.

(Machin & Hoare, 2008; Öz et al., 2010)

The Diamond Model (a comprehensive model of the traffic system)

The *Diamond Model* (Risser, 2004) is another framework that seeks to capture, explain, and predict driver behaviour. In this model, Risser (2004) advocates a holistic approach to traffic psychology. The *Diamond Model* considers not only the driver but also other road users, and takes into account their mutual interactions. It also *incorporates the social dimension*, that is, *the culture and norms of a given society*, whether formal or informal (Chaloupka-Risser, 2005). In addition, the model includes infrastructure and, last but not least, the means of transport and its equipment and technical characteristics (Chaloupka-Risser, 2005). All the above aspects need to be taken into account and integrated with the relevant rules embedded within each part of the system. It is only by considering these interrelated aspects that a traffic psychologist can perform an effective assessment of an individual driver's fitness to drive. The ultimate goal is traffic safety (Chaloupka-Risser, 2005).

In relation to the Diamond Model, Chaloupka-Risser (2005) further emphasises that the role of psychological assessment is to find a balance between protecting society as a whole from dangerous drivers and safeguarding the interests of the individuals who wish to participate in motorised traffic. The model can be used as a theoretical warning system, an aid for identifying problems, and a conceptual framework for finding solutions.

The Diamond Model comprises the following segments:

- *Individual level*: Over time, individuals develop attitudes, values, thoughts, motives, and specific opinions that have been – and will continue to be – influenced by the various levels of the diamond. These factors may be modified throughout the process.
- *Interaction between road users and their experiences*: Individual behaviour in a public space depends on one's current communication history within specific forms of traffic participation. What have these personal experiences been like? (For instance, have there been feelings of inferiority?)
- *Infrastructure*: When road users encounter difficulties, what kind of support does the infrastructure offer to using different modes of transport? How easy and comfortable is it to use these travel modes? What types of behaviour and communication does the infrastructure promote?
- *Mode of transport*: What advantages and perceived disadvantages are associated with various modes of travel? What strain and frustration may a specific mode of travel involve? What needs can be met through particular modes of transport?

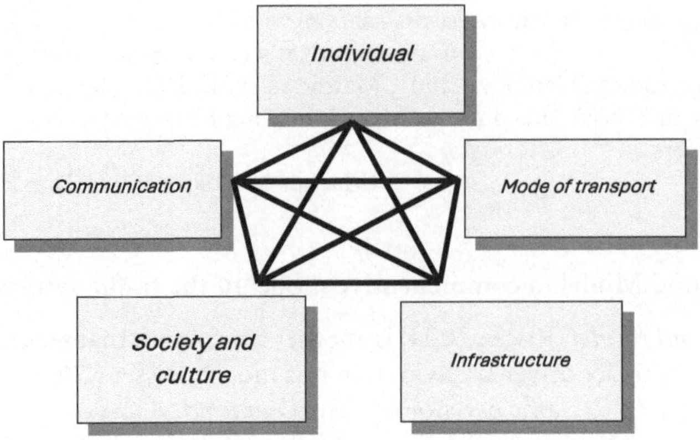

Figure 5.7 Diamond Model (Risser, 2004).

- *Society*: "Society" encompasses the media and the related public discourse – family, friends, one's social environment, and peer groups. How is the role of different forms of traffic participation perceived and discussed across these domains? All these become reflected in legislation, regulations, and informal norms (see Figure 5.7).

The Diamond Model can be seen as an attempt to capture the full range of factors shaping individual behaviour in traffic. As a model striving for a comprehensive description, it is necessarily general in its categories. Nonetheless, it provides the first multifaceted perspective on the individual within the transport system. It serves as a fundamental conceptual framework within which the individual layers can be further elaborated on, yet always with reference to the overarching structure.

References

Aarts, H., & Dijksterhuis, A. (2000). Habits as a means to the automatic pursuit of goals. *Journal of Personality and Social Psychology*, *78*(1), 53–66.

Aarts, H., Dijksterhuis, A., & De Vries, J. (1998). On the automatic pursuit of goals: The case of car use. *Journal of Environmental Psychology*, *18*(2), 177–187.

Åberg, L., Nygård, B., & Wiklund, M. (2008). Young drivers' attitudes toward traffic safety and risky driving: A longitudinal study. *Transportation Research Part F: Traffic Psychology and Behaviour*, *11*(1), 1–13.

Ajzen, I. (1985). From intentions to actions: A theory of planned behavior. In J. Kuhl & J. Beckmann (Eds.), *Action control: From cognition to behavior* (pp. 11–39). Springer.

Ajzen, I. (1991). The theory of planned behavior. *Organizational Behavior and Human Decision Processes*, *50*(2), 179–211.

Ajzen, I., & Fishbein, M. (1980). *Understanding attitudes and predicting social behavior.* Prentice-Hall.

Alloy, L. B., & Abramson, L. Y. (1979). Judgment of contingency in depressed and nondepressed students: Sadder but wiser? *Journal of Experimental Psychology: General*, *108*(4), 441–485.

Bamberg, S., Hunecke, M., & Blöbaum, A. (2007). The impact of beliefs on pro-environmental behavior: A meta-analysis. *Journal of Environmental Psychology*, *27*(4), 287–300.

Bamberg, S., & Möser, G. (2007). Twenty years after: A review of environmental attitude and behavior change. *Journal of Environmental Psychology*, *27*(1), 1–13.

Bamberg, S., & Schmidt, P. (2003). The theory of planned behavior and its relation to behavior change: A meta-analysis. *Journal of Environmental Psychology*, *23*(3), 297–313.

Barradell, S. (2013). The influence of moral and pro-social norms on car use. *Journal of Environmental Psychology*, *33*(1), 1–10.

Biehl, B., Echterhoff, W., & Fischer, K. (1987). *Road safety campaigns: A European perspective*. VTS.

Black, J. S., Stern, P. C., & Elsworth, R. A. (1985). *Personal and contextual influences on household energy use: A literature review*. National Academy Press.

Blöbaum, A., Hunecke, M., & Bamberg, S. (1997). The influence of perceived consequences on pro-environmental behavior. *Journal of Environmental Psychology*, *17*(4), 305–319.

Broughton, J. (2005). The effects of a public transport campaign on car use. *Transportation Research Part F: Traffic Psychology and Behaviour*, *8*(3), 221–234.

Campbell, M., & Stradling, S. G. (2003). The influence of public transport information on travel behaviour. *Transport Policy*, *10*(3), 237–248.

Carrus, G., Steg, L., & Van Der Laan, S. (2008). The role of attitudes and social norms in predicting pro-environmental behavior. *Journal of Environmental Psychology*, *28*(1), 1–11.

Chaloupka-Risser, R. E. (2005). Personality, emotion and risk-taking. *Journal of Environmental Psychology*, *25*(3), 321–331.

Chatzisarantis, N. L. D., & Hagger, M. S. (2009). Effects of self-determination theory and planned behavior theory on physical activity behavior: An integrative model. *Journal of Sport and Exercise Psychology*, *31*(3), 369–387.

Chorlton, K., Forsyth, A., & O'Regan, M. (2012). The effects of a fear-based campaign on young drivers' attitudes and intentions. *Transportation Research Part F: Traffic Psychology and Behaviour*, *15*(1), 1–12.

Deffenbacher, J. L., Oetting, E. R., & Lynch, R. S. (2003). Driving angry: A road rage study. *Journal of Applied Social Psychology*, *33*(2), 261–284.

De Groot, J. I. M., & Steg, L. (2008a). Value orientations to explain beliefs related to environmentally significant behavior: How to measure egoistic, altruistic, and biospheric value orientations. *Environment and Behavior*, *40*(3), 330–354. https://doi.org/10.1177/0013916506297831

De Groot, J. I. M., & Steg, L. (2009). The role of environmental values and norms in the relation between perceived consequences and pro-environmental actions. *Journal of Environmental Psychology*, *29*(4), 438–448.

Donald, I., Cooper, L., & Johnston, D. (2014). Predicting drivers' speed choices: A social cognitive model. *Journal of Environmental Psychology*, *38*(1), 1–10.

Dorn, L., & Matthews, G. (1995). Cognitive failures and driver stress. In *Human factors in traffic safety* (pp. 123–145). Psychology Press.

Drasdo, C. (1969). The effects of a safety campaign on road user behaviour. *Accident Analysis & Prevention*, *1*(1), 1–12.

Elliott, M. A., & Thomson, J. A. (2010). The effects of a speed camera campaign on driver attitudes and behaviour. *Transportation Research Part F: Traffic Psychology and Behaviour*, *13*(2), 89–101.

Elliott, M. A., Thomson, J. A., & Broughton, J. (2007). The effects of a public transport campaign on car use. *Journal of Transport Geography*, *15*(2), 127–138.

Emo, A., Oetting, E., & Deffenbacher, J. L. (2004). Driver anger, aggression, and the theory of planned behavior. *Journal of Applied Social Psychology*, *34*(5), 903–918.

Fishbein, M., & Ajzen, I. (1975). *Belief, attitude, intention, and behavior: An introduction to theory and research*. Addison-Wesley.
Forward, S. E. (2014). The influence of perceived risk and safety on car use. *Journal of Environmental Psychology*, *40*(1), 1–10.
Frey, B. S. (1988). *Modern institutional economics*. Basil Blackwell.
Friedrichsmeier, P., Klöckner, C. A., & Matthies, E. (2013). The role of habit in the theory of planned behavior: A meta-analysis. *Journal of Environmental Psychology*, *35*(1), 1–12.
Fuji, S., Gärling, T., & Kitamura, R. (2002). Time perspective and public transport use. *Transportation Research Part F: Traffic Psychology and Behaviour*, *5*(1), 1–14.
Fuller, R. (2008). *Towards a science of traffic and transport psychology*. Psychology Press.
Geller, E. S., Paterson, L., & Streff, F. M. (1987). Community-based intervention for promoting seat belt use. *Journal of Applied Behavior Analysis*, *20*(4), 387–394.
Harland, P., Staats, H., & Wilke, H. A. M. (1999). Explaining pro-environmental behavior with a model of goal-directed action. *Journal of Environmental Psychology*, *19*(3), 295–311.
Hatakka, M., Keskinen, E., & Katila, E. (2002). The effects of professional driver training on accident involvement. *Accident Analysis & Prevention*, *34*(1), 1–12.
Hunecke, M., Blöbaum, A., & Bamberg, S. (2001). Psychological determinants of pro-environmental behaviour: A review. *Journal of Environmental Psychology*, *21*(1), 1–12.
Jamieson, J., & Kelly, E. (1973). The effects of a safety campaign on driver behaviour. *Accident Analysis & Prevention*, *5*(4), 287–296.
Janssen, W. H. (1994). The effects of speed limits and enforcement on driver behaviour. *Accident Analysis & Prevention*, *26*(4), 481–492.
Klen, R. (1997). The effects of a public transport campaign on car use. *Transportation Research Part F: Traffic Psychology and Behaviour*, *1*(1), 1–12.
Klöckner, C. A., & Blöbaum, A. (2010). The influence of moral and pro-social norms on pro-environmental behavior. *Journal of Environmental Psychology*, *30*(2), 119–129.
Klöckner, C. A., & Friedrichsmeier, T. (2011). A multi-level approach to travel mode choice – How person characteristics and situation specific aspects determine car use in a student sample. *Transportation Research Part F: Traffic Psychology and Behaviour*, *14*(4), 261–277. https://doi.org/10.1016/j.trf.2011.01.006
Klöckner, C. A., & Matthies, E. (2009). The influence of attitudes and habits on environmental behavior. *Journal of Environmental Psychology*, *29*(1), 22–32.
Machin, M. A., & Hoare, P. R. (2008). The effects of driver training on road safety. *Accident Analysis & Prevention*, *40*(3), 1121–1132.
Matthews, G. (2001). *Traffic psychology*. Psychology Press.
Matthews, G. (2002). *Human factors in traffic safety*. Psychology Press.
Matthews, G., Desmond, P. A., & Neubauer, C. (1997). Driver stress, fatigue, and risk. In *Driving behaviour and traffic safety* (pp. 23–45). Psychology Press.
Matthews, G., Desmond, P. A., & Neubauer, C. (1998). Driver stress, fatigue, and risk. *Journal of Environmental Psychology*, *18*(2), 163–175.
Matthews, G., Emo, A. K., & Gilliland, K. (2005). Personality, emotion and risk-taking. *Journal of Environmental Psychology*, *25*(3), 321–331.
Matthews, G., & Wells, A. (1999). *The attention control theory of stress and anxiety*. Guilford Press.
Mayou, R., Ehlers, A., & Bryant, B. (1991). Psychological distress and driving. *Accident Analysis & Prevention*, *23*(3), 205–216.
Menkehorst, T., Janssen, W. H., & O'Regan, M. (1987). The effects of a safety campaign on driver behaviour. *Accident Analysis & Prevention*, *19*(2), 133–145.
Menkehorst, T., Vaa, T., & van der Molen, H. H. (1990). The effects of a safety campaign on driver behaviour. *Accident Analysis & Prevention*, *22*(3), 261–272.

Michon, J. A. (1989). A model of driver behaviour. *Accident Analysis & Prevention*, *21*(4), 311–322.
Mirman, S., & Kay, D. (2012). The effects of a fear-based campaign on young drivers' attitudes and intentions. *Transportation Research Part F: Traffic Psychology and Behaviour*, *15*(1), 1–12.
Mohr, D. L., & Clemmer, T. A. (1989). *Road safety campaigns: A critical review*. VTS.
Murtagh, N., Gatersleben, B., & Jackson, E. (2012). The role of habits in the theory of planned behavior: A meta-analysis. *Journal of Environmental Psychology*, *32*(1), 1–12.
Nordlund, A. M., & Garvill, J. (2003). The role of values and attitudes in pro-environmental behaviour. *Environment and Behavior*, *35*(6), 797–812.
O'Neill, B. (1977). The human factor in traffic safety. *Traffic Safety*, *77*(1), 1–10.
Oreg, S., & Katz-Gerro, T. (2006). Predicting pro-environmental behavior with a model of goal-directed action. *Journal of Environmental Psychology*, *26*(3), 223–236.
Öz, B., Van der Molen, H. H., & Buse, J. (2010). The effects of driver training on road safety. *Accident Analysis & Prevention*, *42*(1), 11–22.
Paris, J., & Van den Broucke, S. (2008). The effects of a public transport campaign on car use. *Journal of Transport Geography*, *16*(2), 127–138.
Passafaro, P., Cacciabue, P. C., & Borrion, H. A. (2014). The effects of a public transport campaign on car use. *Journal of Transport Geography*, *40*(1), 1–10.
Poortinga, W., Steg, L., & Vlek, C. A. J. (2004). The influence of perceived consequences on pro-environmental behavior. *Journal of Environmental Psychology*, *24*(3), 293–306.
Prochaska, J. O., & DiClemente, C. C. (1983). Stages and processes of self-change of smoking: Toward an integrative model of change. *Journal of Consulting and Clinical Psychology*, *51*(3), 390–395. https://doi.org/10.1037/0022-006X.51.3.390
Risser, R. (2004). Personality, emotion and risk-taking. *Journal of Environmental Psychology*, *24*(3), 321–331.
Robinson, P. A. (1996). The effects of speed limits and enforcement on driver behaviour. *Accident Analysis & Prevention*, *28*(1), 1–10.
Rodwell, D., Watson-Brown, H., & Mirman, S. (2018). The effects of a fear-based campaign on young drivers' attitudes and intentions. *Transportation Research Part F: Traffic Psychology and Behaviour*, *58*(1), 1–12.
Rumar, K., Hedman, L., & Bäckström, G. (1976). The effects of a public transport campaign on car use. *Transportation Research Part F: Traffic Psychology and Behaviour*, *10*(1), 1–12.
Schwartz, S. H. (1977). Normative influences on altruism. In L. Berkowitz (Ed.), *Advances in experimental social psychology* (Vol. 10, pp. 221–279). Academic Press.
Sørensen, M., & Mosslemi, M. (2009). The effects of driver training on road safety. *Accident Analysis & Prevention*, *41*(1), 1–12.
Steg, L., & De Groot, J. I. M. (2010). The role of environmental values and norms in the relation between perceived consequences and pro-environmental actions. *Journal of Environmental Psychology*, *30*(1), 1–12.
Steg, L., Dreijerink, L., & Abrahamse, W. (2005). The role of values, attitudes and norms in the relation between perceived consequences and pro-environmental actions. *Journal of Environmental Psychology*, *25*(3), 347–358.
Stern, P. C. (2000). New environmental theories: Toward a coherent theory of environmentally significant behavior. *Journal of Social Issues*, *56*(3), 407–424.
Stern, P. C., & Dietz, T. (1994). The value-basis of environmental concern. *Journal of Environmental Psychology*, *14*(1), 1–16.
Stern, P. C., Dietz, T., Abel, T., Guagnano, G. A., & Kalof, L. (1999). A value-belief-norm theory of support for social movements: The case of environmentalism. *Human Ecology Review*, *6*(2), 81–97.

Streff, F. M., Geller, E. S., & Fincher, J. M. (1987). The effects of a public transport campaign on car use. *Journal of Applied Behavior Analysis*, *20*(4), 387–394.
Summala, H., & Näätänen, R. (1988). *The psychology of the driver*. North-Holland.
Sutton, S. R. (1998). A critical review of the theory of planned behavior. *Journal of Applied Psychology*, *83*(6), 843–850.
Tolppanen, S., & Kang, J. (2021). The influence of moral and pro-social norms on car use. *Journal of Environmental Psychology*, *73*(1), 1–12.
Vaa, T., Menkehorst, T., & van der Molen, H. H. (2015). The effects of a safety campaign on driver behaviour. *Accident Analysis & Prevention*, *74*(1), 1–12.
Van der Molen, H. H., & Bötticher, A. (1988). The effects of a public transport campaign on car use. *Journal of Transport Geography*, *10*(1), 1–12.
Van Haaf, T. M. (2002). The effects of speed limits and enforcement on driver behaviour. *Accident Analysis & Prevention*, *34*(3), 321–331.
Verplanken, B., Aarts, H., & van Knippenberg, A. (1997). The role of habit in the theory of planned behavior: A meta-analysis. *Journal of Applied Social Psychology*, *27*(13), 1184–1205.
Vlakveld, W. P. (2008). The effects of driver training on road safety. *Accident Analysis & Prevention*, *40*(3), 1121–1132.
Wagenaar, W. A. (1992). *The driver and the road*. North-Holland.
Watson-Brown, N., Rodwell, D., & Mirman, S. (2018). The effects of a fear-based campaign on young drivers' attitudes and intentions. *Transportation Research Part F: Traffic Psychology and Behaviour*, *58*(1), 1–12.
Weinstein, N. D. (1988). The theory of optimistic bias. *Psychological Bulletin*, *103*(1), 31–43.
Wells, A., & Matthews, G. (1996). *Attention and emotion: A clinical perspective*. Erlbaum.
Wilde, G. J. S. (1982). *The theory of risk homeostasis*. North-Holland.
Wilde, G. J. S. (1988). *Risk homeostasis theory and traffic accidents*. VTRC.
Wilde, G. J. S. (1994). *Target risk: Dealing with the danger of death, disease, and damage in everyday decisions*. PDE Publications.
Wilde, G. J. S. (2014). The theory of risk homeostasis. *Journal of Safety Research*, *49*(1), 1–10.
Wilde, G. J. S., & Murdoch, S. (1982). *The theory of risk homeostasis*. North-Holland.
Wood, W., Quinn, J. M., & Kashy, D. A. (2002). Habits and the theory of planned behavior. *Journal of Personality and Social Psychology*, *83*(6), 1283–1296.

6 Risk-taking in traffic

Ralf Risser

Introduction

Road traffic safety remains a critical global health issue, with accidents standing out as a leading cause of death, particularly among young individuals. While technological advancements and engineering solutions are crucial for enhancing safety, human behaviour is widely acknowledged as the primary factor in the vast majority of traffic accidents. A central concept for understanding driver behaviour and accident causation is risk-taking. Risk-taking in traffic is not a simple concept; it is shaped by a complex interplay of individual psychological factors, the immediate traffic environment, broader societal and cultural contexts, and the characteristics inherent in the chosen transport mode (Risser & Šucha, 2020a). This chapter, which draws upon the source materials provided, delves into the multifaceted nature of risk-taking in traffic, examining how communication among road users, societal preconditions, infrastructure, and the chosen transport mode contribute to or mitigate risky behaviours.

Understanding risk-taking in traffic behaviour

In traffic psychology, risk-taking refers to engaging in actions while perceiving elements of uncertainty or the possibility of loss (Lund & Rundmo, 2009). This concept is strongly linked to accidents, especially among young drivers, on the basis of the premise that drivers who act as they do despite being aware of the potential risks are more likely to cause accidents. However, the sources highlight that individuals may indulge in outwardly similar "risky" behaviours for different underlying reasons. Some might simply be unaware of the danger, potentially because of issues with hazard perception. Others deliberately take risks for various motives, such as seeking thrills (Lund & Rundmo, 2009).

Nearly all traffic behaviours involve some level of risk-taking because of the inherent uncertainties compared to other daily activities (de-Juan-Ripoll et al., 2021). A significant characteristic of traffic situations is high time pressure, which influences risk-taking. Hazard perception, the ability to foresee

DOI: 10.4324/9781003608394-6

and detect potentially dangerous objects or situations, is essential for safe driving. However, even with good hazard perception, drivers evaluate the risk on the basis of their perceived driving skills in relation to the perceived hazards. If drivers believe their skills are sufficient to handle a challenging situation, the risk might be judged as low, even if the hazards are perceived correctly. Crucially, risk-taking behaviour is also significantly influenced by risk utility. Even when a risk is perceived as high, if the potential gain or "utility" – such as saving time or gaining admiration from others – is judged more important, the risk is likely to be accepted. Positive reinforcement, such as shortening travel time by speeding, rewards high-risk behaviour directly, making it more likely to be repeated. Human motivations for seeking risk utility are varied and include stress release, improved travel efficiency, seeking arousal, expressing independence, rebellion, and gaining peer admiration. Sensation-seeking, defined as an individual trait involving the pursuit of varied, novel, complex, and intense sensations and a willingness to take risks for such experiences, is particularly relevant to traffic behaviour (Grace et al., 2019). It has been explicitly linked to risky driving, especially among novice drivers. Studies indicate that a high percentage of novice drivers exhibit high levels of thrill- and adventure-seeking (Lund & Rundmo, 2009). Understanding risk utility is of paramount importance because improving risk perception alone may not be sufficient to reduce risk-taking if the perceived benefits of the risky behaviour outweigh the perceived risks (de-Juan-Ripoll et al., 2021). According to these authors, risk-taking is a component of the decision-making process in situations involving uncertainty where the probability of outcomes (rewards/consequences) is known. Theories about risk-taking distinguish between perceptual bias, which involves inaccurate assessments of situations, and proneness bias, also known as risk proneness, which is an individual's propensity to be attracted to potentially risky activities. Both risk perception and risk proneness affect risk-taking behaviours. Personality, sensation-seeking, and impulsivity traits are identified as proneness biases influencing risk-taking, as are locus of control, emotion regulation, and executive control.

Other theoretical models attempt to explain how broader safety measures interact with driver behaviour. Risk compensation suggests that participants in traffic might counteract the benefits of safety improvements by engaging in riskier behaviour, such as increasing speed on an improved road. Wilde's Risk Homeostasis Theory (RHT) posits that individuals maintain a target level of risk for activities and adjust their behaviour to keep this level constant (Wilde, 1989). According to this theory, accident numbers would only decrease if safety measures lowered this target risk level itself. However, RHT is often criticised for assuming that drivers can accurately estimate the safety merits of measures, which is unlikely (Robertson & Pless, 2002).

The influence of communication among road users

Effective communication is fundamental for the smooth, safe, and comfortable cooperation of all actors in traffic (Chaloupka & Risser, 2020). Accidents are frequently the result of problematic behaviour and interactions between different road users, rather than technical defects. Road user behaviour should be viewed as social behaviour within a complex social context, not just as isolated individual actions. Communication in traffic serves multiple essential functions: allowing users to interpret others' behaviour, making their own intentions clear, and using signals to improve safety and flow. This includes adherence to both formal traffic rules (laws) and informal rules or signals (actions and gestures) (Färber, 2016; Risser, 1985, 1988). Problems arise when different road users operate on the basis of conflicting formal or informal rules. Misunderstood or misleading communication, whether official or non-official, is a risk factor for accidents and can lead to errors or conflicts and potentially dangerous situations (Hydén, 1987). For example, a car driver might expect a pedestrian to yield on the basis of an informal rule ("pedestrians usually let car drivers pass here"), while the pedestrian expects the driver to yield on the basis of a formal rule ("car drivers have to stop and let pedestrians pass at a pedestrian crossing"), creating potential conflict (Šucha et al., 2017).

Vehicle speeds have a significant impact on situations requiring communication and coordination; quick and appropriate reactions to unexpected moves are easier at lower speeds (Risser, 2004). The traffic environment has unique characteristics that affect communication, including speed, anonymity, wide discrepancies in user experience and skill, subtle forms of communication open to multiple interpretations, and a mix of written and unwritten rules. Compared to everyday communication, traffic much more often lacks immediate social feedback, which is particularly important for vulnerable road users (VRUs) such as children and seniors, who rely on unambiguous communication (Chaloupka & Risser, 2020). The rise of increased car automation introduces new communication challenges. Generally, with increasing vehicle automation, interpersonal communication is expected to decrease, while digital communication will increase. While this could potentially improve safety and efficiency, experts also identify risks, particularly misunderstandings between automated vehicles and VRUs, especially during the transition phase with mixed traffic. In manual driving, informal communication, such as eye contact between drivers and pedestrians and cyclists, is crucial for coordinating actions and ensuring safety in ambiguous situations (Chaloupka-Risser & Füssl, 2017). With distracted or absent drivers in automated vehicles, this vital source of information is limited. This necessitates a transformation in how vehicles and VRUs communicate. External human–machine interfaces (eHMIs) are being explored as a way to compensate for this lack of implicit communication and enhance the feeling of safety for VRUs interacting with automated vehicles (Hensch et al., 2020).

Another type of communication is meant and represented by Road Safety Communication Campaigns. They are a key tool for influencing road user behaviour, with the goal being to reduce risk-taking, among other things (Ulleberg et al., 2009; Redshaw et al., 2008). The core assumption is that improving communication with road users is necessary to change their risk-taking behaviour. This requires understanding and considering the motives, needs, and interests that steer (risk-taking) behaviour and act as barriers to change. Effective campaigns need to tailor messages to target groups on the basis of this understanding to enhance personal relevance and promote mental processing of the message. Strategies can include using personal communication, addressing the risk of harm, highlighting cognitive dissonance, and employing emotional appeals. The "Yellow-Card" campaign in Victoria, Australia, is cited as an example using social communication, allowing people to express disappointment about dangerous driving to friends, rather than relying solely on official sources (Faulks, 2011).

Societal preconditions and social context

Road users are members of society, and their behaviour is profoundly shaped by a complex social context. Societal preconditions and structures influence how individuals interact within the traffic system and perceive risks (Rothe, 1992; Šucha, 2019). Risk perception is influenced by media focus, topics of discussion, cultural norms, and the technical/legal opportunities for risk control (Maurer et al., 2016). Social factors, such as peer pressure and subcultures, can have a significant impact on risk-taking behaviour. Peer passenger influence is also specifically linked to risky driving among novice drivers (Lewis, 1985). Worldwide, traffic accidents are the leading cause of death for young people aged 15–29 years. This highlights a specific demographic whose behaviour is particularly relevant and often associated with risk-taking. For instance, in some subcultures, drink-driving might be viewed as acceptable, providing a strong extrinsic reward (social respect) that makes changing behaviour more difficult (Porter, 2011; Vankov & Kaufmann, 2017).

Cultural differences also play a significant role in shaping perceptions of risk and attitudes towards traffic safety (Lund & Rundmo, 2009). Studies comparing risk perception in different countries (e.g. Norway and Ghana) have found differences that are potentially linked to varying levels of exposure to daily risks and the struggle for survival. People in developing countries, who are often more exposed to threats such as disease, poverty, or food insecurity, may be more sensitive to risks in general. Conversely, familiarity with a hazard reduces the perception of risk, meaning something seen as highly risky in a high-income country might be viewed with equanimity in a developing one, and vice versa. It is noted that low-income countries account for a disproportionately high percentage of global traffic fatalities, and the types of accidents differ, with many fatalities involving pedestrians and passengers in public transport (Porter, 2011). This strongly suggests that effective road safety strategies, including campaigns, must be adapted to specific cultural and societal contexts (Lagarde, 2007; Muhlrad, 1987).

Dialogue with decision-makers is crucial for enhancing public acceptance of safety measures and avoiding misleading information. The participation of road users in a systematic communication process is highlighted as a key issue for behaviour change and the acceptance of measures that should reduce risk-taking (Ulleberg et al., 2009). Car drivers' negative reactions to speed limit reductions indicate the need to understand community acceptance before implementation (Fink, 2010). In this sense, road safety strategies can potentially be developed jointly by government and the public through collaboration tools, among other things by making use of the concept of social marketing (e.g. Lee & Kotler, 2015). This approach applies marketing principles, viewing road safety as a "product" to be "sold" to citizens. This requires understanding citizens' motivations, preconditions, and the perceived benefits and costs associated with adopting safer or more sustainable behaviours. A holistic approach combining information, product design (making avoiding risks feasible), communication, incentives, and effective distribution of product and communication elements is considered necessary for success. Examples include insurance bonuses for accident-free periods of driving.

In a "reflexive society", citizens may seek information from various sectors to form their own opinions on road safety, rather than simply accepting explanations from authorities (Fink, 2010). The public's sensitivity to accidents involving new technologies such as autonomous driving underscores the need for trust-building through open and early communication about potential risks (Hensch et al., 2020). Stakeholder involvement (drivers' associations, media, and consumer groups) is vital for societal acceptance (Litman, 2007; Risser & Šucha, 2020a).

The role of infrastructure

Road infrastructure significantly influences driver behaviour and safety, and it plays a direct role in the communication between road users. The design of roads, intersections, and pedestrian crossings affects how road users interact and perceive risk. For example, a good intersection design can enhance the visibility of cyclists, improving communication and potentially reducing critical events involving VRUs (Chaloupka-Risser & Füssl, 2017). Changing the immediate environment can be an effective way to reduce risky behaviour (Šucha et al., 2017). However, changes in infrastructure can and should be deliberate actions intended to influence road user behaviour in this sense. Narrowing roads, for instance, aims to reduce car speeds, effectively acting as a form of communication to drivers (Theeuwes & Van Der Horst, 2017). At the same time, infrastructure measures meant to improve safety can trigger risk behaviour, such as drivers increasing their speed on a seemingly safer road that is straight, broad, and smooth. A road that looks like a motorway will be used like a motorway, and low speed limits will not be respected if the infrastructure does not credibly reflect the necessity for lower speeds. This highlights the complex interaction between infrastructure design and behavioural responses. Specific road characteristics have been empirically linked to accident rates (Andersen et al., 2013). These include factors such as annual daily traffic, carriageway width, the presence of hard or soft shoulders, the density of intersections and road access points, and the extent of road markings. Studies analyse the relationship between these characteristics and accident density to identify hazardous locations. Understanding these links is essential for designing infrastructure that promotes safer behaviour and reduces opportunities for risk-taking.

Infrastructure near pedestrian crossings should be designed to meet pedestrians' needs for feeling safe and comfortable, in addition to ensuring objective safety (Risser & Šucha, 2020b). It is argued that adjusting the public space for easy use by the weakest groups (such as children and seniors) may benefit the whole population. They will be less at risk. Most of all, reducing risky behaviour by car drivers will benefit them (Porter, 2011; Ulleberg et al., 2009; Risser & Šucha, 2020b).

Transport mode and risk

The chosen transport mode inherently involves different types and levels of risk, both for the individual user and for other road users (Nilsson, 2004; Shinar, 2017). This risk can be differentiated between the risk one undergoes oneself and the risk one constitutes for others. A significant theme when discussing risks should be a shift towards more sustainable transport modes, particularly walking and cycling, to address global warming and mitigate associated ecological risks (Toth-Szabo et al., 2011). This transition often involves moving away from individual car use, thus reducing environmental damage. In the course of such attempts, promoting walking and cycling requires an understanding of the psychological factors (motivations, barriers, and preconditions) that influence mode choice (Risser & Šucha, 2020b). The risk that car drivers constitute for persons not sitting in cars plays an important role, there. As well, personal and environmental characteristics are associated with the choice between active transport modes and car use (Scheepers et al., 2013).

Different modes of transport involve different vulnerabilities. Pedestrians and cyclists are classified as vulnerable road users (VRUs). Over one-third of road traffic deaths globally are of pedestrians and cyclists (Asian Biomedicine, 2017). Their interactions with other, more protected modes such as cars are crucial for safety. Well-functioning communication between VRUs and drivers is vital in order to avoid risks. As discussed earlier, the communication dynamics, particularly the loss of informal cues such as eye contact with the rise of automated vehicles, has a disproportionate impact on the safety of VRUs (de-Juan-Ripoll et al., 2021). Risk perception can also vary depending on the transport mode and role (e.g. pedestrian vs. driver; Dinh et al., 2020). Promoting a shift to modes such as walking requires not only the improvement of preconditions (such as the infrastructure for walking) and providing incentives but also the effective provision of information and the addressing of potential arguments for and against these modes, not least with respect to risk aspects. The perceived costs and benefits of using a particular mode and the perceived risks associated with it are central to an individual's mode choice and willingness to engage in specific behaviours while using that mode (Maurer et al., 2016). Promoting a shift in transport mode requires psychology-based measures. For example, encouraging walking involves understanding what discourages people from walking and what motivates those who do, identifying the barriers and necessary preconditions, and recognising that these are perceived differently across individuals. Safety needs – that is, the needs of not being at risk – are an important element in this context. Improving the perceived safety of pedestrians can make walking a more attractive alternative for short trips (Risser, 1993).

Interplay of factors and implications for intervention

The factors discussed earlier – communication, societal preconditions, infrastructure, and transport mode – do not operate in isolation. They interact dynamically to influence road user behaviour and risk-taking. As already underlined further earlier, infrastructure influences communication opportunities, societal norms shape perceptions and weighting of risk and acceptable behaviour within different modes, and communication campaigns draw on social marketing principles to promote safer practices across various modes (Ulleberg et al., 2009). Effective road safety interventions must adopt a comprehensive, holistic approach that considers this complex interplay.

The Diamond Model by Ralf Risser is proposed as a framework for analysing this interplay and summarises what has been said earlier (Risser, 2004; Chaloupka & Risser, 2020). It suggests analysing both the physical and social aspects of the public space and the individual's position, considering five interacting areas: individual characteristics, communication, socialisation agents/culture/social establishment/media (social aspects), infrastructure, and vehicle, or mode, characteristics. The model posits that individuals develop attitudes influenced by these other levels. Interactions produce a communication history, infrastructure provides physical conditions, modes have varying advantages and disadvantages, and society, that is, societal structures (media, family, friends, peer groups, laws, and informal norms), shape views on participation in traffic. This model suggests that a comprehensive approach is needed to influence road user behaviour, including risk-taking. It is argued that influencing behaviour by targeting factors "from the outside" (situational factors such as infrastructure or laws) might be more promising, quicker, and more efficient than solely targeting internal factors such as attitudes. One may assume, though, that targeting internal factors such as attitudes through education and campaigns remains important for medium- and long-term change, because a lasting change of attitudes will arguably have long-term behavioural effects.

Some psychological theories as a basis for interventions

Psychological theories provide valuable insights into how risk-taking behaviour could be influenced. Festinger's *cognitive dissonance theory* (1957) explains that discomfort arises when beliefs or actions conflict (e.g. wanting to be safe but speeding). Strategies to reduce this dissonance include changing beliefs, introducing new ones, or changing behaviour. Often, this leads to rationalisations rather than to changes reflecting reality ("speeding is risky but I am a very skilful driver"). However, authorities can use campaigns to change the "concert of arguments" or provide incentives to encourage new behaviour that creates cognitive consonance (e.g. highlighting comfort and relaxation as a personal benefit of safe driving).

Reactance theory (Brehm, 1966) suggests that restrictions on freedom of choice are unpleasant and provoke resistance. Being forced to respect speed limits, for instance, might be felt as a restriction on one's autonomy, potentially leading to intentional neglect of the rule. Communication is needed to help citizens understand the reasons for law enforcement, so that they avoid speeding because they understand the necessity, not just because they are forced to – which supports the assumptions that campaigns should be useful.

Attribution theory (Meyer, 2003) states that our reaction to others' behaviour depends on the reasons we attribute to them. Attributing negative intentions leads to negative feelings, while attributing unintentional reasons generates fewer problems. Misattributions can derail communication, and attributing negative intentions can provoke aggressive and thus risky behaviour. Assuming that there are positive attitudes underlying others' actions can support positive attitudes towards others. Campaigns can aim to change attitudes in the sense that we assume negative intentions less often when we are confronted with other road users' behaviour. For instance, we need to understand that we tend to attribute our own violations to situational factors (explaining "everything"), while attributing others' violations to their personal dispositions, labelling them as "idiots" or similar things (the actor-observer effect).

The *theory of planned behaviour* (Ajzen, 1991) posits that behavioural intentions are influenced by attitudes towards behaviour, subjective norms (expectations of significant others), and perceived behavioural control. A change in behaviour is more likely if attitudes are positive, if significant others adhere to the behaviour, and if the behaviour is perceived as easy to perform. To "market" risk-avoiding behaviour should, thus, include arguments that support wished-for attitudes, information that shows that significant others adhere to risk-avoiding behaviour, and demonstrations that it is easy to behave in a risk-avoiding way.

The *situational model* of behaviour by Rosenstiel (2003) proposes that both personal factors (desire and ability) and situational variables (obligations and permissions) affect behaviour. An individual's value system develops on the basis of societal attitudes and values, and behaviour aligned with this system is accepted and internalised, but only if preconditions allow it. For example, respecting speed limits requires a public discussion that makes the relationship between speed and risk clear, but at the same time infrastructure that credibly reflects the need for lower speeds.

The *theory of social comparison processes* (Festinger, 1954) suggests that people evaluate their own thoughts and actions by comparing themselves to others. This comparison is more important when it involves significant others and provides a benchmark, especially in new situations. Behaving differently from others can cause discomfort or conflict. If a large majority of the other road users exhibit risky behaviour, it will be difficult for an individual to behave differently. At the same time, involving media, including social media, in traffic safety activities could provide the possibility of demonstrating increasing public proneness to act in a considerate, risk-avoiding way by selecting and presenting cases of good practice.

The *model of ecological behaviour* (Fietkau & Kessel, 1981) identifies five factors for enhancing pro-environmental behaviour: environmental knowledge, pro-environmental attitudes, opportunities to act, incentives, and perceived consequences. This model can also be applied to enhance risk-avoiding behaviour. Knowledge is needed to perform the behaviour, attitudes can be enhanced by education and campaigns, opportunities require appropriate preconditions (e.g. infrastructure), incentives support using opportunities, and positive feedback is crucial (e.g. bonuses provided by insurance companies as mentioned earlier).

The *transtheoretical model* (TTM) (Prochaska & DiClemente, 1982) proposes segmenting the population according to their stage of awareness regarding problematic behaviour, in this case risk-taking behaviour. The stages are pre-contemplation, contemplation, preparation, action, maintenance, change adopted, and relapse. Interventions should be tailored to these stages. For instance, causing dissonance ("my behaviour today is actually not very clever, but I would like to see myself as a clever person") can move someone from pre-contemplation to contemplation. The model views behaviour change as a learning process that may involve several rounds before becoming a habit. Thereby, arguments and information become more specific and detailed step by step.

Finally, Schneider (2013) combines several theories into a strategy for enhancing sustainable transport modes in his theory of routine mode choice decisions. Campaigns should target behaviours empirically linked to accidents and injuries, such as speeding, drink-driving, and mobile phone use while driving. It is recommended that campaigns aim to influence road users both within the immediate context of the behaviour (e.g. feedback on speed) and outside (e.g. addressing underlying beliefs). In-context measures can cause temporary change, while off-context measures are important for beliefs, which suggests that a combination is effective. Understanding the motives, needs, and interests of the target group is of paramount importance for tailoring messages and enhancing personal relevance (Lee & Kotler, 2015; Chaloupka & Risser, 2020). Psychological concepts such as cognitive dissonance, the actor-observer bias, and emotional appeals can enhance persuasion.

Beyond traditional campaigns, involving the public through participation and dialogue, potentially using Web 2.0 technologies, can facilitate a dynamic exchange on risk-avoiding behaviour of road users and enhance acceptance of measures (Dobranskyte-Niskota et al., 2007).

Ultimately, reducing risk-taking requires a concerted effort across various domains, involving experts, the public, decision-makers, practitioners, researchers, and policymakers collaborating to design, implement, and evaluate interventions (Boulanger et al., 2007). Systematic evaluation and reporting of both analyses and measures are essential for future learning. The reasons cited for why problematic behaviour such as speeding persists include the lack of immediate negative feedback and the fact that drivers often attribute their own faults to situational factors while blaming others' infringements on intentional acts (the actor-observer effect). Unclear public and institutional attitudes will also make it difficult to develop a sense of right and wrong in traffic (Maurer et al., 2016).

Conclusion

Risk-taking behaviour is a primary factor contributing to road accidents and injuries worldwide. Understanding this behaviour necessitates looking beyond individual psychological traits to the dynamic context in which road users operate. Communication among road users, involving formal rules and informal interactions, is vital for coordination but can also be a source of error, particularly with new technologies such as automated vehicles. Societal preconditions and social context shape norms, influence risk perception through cultural factors and media, and provide the environment within which interventions must function. Infrastructure design has a direct impact on safety and behaviour, setting the physical stage for interactions and potentially triggering behavioural adaptations such as risk compensation. Finally, the transport mode itself dictates the nature of the risks faced and posed, influencing vulnerabilities and interaction dynamics between different user groups. Effective road safety strategies must recognise and address the complex interplay of these factors. Integrating insights from psychological models, communication theories, societal influences, and infrastructure impacts is essential. A holistic approach combining targeted communication campaigns informed by psychological principles, infrastructure improvements designed to promote safety and clarity, incentives for safer and more sustainable modes, and genuine public participation, all informed by rigorous research and evaluation, is necessary to meaningfully influence road user behaviour, reduce risk-taking, and enhance safety for all. Leveraging psychological theories such as cognitive dissonance, reactance, attribution, and the transtheoretical model, alongside frameworks such as social marketing and the Diamond Model, can provide a robust foundation for developing interventions tailored to different contexts and target groups.

References

Ajzen, I. (1991). The theory of planned behavior. *Organizational Behavior and Human Decision Processes*, *50*, 179–211.

Andersen, C. S., Olesen, A. V., & Bolet, L. (2013). Influence of road characteristics on density of accidents on secondary rural roads in Denmark. In *Sustainable and safe road design from a human behaviour point of view*. Aalborg University, Traffic Research Group.

Asian Biomedicine, T. W. (2017). Evidence-based policy, legislation, and enforcement: A need for concerted actions toward road safety. *Asian Biomedicine*, *11*(4), 299–330.

Boulanger, A., Daniels, S., Delhomme, P., Deugnier, M., Divjak, M., & Eyssarteir, C. (2007). *Campaigns and awareness-raising strategies in traffic safety*. Deliverable 2.2 of the EU-Project CAST: Comparison of Research Designs.

Brehm, J. W. (1966). *A theory of psychological reactance* (pp. 47–48). Academic Press.

Chaloupka, C., & Risser, R. (2020). Communication between road users and the influence of increased car automation. *Transactions on Transport Sciences*, *10*(2), 5–17.

Chaloupka-Risser, C., & Füssl, E. (2017). The importance of communication between cyclists and other traffic participants and its potential in reducing traffic safety-critical events. *Transactions on Transport Sciences*, *8*(1), 24–30.

de-Juan-Ripoll, C., Chicchi Giglioli, I. A., Llanes-Jurado, J., Marín-Morales, J., & Alcañiz, M. (2021). Why do we take risks? Perception of the situation and risk proneness predict domain-specific risk taking. *Frontiers in Psychology*, *12*, 562381.

Dinh, D. D., Vũ, N. H., McIlroy, R. C., Plant, K. A., & Stanton, N. A. (2020). Effect of attitudes towards traffic safety and risk perceptions on pedestrian behaviours in Vietnam. *IATSS Research*, *44*(3), 238–247.

Dobranskyte-Niskota, A., Perujo, A., & Pregl, M. (2007). *Indicators to assess sustainability of transport activities*. European Commission, Joint Research Centre.

Färber, B. (2016). Human factors in automated driving. In M. Maurer, J. C. Gerdes, B. Lenz, &, H. Winner (Eds.), *Autonomous driving* (pp. 125–136). Springer.

Faulks, I. (2011). Road safety advertising and social marketing. *Journal of the Australasian College of Road Safety*, 22, 34–40.

Festinger, L. (1954). A theory of social comparison processes. *Human Relations*, 7, 117–140.

Festinger, L. (1957). *Theory of cognitive dissonance*. Stanford University Press.

Fietkau, H. J., & Kessel, H. (1981). *Umweltlernen: Veränderungsmöglichkeiten des Umweltbe-wußtseins*. Hain.

Fink, D. (2010, June 20–23). *Road safety 2.0: Insights and implications for government*. 23rd Bled eConference & eTrust: Implications for the Individual, Enterprises and Society, Bled, Slovenia.

Grace, P. S., Sumit, K., & Chakraborty, N. (2019). Sensation seeking and peer passenger influence on risky driving among novice drivers in Udupi Taluk, India. *Clinical Epidemiology and Global Health*, *8*(1), 221–225.

Hensch, A. C., Neumann, I., Beggiato, M., Halama, J., & Krems, J. F. (2020). How should automated vehicles communicate? Effects of a light-based communication approach in a Wizard-of-Oz study. In *Advances in human factors of transportation – proceedings of the AHFE 2019 international conference on human factors in transportation, July 24–28, 2019, Washington DC, USA* (Vol. 10, pp. 79–91). Springer International Publishing.

Hydén, C. (1987). The development of a method for traffic safety evaluation: The Swedish traffic conflicts technique. *Lund Institute of Technology Bulletin*, *70*.

Lagarde, E. (2007). Road traffic injury is an escalating burden in Africa and deserves proportionate research efforts. *PLOS Medicine*, *4*(6), e170.

Lee, N. R., & Kotler, P. (2015). *Social marketing: Changing behaviors for good*. Sage Publications.
Lewis, C. (1985). Adolescents' traffic casualties: Causes and interventions. In L. Evans & R. Schwing (Eds.), *Human behavior and traffic safety* (pp. 67–85). Plenum.
Litman, T. (2007). *Well measured: Developing indicators for comprehensive and sustainable transport planning*. Victoria Transport Policy Institute.
Lund, I. O., & Rundmo, T. (2009). Cross-cultural comparisons of traffic safety, risk-perception, attitudes and behaviour. *Safety Science*, *47*(4), 547–553.
Maurer, M., Gerdes, J. C., Lenz, B., & Winner, H. (2016). *Autonomous driving: Technical, legal and social aspects*. Springer Nature.
Meyer, W. U. (2003). *Einige grundlegende Annahmen und Konzepte der Attributionstheorie*. Universität Bielefeld.
Muhlrad, N. (1987). *Traffic safety research for developing countries: Methodologies and first results* (No. 7). INRETS.
Nilsson, G. (2004). *Traffic safety dimensions and the power model to describe the effect of speed on safety* [Doctoral dissertation, Lund University].
Porter, B. E. (Ed.). (2011). *Handbook of traffic psychology*. Academic Press.
Prochaska, J. O., & DiClemente, C. C. (1982). Transtheoretical therapy: Toward a more integrative model of change. *Psychotherapy: Theory, Research & Practice*, *19*(3), 276–288.
Redshaw, S., Irwin, J. D., & Faulks, I. J. (Eds.). (2008, July 10). *Getting the road safety message across: What can we learn from public health advertising?* Proceedings of a Seminar held at the George Institute for International Health, Sydney.
Risser, R. (1985). Behavior in traffic conflict situations. *Accident Analysis & Prevention*, *17*(2), 179–197.
Risser, R. (1988). *Kommunikation und Kultur des Straßenverkehrs* [Habilitation thesis, University of Vienna].
Risser, R. (1993). What is sustainable traffic? Are safety, mobility and life quality competing with each other. In Austrian Road Safety Board & ICTCT (Eds.), *Safety evaluation of traffic systems: Traffic conflicts and other measures*. Proceedings of the 6th ICTCT Workshop, Salzburg.
Risser, R. (2004). Philosophy of traffic calming. *The Asian Journal: Journal of Transport and Infrastructure*, *11*(1), 1–9.
Risser, R., & Šucha, M. (2020a). Start walking! How to boost sustainable mode choice – psychological measures to support a shift from individual car use to more sustainable traffic modes. *Sustainability*, *12*(1), 554.
Risser, R., & Šucha, M. (2020b). *Psychological perspectives on walking: Interventions for achieving change*. Routledge.
Robertson, L. S., & Pless, I. B. (2002). Does risk homoeostasis theory have implications for road safety: Against. *BMJ (Clinical Research ed.)*, *324*(7346), 1149–1152.
Rosenstiel, L. (2003). *Grundlagen der Organisationspsychologie. Basiswissen und Anwendungshinweise* (5th ed.). C. E. Poeschel Verlag.
Rothe, J. P. (1992). Traffic sociology: Social patterns of risk. *International Journal of Adolescent Medicine and Health*, *5*(3–4), 187–198.
Scheepers, E., Wendel-Vos, W., van Kempen, E., Panis, L. I., Maas, J., Stipdonk, H., & Schuit, J. (2013). Personal and environmental characteristics associated with choice of active transport modes versus car use for different trip purposes of trips up to 7.5 kilometers in The Netherlands. *PLOS One*, *8*(9), e73105.
Schneider, R. J. (2013). Theory of routine mode choice decisions: An operational framework to increase sustainable transportation. *Transport Policy*, *25*, 128–137.
Shinar, D. (2017). *Traffic safety and human behavior*. Emerald Publishing Limited.
Šucha, M. (2019). *The human factor in traffic – possible ways of influencing people's behaviour* [Habilitation thesis, Charles University].

Šucha, M., Dostal, D., & Risser, R. (2017). Pedestrian-driver communication and decision strategies at marked crossings. *Accident Analysis & Prevention*, *102*, 41–50.

Theeuwes, J., & Van Der Horst, R. (2017). *Designing safe road systems: A human factors perspective*. CRC Press.

Toth-Szabo, Z., Várhelyi, A., Koglin, T., & Angjelevska, B. (2011). *Measuring sustainability of transport in the city: A development of an indicator-set*. Department of Technology and Society, Lund University.

Ulleberg, P., Vaa, T., Ausserer, K., Carstensen, G., Forward, S., Krol, B., Malasek, J., Meng, A., Møller, M., Moan, I. S., Phillips, R., Risser, R., Sardi, G. M., Sedà, E., & Simoes, A. (2009). *Road user model and persuasion technique: Final version* (Uden NAVN No. D 1.4). DTU-Library, Copenhagen.

Vankov, D., & Kaufmann, C. (2017). IMPACT-traffic safety campaigns for young people. *Transactions on Transport Sciences*, *8*(1), 11–23.

Wilde, G. J. S. (1989). Accident countermeasures and behavioural compensation: The position of risk homeostasis theory. *Journal of Occupational Accidents*, *10*(4), 267–292.

7 Novice drivers and risk

Kristýna Josrová

Introduction

Drivers are the most frequent cause of road accidents. The most common contributing factors include excessive speed, driving under the influence of alcohol or other psychoactive substances, distraction, and failure to use safety features. As it is ultimately drivers who decide what driving style they choose, the vast majority of road crashes could be avoided if drivers were more cautious and attentive.

Young people are among the most risky and vulnerable driver groups. Road traffic crashes are the leading cause of death for young people. Young men, in particular, are prone to risky driving behaviour, with a stronger tendency towards fast and dangerous driving. Adolescence is associated with increased sensation-seeking inclinations, while the capacity for self-regulation – necessary to curb impulsive behaviour – is still underdeveloped. This is largely due to brain immaturity, as certain cerebral regions continue to develop until the late 20s. The prefrontal cortex, which plays a key role in decision-making and self-regulation, matures around the age of 25. Additionally, young people have a more active reward system in the brain, which makes them more susceptible to seeking short-term gratification, such as the thrill of speed and peer recognition.

Young drivers are typically at an age when social relationships are especially important. As a result, they often behave in ways that will enhance their reputation and prevent social exclusion. They tend to take more risks in the company of their peers and are more responsive to peer influence, which often encourages faster driving. However, when driving alone, they generally drive more safely. Peer presence is a significant risk factor, whereas the presence of an adult passenger is conducive to safer driving.

Novice drivers, those with less than three years of experience, are much more likely to cause an accident than experienced drivers. Their driving presents the greatest risk within their first month on the road, but rates of crashes involving newly qualified drivers tend to decline with each passing month of their driving practice. Their inexperience makes it difficult for them to cope with complex traffic situations. Driving requires intense concentration,

DOI: 10.4324/9781003608394-7

which they find difficult to sustain, and distraction has a particularly negative impact on their performance.

Beginners also differ significantly from experienced drivers in terms of visual patterns. They have narrower peripheral vision and make fewer eye movements while driving. They find it particularly challenging to predict and detect potential hazards, particularly hidden risks. Lacking experience, they do not look towards areas where dangers may emerge, and they also tend to underestimate how dangerous certain situations may be. Their responses are slower, as they take longer to process input information than more experienced drivers. All of these risk-related abilities and skills improve with experience. At the same time, novice drivers often overestimate their driving competencies, particularly as regards detecting and anticipating risk.

Prevention strategies for young and inexperienced drivers can take many forms. One approach, widely used globally, is the graduated driver licensing (GDL) system. Learners must progress through several stages of driving instruction, each involving specific restrictions, such as bans on carrying young passengers or driving at night. These restrictions reduce crash rates significantly by addressing the most high-risk factors for this driver group. Another common and effective measure is accompanied driving, where the novice is not allowed to drive alone but must be accompanied and supervised by an older, more experienced driver, who ensures that their driving behaviour is safe.

The presence of an older and experienced passenger contributes to safer driving. In addition, such a passenger can guide the novice towards an appropriate driving style and draw attention to potential hazards, which will help them better perceive emerging dangers for themselves. Contrarily, the presence of peer passengers greatly increases the likelihood of an accident and should therefore be limited. Cognitive training is highly useful for inexperienced drivers, as it enhances specific skills. Such educational programmes can improve their ability to perceive and respond to risks.

Road traffic crash rates and contributing factors

Road traffic crashes claim nearly 1.2 million lives globally each year (World Health Organization (WHO), 2023). Within the European Union, approximately 20,000 people are killed in road crashes annually, with over one million injured (European Commission, 2025). However, the risk of being involved in a crash is not distributed evenly – children and young people are disproportionately affected. Males are also at higher risk, being three times more likely to die on the road than females (WHO, 2023).

In particular, improving road safety requires interventions addressing the human factor, which is responsible for up to 95% of crashes. Contributing to approximately 30% of road fatalities, excessive or inappropriate speed remains a critical risk factor (European Commission, 2024). Other frequent causes include driving under the influence of alcohol or other psychoactive substances, driver distraction (e.g. mobile device use), and the failure to use protective safety features such as seat belts, helmets, and child restraints. These high-risk behaviours are commonly associated with both fatal and serious non-fatal crashes (European Commission, 2024). Accordingly, safety recommendations emphasise the implementation of legislative frameworks and supporting measures aimed at reducing such hazardous practices worldwide (WHO, 2023).

Young drivers

Crash involvement of young drivers

Road traffic crashes are a major threat to young people; they represent the leading cause of death among individuals aged 15–29 (WHO, 2022). Young drivers are the most risky and vulnerable group of motorists. They not only endanger themselves but also pose a threat to other road users. While comprising a relatively small segment of all drivers, they are responsible for a disproportionately high number of crashes. Age, driving experience, and gender all play significant roles. Young males, in particular, are more prone to engaging in risky driving behaviours. They also account for the majority of deaths among young drivers. In the EU, males represent 82% of road traffic fatalities involving young drivers (European Commission, 2023).

Sensation-seeking

In youth, risk can be actively sought. This tendency is linked to the concept of sensation-seeking, which refers to the tendency to seek novel, intense, and thrilling experiences. Sensation-seekers typically prefer excitement and stimulation while downplaying risks and showing lowered inhibitions (Zuckerman, 1994). Fast and risky driving can provide exactly the kind of experience they seek. Drivers with high sensation-seeking scores are more likely to engage in risky and aggressive driving, make more driving errors, and consequently face more crashes and traffic fines (Zhang et al., 2019a). Young drivers who score high in sensation-seeking are prone to speeding (Delhomme et al., 2012), with this tendency being more pronounced among young males (Cestac et al., 2011).

Young people are more susceptible to risky behaviours partly because sensation-seeking increases during adolescence. It peaks around the age of 19 and gradually declines over the following decade. As a result, young drivers seek out stimulating experiences far more often than drivers in their 30s. Although the tendency remains noticeable among drivers in their late 20s, additional biological factors, such as brain maturation and improved self-regulation, come into play at that stage (Steinberg et al., 2018).

Brain development and functioning in young drivers

When young people begin to drive, their brains are still developing. Certain brain areas continue to mature beyond the age of 20, with some still developing into the late 20s. Significantly, these regions are relevant to driving and influence driving behaviour (Somerville, 2016). Brain maturation proceeds from the back of the brain to the front, meaning that the prefrontal cortex – responsible for cognitive functions – matures last. This area, which plays a key role in stress regulation, decision-making processes, inhibitory control, and memory, does not fully develop until around the age of 25. During this period, white matter in the prefrontal cortex increases. In young drivers, this region is less active than in older drivers, which can lead to more risk-taking because of their difficulty in suppressing impulsive and inappropriate behaviour (Foy et al., 2016).

The brain's reward system also plays a significant role. It provides pleasurable feelings and encourages repetition of behaviour that previously led to pleasant experiences. In the adolescent brain, this system is more active, whereas sensitivity to aversive stimuli may be attenuated (Spear, 2013). When adolescents consider taking a risk, their subcortical regions show elevated activity. The nucleus accumbens, known as the "pleasure centre", is also engaged. As a result, short-term rewards, positive feelings, or peer recognition are prioritised, while long-term consequences are sidelined, a situation conducive to risk-taking behaviour (Glendon, 2014).

The reward system is easily activated in young people, making them more prone to sensation-seeking and engaging in risky activities such as speeding or tailgating. However, they lack the self-regulation needed to moderate these tendencies. Self-regulation, that is, the ability to control emotions and behaviour and to resist temptation, peaks at around the age of 23. Therefore, young people are more likely to succumb to immediate impulses, and their emotions may override inhibitory control (Steinberg et al., 2018).

The amygdala, the structure with the main responsibility for emotional regulation, matures later in males, making young men particularly prone to aggressive behaviour, which can manifest itself in dangerous driving. The cerebellum, involved in motor coordination, also continues to mature and can affect the physical aspects of driving. Overall, brain function in young drivers is shaped by the asynchronous development of different brain areas. This can result in one region dominating over another, less developed one (Glendon, 2014).

Social influences and the role of passengers

Young people are strongly influenced by their peers, which corresponds with their developmental task: to establish themselves and gain support from their social environment. At this stage, their primary goal is to integrate into their social group. Acceptance by the group is often more important to them than other risks, including threats to their own health. During that period, young people are especially sensitive to social risks such as embarrassment, a drop in social status, or exclusion. Rejection by peers is perceived as particularly threatening in adolescence. As a result, they are willing to do whatever it takes to avoid ostracism. Finding the opinions of others crucial, they tend to imitate what they believe their peers do. In the presence of peers, young people are more likely to engage in risky behaviour – less so when they are alone. Their risk perception is shaped by the views of their peers, and they adjust their own judgement accordingly (Blakemore, 2018).

Driving behaviour is significantly influenced by passengers. When driving with adults, young people tend to drive more safely. In contrast, when accompanied by peers, they are more inclined to risk behaviour behind the wheel. The presence of risk-taking friends in the car significantly increases the likelihood of dangerous driving (Simons-Morton et al., 2011). Young men are particularly susceptible to direct peer pressure encouraging risky driving behaviour (Trógolo et al., 2022). Young drivers are more likely to succumb to peer pressure and break traffic rules in order to gain the respect of others (Weston & Hellier, 2018). The presence of one or more peer passengers is a significant risk factor for young drivers. It dramatically increases the likelihood of them causing a crash resulting in injury or even death (Ouimet et al., 2015).

Novice drivers

Novice driver crash rates

Novice drivers is a term used in reference to individuals with less than three years of driving experience. Although they represent a relatively small group, they account for a disproportionately high number of crashes. In addition, novice drivers under 30 or over 55 years of age face a higher risk of being fatally injured in a crash (Chen et al., 2021).

Especially during the early stages, the length of driving experience significantly affects driving safety. The first six months are the most hazardous, with newly licensed drivers recording the highest crash rates during this period (Rowe et al., 2022). The first month is particularly critical, but drivers gain experience rapidly once on the road. Each month after obtaining a licence sees a decline in the crash risk. The crash rate continues to drop steadily over the first few months and decreases substantially after two years of driving (Mayhew et al., 2003).

Over time, novice drivers typically become more experienced, showing lower crash involvement. Those most at risk are the ones with fewer than six months of driving experience or less than 2,000 kilometres driven (Fisher et al., 2006). After covering the first 800 kilometres, their crash risk declines by two-thirds (McKnight & McKnight, 2003).

Automaticity of driving fundamentals

In the initial phase of driver training, the objective is to develop automaticity in basic skills, such as starting the engine, changing gear, or checking rear-view mirrors, so that these tasks become routine and effortless. However, mastering the fundamentals reliably takes time. Limited experience with vehicle control means that novice drivers need more time to perform intended actions. This delay can lead to crashes (Zhang et al., 2019a). Intersections may pose significant difficulties, as they require quick and effective assessment of the situation, which is something novice drivers find challenging because of a lack of practice in responding to unexpected events (Borowsky & Oron-Gilad, 2013).

It is not until they have securely mastered the basics that drivers have the mental capacity to respond to more complex and unexpected situations. A common issue is that driver training tends to focus on vehicle control and understanding of traffic rules while dedicating minimal attention to aspects essential for safe driving (Hatakka et al., 2002). Depending on the country, learner drivers are required to spend about ten to several dozen hours behind the wheel. This is far too little time to build sufficient experience across different driving contexts. It is through repeated exposure to diverse traffic situations that drivers learn to avoid potential hazards.

The high number of crashes among newly licensed drivers suggests that they are not sufficiently prepared for real-life traffic. In driving schools, they learn the technical basics needed to control the car within a few hours. However, learning to operate a vehicle and learning how to behave in traffic are two distinct tasks. While basic vehicle control is a prerequisite for safe driving, it is by no means sufficient. Drivers also need to deal with complex cognitive challenges. Driving is a multifaceted task requiring dozens of simultaneous activities: in addition to steering and managing speed, drivers must monitor their surroundings, assess information, anticipate situations, and communicate with other road users (Groeger, 2001). Operations that may seem easy for seasoned drivers pose greater difficulties for less experienced ones. Even routine activities such as parking can cause heightened heart activity in novice drivers (Wang et al., 2018). As experience accumulates, fundamental skills become ingrained and drivers can focus their attention on observing the road environment.

Acquiring essential driving skills and developing automaticity is a time-consuming process. Once novices have achieved a level of technical competence, they must continue learning to function in everyday traffic, which poses far more demands than the controlled conditions of a training ground. Real-life driving practice exposes road users to a wide range of traffic situations, helping them to discern what to watch out for and what constitutes a genuine risk. This ongoing process contributes to driving ability improving progressively.

Risk perception in novice and experienced drivers

The crucial difference between novice and experienced drivers lies in how they perceive and assess risk. These differences are particularly evident in visual perception. Each group scans the environment differently when driving: novice drivers tend to show more naïve gaze behaviour, as if expecting a smooth and uneventful journey.

As novice drivers improve and gain experience, their visual fixation patterns, that is, where they direct their gaze, change systematically over time (Shinar, 2008). Experienced drivers show better peripheral vision, whereas novices tend to focus more on keeping the vehicle on course, with their peripheral awareness developing gradually with increased practice (Deng et al., 2021). In comparison with their more experienced counterparts, novice drivers have a narrower horizontal spread of visual search, meaning they do not gain as broad a view of their surroundings as they could. As a result, they neither expect nor search for potential hazards that may lie at the outer edges of their visual field (Robbins & Chapman, 2019). They also exhibit fewer eye movements during driving (Wang et al., 2018).

While all drivers are generally aware of obvious dangers, novice drivers may often fail to detect potential or hidden threats. Even when such hazards are recognised, their reactions tend to be slower than those of more advanced drivers (Zhang et al., 2019b). The more experienced a driver becomes, the more sensitive they are to emerging risks. Seasoned drivers tend to identify potentially dangerous situations early. They are often able to detect subtle cues more than a second before the actual event occurs. Inexperienced drivers, on the contrary, typically do not anticipate hidden threats and may often miss the subtle indicators of a developing emergency (Crundall, 2016; Borowsky & Oron-Gilad, 2013). They take longer to shift their gaze to hazardous spots and are particularly prone to overlooking risks obscured by visual barriers, such as a pedestrian stepping out onto a zebra crossing from behind a parked car. Experienced drivers, in contrast, fixate such hazards more quickly and maintain visual attention on them for longer (Crundall et al., 2012). They also scan their surroundings more thoroughly while driving (Huestegge et al., 2010).

Novice drivers need more time to process potentially dangerous situations, while experienced drivers benefit from shorter reaction times, as they are able to interpret risk-specific information more quickly. They can assess more rapidly whether a situation is hazardous and brake in time if needed (Huestegge et al., 2010). Their brain can process more driving information and their visual nerve processes such information faster. Meanwhile, their vagus nerve and heart show less activity (Wang et al., 2018).

Risk perception manifests itself on both the psychological and physical levels. A given traffic situation may be perceived as hazardous by experienced drivers, while novices may not see it as risky. The latter tend to be less sensitive to danger and find certain situations not so threatening. As a result, they may react too late, only registering the risk when the obstacle is already too close (Zhang et al., 2019b). In contrast, experienced drivers have been found to produce up to three times more skin conductance responses to a potential hazard, which demonstrates that they are aware of the danger and preparing to avoid it. Novices, on the other hand, remain unaware of the risks at that point and hazardous situations thus have little emotional effect on them. However, as their driving experience grows, their psychophysiological systems learn to respond accordingly (Kinnear et al., 2013).

Attention in novice drivers

Novice drivers are quickly and easily overwhelmed while driving. Compared to more experienced drivers, they tend to feel more overloaded and under greater stress behind the wheel (Paxion et al., 2014). They also become fatigued more quickly when driving in monotonous environments (Li et al., 2017). For newly qualified drivers, driving itself is highly demanding in terms of attention. When they are engaged in secondary tasks while driving, they struggle to maintain the vehicle in the lane and regulate their speed. More experienced drivers are better at coping with such distractions (Freydier et al., 2016). They also demonstrate better accuracy in sustained- and divided-attention tasks (Kerruish et al., 2022). Furthermore, they are capable of maintaining concentration for longer periods, meaning that they require less effort to manage cognitively demanding situations. For novices, monitoring the environment and staying attentive are more challenging (Crundall, 2016).

While driving, both novice and experienced drivers engage in secondary tasks such as reaching for a mobile phone or looking up a contact number. However, novice drivers are significantly more likely to cause a crash or near-crash in such situations. Distracted driving is risky for all driver groups, but more experienced drivers are better equipped to manage such distractions and can partially compensate for the increased risk (Siebe, 2014).

Overestimation of driving abilities

Drivers in general tend to be overly optimistic about their own driving abilities, and the less experience they have, the greater the discrepancy between their self-assessment and the reality. Inexperienced drivers are particularly prone to overestimating their ability to predict and detect hazards, as well as to maintain a safe distance (Martinussen et al., 2017).

One reason may be that novices do not yet fully understand the range of skills required for safe driving. While they may have mastered the basic controls, this is often the full extent of their knowledge at that point. They may be totally unaware that safe driving involves a range of additional skills. Without the insight that there are still areas in which they need to improve, they may feel sufficiently competent and thus overlook various risks (Martinussen et al., 2017).

Prevention in young and novice drivers

Graduated driver licensing

Young drivers typically lack sufficient experience and are also more vulnerable to certain driving conditions than their older and more experienced counterparts. It is therefore useful to strive to reduce the influence of the specific risk factors to which young drivers are most exposed. The process of obtaining a driving licence varies from country to country. In many jurisdictions, learners complete a driving course within a few months, pass the tests, and become fully licensed drivers. A well-established and widely used alternative in many countries is the graduated driver licensing (GDL) system, which has been shown to significantly reduce crash rates and fatalities among young drivers (McCartt & Teoh, 2015).

GDL is a process involving several stages of driver training. At the initial stage, learner drivers are subject to the strictest limitations, which are gradually relaxed as they progress through the system. Once the final stage is completed, the restrictions are lifted entirely. Common restrictions include driving bans during night-time hours, limits on the number of passengers, bans on mobile phone and electronic device use, and zero-tolerance policies on the alcohol blood level (Williams et al., 2016). In some jurisdictions, additional rules apply to novice drivers, such as reduced maximum speeds or stricter penalties for traffic violations, for example, incurring more demerit points.

The GDL process acknowledges that novice drivers are still in training and should behave accordingly on the road. As long as they are still learning, they are more likely to make mistakes, and since many of these mistakes are predictable, they can be mitigated. GDL thus makes it possible for young drivers to gain driving experience step by step while avoiding high-risk conditions that are especially dangerous for this driver group.

Accompanied driving

Accompanied driving is another measure which is widely used internationally. During the early stages of training, learner drivers are not allowed to drive alone but must be accompanied by an older, more experienced driver, typically one who meets specific age and driving experience requirements. These accompanying drivers are usually laypeople who supervise and support the novice to drive more safely. Accompanied driving is often integrated into the GDL system. In some jurisdictions, it is required throughout some or all of the GDL stages, while elsewhere the mentor's role is limited to helping the learner accumulate the required driving practice, which is often quantified in terms of a minimum number of hours. Specific conditions or driving environments, such as night-time driving, in which the practice must occur are frequently determined to ensure that the learner gains experience under more demanding circumstances.

Social influences and the role of passengers

The presence of adults, typically parents, tends to contribute to safer driving. The mere presence of an adult in the car is not necessarily beneficial. What matters most is that the adult passenger comments on the driver's behaviour. Young drivers then often apply safety-related advice they hear during accompanied driving even when driving alone (Chung et al., 2014). Parental comments usually focus on vehicle control, whereas other important areas relevant to safe driving may be neglected, and yet it is precisely comments on driving style and risk awareness that would be most valuable to novice drivers. Better-targeted feedback from parents could help young drivers develop the ability to identify and anticipate hazards more effectively in the future (Goodwin et al., 2014).

Younger passengers, on the other hand, present a risk: they have been found to double the likelihood of a crash. Therefore, many countries have adopted restrictions that limit whether novice drivers are allowed to carry peer passengers. These restrictions are usually lifted only after a certain period, once the driver has gained sufficient experience or advanced to a higher licensing stage. Banning young passengers helps prevent negative influences such as distraction and encouragement towards risky driving behaviour. As young people's driving is strongly influenced by their social group, when alone in a car, a young driver will not be exposed to the pressure of their peers and will stay fully focused on the driving task (Cassarino & Murphy, 2018).

Given their strong influence on the driver, young passengers also have a significant potential to exert a positive effect. They may thus become a protective factor. When a driver's peers value safe driving and consider it a social norm, they actually encourage the driver to adopt a safer driving style. In such cases, the driver is more likely to conform to the group's pressure and drive more cautiously (Sutherland et al., 2022). Passengers may even comment that they would prefer a slower ride, prompting the driver to comply (Shepherd et al., 2011). When friends or family disapprove of risky driving, the young driver may fear social criticism and modify their behaviour accordingly. This principle works both ways: peer influence can promote either safe or risky driving styles (Scott-Parker et al., 2015).

Training in cognitive skills

Novice drivers must first master the basic skills of vehicle control, but these alone are far from sufficient for safe driving. In this respect, the training of cognitive skills, which has been shown to reduce crash rates, is the most beneficial approach (Cassarino & Murphy, 2018). Targeted programmes help young drivers develop essential abilities, such as hazard anticipation. Computers, simulators, or video-based methods are used, allowing the training sessions to take place in a safe environment (McDonald et al., 2015).

Cognitive training programmes are generally focused on specific abilities or skills, such as responding to risk, anticipating hazards, or maintaining attention. However, these elements can also be combined within a single programme. Feedback is often used as part of the training process (Yamani et al., 2016). The majority of the studies in this field evaluate short-term outcomes. Only a limited amount of research has been concerned with the long-term effects, but it has been shown that some programmes remain effective even after several months (McDonald et al., 2015). As risk management is a topic of paramount importance, it is advisable to use evidence-based cognitive training programmes with proven long-term effectiveness in order to support young drivers in developing safer driving practices.

Another effective technique is road commentary training, where novice drivers describe aloud what they do and see while driving. This practice helps them become more aware of potential hazards and directs their attention to high-risk areas, thereby broadening their visual field (Cantwell et al., 2013).

References

Blakemore, S. J. (2018). Avoiding social risk in adolescence. *Current Directions in Psychological Science*, *27*(2), 116–122.

Borowsky, A., & Oron-Gilad, T. (2013). Exploring the effects of driving experience on hazard awareness and risk perception via real-time hazard identification, hazard classification, and rating tasks. *Accident Analysis & Prevention*, *59*, 548–565.

Cantwell, S. J., Isler, R. B., & Starkey, N. J. (2013). *The effects of road commentary training on novice drivers' visual search behaviour: A preliminary investigation*. ACRS.

Cassarino, M., & Murphy, G. (2018). Reducing young drivers' crash risk: are we there yet? An ecological systems-based review of the last decade of research. *Transportation Research Part F: Traffic Psychology and Behaviour*, *56*, 54–73.

Cestac, J., Paran, F., & Delhomme, P. (2011). Young drivers' sensation seeking, subjective norms, and perceived behavioral control and their roles in predicting speeding intention: How risk-taking motivations evolve with gender and driving experience. *Safety Science*, *49*(3), 424–432.

Chen, S., Shao, H., & Ji, X. (2021). Insights into factors affecting traffic accident severity of novice and experienced drivers: A machine learning approach. *International Journal of Environmental Research and Public Health*, *18*(23), 12725.

Chung, E. K., Choe, B., Lee, J. E., Lee, J. I., & Sohn, Y. W. (2014). Effects of an adult passenger on young adult drivers' driving speed: Roles of an adult passenger's presence and driving tips from the passenger. *Accident Analysis & Prevention*, *67*, 14–20.

Crundall, D. (2016). Hazard prediction discriminates between novice and experienced drivers. *Accident Analysis & Prevention*, *86*, 47–58.

Crundall, D., Chapman, P., Trawley, S., Collins, L., Van Loon, E., Andrews, B., & Underwood, G. (2012). Some hazards are more attractive than others: Drivers of varying experience respond differently to different types of hazard. *Accident Analysis & Prevention*, *45*, 600–609.

Delhomme, P., Chaurand, N., & Paran, F. (2012). Personality predictors of speeding in young drivers: Anger vs. sensation seeking. *Transportation Research Part F: Traffic Psychology and Behaviour*, *15*(6), 654–666.

Deng, M., Wu, F., Gu, X., & Xu, L. (2021). A comparison of visual ability and its importance awareness between novice and experienced drivers. *International Journal of Industrial Ergonomics*, *83*, 103141.

European Commission. (2023). *Road safety thematic report – young novice drivers*. European Road Safety Observatory. Brussels, European Commission, Directorate General for Transport. https://road-safety.transport.ec.europa.eu/document/download/8665e63d-7dbf-40a4-be32-0bf86accb35a_en?filename=ERSO-TR-NoviceDrivers-20231512.pdf&prefLang=fi

European Commission. (2024). *Road safety thematic report – main factors causing fatal crashes*. European Road Safety Observatory. European Commission, Directorate General for Transport. https://road-safety.transport.ec.europa.eu/document/download/a7428369-8eaf-4032-806e-ea08b46028c0_en?filename=ERSO-TR-MainCauses.pdf

European Commission. (2025). *Annual statistical report on road safety in the EU, 2025*. European Road Safety Observatory. European Commission, Directorate General for Transport. https://road-safety.transport.ec.europa.eu/document/download/17d70e9c-d9c4-4273-b497-41b61194e808_en?filename=ERSOnext_AnnualReport_20250227.pdf

Fisher, D. L., Pollatsek, A. P., & Pradhan, A. (2006). Can novice drivers be trained to scan for information that will reduce their likelihood of a crash? *Injury Prevention*, *12*(Suppl. 1), i25–i29.

Foy, H. J., Runham, P., & Chapman, P. (2016). Prefrontal cortex activation and young driver behaviour: A fNIRS study. *PLOS One*, *11*(5), e0156512.

Freydier, C., Berthelon, C., & Bastien-Toniazzo, M. (2016). Does early training improve driving skills of young novice French drivers? *Accident Analysis & Prevention*, *96*, 228–236.

Glendon, A. I. (2014). An approach to novice driver training. *European Review of Applied Psychology*, *64*(3), 111–122.

Goodwin, A. H., Foss, R. D., Margolis, L. H., & Harrell, S. (2014). Parent comments and instruction during the first four months of supervised driving: An opportunity missed? *Accident Analysis & Prevention*, *69*, 15–22.

Groeger, J. A. (2001). *Understanding driving: Applying cognitive psychology to a complex everyday task*. Routledge.

Hatakka, M., Keskinen, E., Gregersen, N. P., Glad, A., & Hernetkoski, K. (2002). From control of the vehicle to personal self-control; broadening the perspectives to driver education. *Transportation Research Part F: Traffic Psychology and Behaviour*, *5*(3), 201–215.

Huestegge, L., Skottke, E. M., Anders, S., Müsseler, J., & Debus, G. (2010). The development of hazard perception: Dissociation of visual orientation and hazard processing. *Transportation Research Part F: Traffic Psychology and Behaviour*, *13*(1), 1–8.

Kerruish, L., Cheng, A. S., Ting, K. H., & Liu, K. P. (2022). Exploring the sustained and divided attention of novice versus experienced drivers. *Transportation Research Interdisciplinary Perspectives*, *16*, 100717.

Kinnear, N., Kelly, S. W., Stradling, S., & Thomson, J. (2013). Understanding how drivers learn to anticipate risk on the road: A laboratory experiment of affective anticipation of road hazards. *Accident Analysis & Prevention*, *50*, 1025–1033.

Li, R., Su, W., & Lu, Z. (2017). Physiological signal analysis for fatigue level of experienced and inexperienced drivers. *Traffic Injury Prevention*, *18*(2), 139–144.

Martinussen, L. M., Møller, M., & Prato, C. G. (2017). Accuracy of young male drivers' self-assessments of driving skill. *Transportation Research Part F: Traffic Psychology and Behaviour*, *46*, 228–235.

Mayhew, D. R., Simpson, H. M., & Pak, A. (2003). Changes in collision rates among novice drivers during the first months of driving. *Accident Analysis & Prevention*, *35*(5), 683–691.

McCartt, A. T., & Teoh, E. R. (2015). Tracking progress in teenage driver crash risk in the United States since the advent of graduated driver licensing programs. *Journal of Safety Research*, *53*, 1–9.
McDonald, C. C., Goodwin, A. H., Pradhan, A. K., Romoser, M. R., & Williams, A. F. (2015). A review of hazard anticipation training programs for young drivers. *Journal of Adolescent Health*, *57*(1), S15–S23.
McKnight, A. J., & McKnight, A. S. (2003). Young novice drivers: Careless or clueless? *Accident Analysis & Prevention*, *35*(6), 921–925.
Ouimet, M. C., Pradhan, A. K., Brooks-Russell, A., Ehsani, J. P., Berbiche, D., & Simons-Morton, B. G. (2015). Young drivers and their passengers: A systematic review of epidemiological studies on crash risk. *Journal of Adolescent Health*, *57*(1), S9–S14.
Paxion, J., Galy, E., & Berthelon, C. (2014). Mental workload and driving. *Frontiers in Psychology*, *5*, 1344.
Robbins, C., & Chapman, P. (2019). How does drivers' visual search change as a function of experience? A systematic review and meta-analysis. *Accident Analysis & Prevention*, *132*, 105266.
Rowe, R., Stride, C. B., Day, M. R., Thompson, A. R., McKenna, F. P., & Poulter, D. R. (2022). Why are newly qualified motorists at high crash risk? Modelling driving behaviours across the first six months of driving. *Accident Analysis & Prevention*, *177*, 106832.
Scott-Parker, B., Watson, B., King, M. J., & Hyde, M. K. (2015). "I would have lost the respect of my friends and family if they knew I had bent the road rules": Parents, peers, and the perilous behaviour of young drivers. *Transportation Research Part F: Traffic Psychology and Behaviour*, *28*, 1–13.
Shepherd, J. L., Lane, D. J., Tapscott, R. L., & Gentile, D. A. (2011). Susceptible to social influence: Risky "driving" in response to peer pressure. *Journal of Applied Social Psychology*, *41*(4), 773–797.
Shinar, D. (2008). Looks are (almost) everything: Where drivers look to get information. *Human Factors*, *50*(3), 380–384.
Siebe, C. (2014). Distracted driving and risk of road crashes among novice and experienced drivers: Klauer SG, Guo F, Simons-Morton BG, et al. The New England Journal of Medicine 2013; 370: 54–9. *Journal of Emergency Medicine*, *46*(4), 600–601.
Simons-Morton, B. G., Ouimet, M. C., Zhang, Z., Klauer, S. E., Lee, S. E., Wang, J., Chen, R., Albert, P., & Dingus, T. A. (2011). The effect of passengers and risk-taking friends on risky driving and crashes/near crashes among novice teenagers. *Journal of Adolescent Health*, *49*(6), 587–593.
Somerville, L. H. (2016). Searching for signatures of brain maturity: What are we searching for? *Neuron*, *92*(6), 1164–1167.
Spear, L. P. (2013). Adolescent neurodevelopment. *Journal of Adolescent Health*, *52*(2), S7–S13.
Steinberg, L., Icenogle, G., Shulman, E. P., Breiner, K., Chein, J., Bacchini, D., Chang, L., Chaudhary, N., Giunta, L. D., Dodge, K. A., & Takash, H. M. (2018). Around the world, adolescence is a time of heightened sensation seeking and immature self-regulation. *Developmental Science*, *21*(2), e12532.
Sutherland, J. E., Hassein, U., Day, D. M., & Easa, S. M. (2022). Modeling social rejection, physiological arousal, and peer influence on risky driving among adolescents and young adults. *Transportation Research Part F: Traffic Psychology and Behaviour*, *84*, 114–138.
Trógolo, M. A., Ledesma, R., Medrano, L. A., & Dominguez-Lara, S. (2022). Peer pressure and risky driving: Development of a new scale. *Journal of Safety Research*, *82*, 48–56.

Wang, F., Xu, Q., Fu, R., & Sun, G. (2018). Study of driving skill level discrimination based on human physiological signal characteristics. *RSC Advances*, *8*(73), 42160–42169.
Weston, L., & Hellier, E. (2018). Designing road safety interventions for young drivers – the power of peer influence. *Transportation Research Part F: Traffic Psychology and Behaviour*, *55*, 262–271.
Williams, A. F., McCartt, A. T., & Sims, L. B. (2016). History and current status of state graduated driver licensing (GDL) laws in the United States. *Journal of Safety Research*, *56*, 9–15.
World Health Organization. (2022). *Preventing injuries and violence: An overview*. https://iris.who.int/bitstream/handle/10665/361331/9789240047136-eng.pdf?sequence=1
World Health Organisation. (2023). *Global status report on road safety*. https://iris.who.int/bitstream/handle/10665/375016/9789240086517-eng.pdf?sequence=1
Yamani, Y., Samuel, S., Knodler, M. A., & Fisher, D. L. (2016). Evaluation of the effectiveness of a multi-skill program for training younger drivers on higher cognitive skills. *Applied Ergonomics*, *52*, 135–141.
Zhang, J., Guo, G., Wang, P., Tang, Q., & Zhang, Z. (2019a, July). Research on the differences of risk perception ability between novice and experienced drivers. In *International conference on human-computer interaction* (pp. 433–447). Springer.
Zhang, X., Qu, X., Tao, D., & Xue, H. (2019b). The association between sensation seeking and driving outcomes: A systematic review and meta-analysis. *Accident Analysis & Prevention*, *123*, 222–234.
Zuckerman, M. (1994). *Behavioral expressions and biosocial bases of sensation seeking*. Cambridge University Press.

8 Elderly drivers

Eva Adamovská

Introduction

We have been experiencing a significant global increase in the proportion of older drivers for whom driving is an integral part of their lifestyle and a necessary mode of transportation to fulfil their mobility needs. This trend reflects broader demographic shifts such as population ageing, more active ways of living among the elderly, and the increasing number of female drivers. Gerontological research highlights the vital role of the automobile for the elderly population and their quality of life. Numerous studies highlight the critical importance that older adults attribute to driving and point to a substantial decline in mobility. Additionally, a significant deterioration in mental and physical well-being that accompanies the involuntary cessation of driving has often been detected. On the other hand, there are growing concerns regarding traffic safety and a rise in accidents among older drivers. How can we ensure the safety of all road users while also preserving the independence that driving offers to older individuals? This ongoing tension between safety and personal autonomy lies at the heart of the current debates.

This chapter aims to provide insight into the topic of older drivers. It starts by presenting the most up-to-date statistics on older drivers, highlighting the importance and relevance of this topic (1). Next, it addresses age-related changes in sensory, psychomotor, and cognitive abilities that affect driving skills (2). When discussing older drivers, the focus has increasingly shifted to those aged 75 and above. This contrasts with past trends, in which the category typically included individuals aged 65 and older. This change reflects the growing vitality of older populations. However, using fixed age thresholds can be misleading. Ageing is a highly individual process, and there can be significant differences in driving ability, physical health, and mental capacity among individuals of the same age. Additionally, specific diseases and medications can have a significant impact on driving ability, which is addressed in another section of this chapter (3). The final section addresses the psychosocial aspects and consequences of driving cessation (4).

DOI: 10.4324/9781003608394-8

Increase in older drivers – illustrative trends and statistics

Although the percentage of licensed drivers in older age groups varies significantly across countries, the data shows a similar trend of a substantial increase in the number and proportion of older drivers. In particular, licensing rates for the oldest group, aged 75 years and older, and for women have increased. Aggregated EU statistics on older drivers which would provide a broader picture of the current situation on European roads are lacking. Neither the updated share of driving licence holders by age category nor the average age at which people decide to give up their driving licence is available. To obtain the most recent figures, we must consult individual country data sources.

For instance, in *Czechia,* there were more than 1.5 million driving licence holders aged 65 and over in 2023, representing approximately a quarter of the population eligible to drive. Men consistently outnumbered women, with the gender gap widening as age increased. The number of people aged 80 and above holding a driving licence (type B) shows a clear upward trend. Between 2018 and 2023, the number of drivers aged 80 and over increased by 41%, from approximately 96,000 to 135,000. While male drivers dominated significantly, constituting over 75% of the group and showing an increase of approximately 30% over the observed period, the number of female drivers increased at a much faster rate, nearly doubling the initial value, rising from about 16,500 to approximately 31,500 (CSU, 2023a, 2023b).

In the *United Kingdom*, the proportion of full driving licence holders aged 70 and over rose from 44% in 2003 to 73% in 2023. Additionally, growth is particularly notable among older women: in 2023, 63% of women aged 70 or above held a driver's licence, compared to just 26% in 2003. According to the latest figures from the National Travel Survey, there are now 5.8 million driving licence holders aged 70 or older, accounting for nearly 17% of all licensed drivers in the United Kingdom. Within this age group, 84% of men and 63% of women hold a valid licence (DfT, 2024).

It is not a surprising fact that in the *United States*, the epitome of car-centric culture, the percentage of driving licence holders is even higher than in European countries. Among the population aged 65 and older, 93% of men and 86% of women are licensed drivers. Especially among men, the percentage decreases only after the age of 85 and over is reached, from 95% to 88% (FHA, 2023). In 2023, 19% of all fatal crashes involved drivers aged 65 years or above. However, the older population has grown at a rate faster than the increase in fatalities involving older drivers (NHTSA, 2023).

This global trend of an ageing driver population is also observable in EU road safety data. According to data from the European Road Safety Observatory, there were 5,180 senior fatalities (aged 65 or older) in 2020 in Europe, accounting for 26% of all road fatalities (EC, 2021). Older adults have the second-highest traffic fatality rate, with only young adults aged 18–24 experiencing a slightly higher number of deaths per million people. This is partly due to the increased physical vulnerability of older individuals. When involved in traffic collisions, they are more likely to succumb to their injuries compared to other age groups. Although the fatality rate is higher, the injury rate is often lower, indicating that they may be involved in fewer crashes overall but experience more severe outcomes when accidents do happen (Carson et al., 2023; EC, 2024).

Age-related changes in sensory, psychomotor, and cognitive abilities affecting driving skills

Driving is one of the most complex instrumental activities of daily living, involving both acquired skills and the integration of cognitive and physical functions. Challenges in managing this complexity – whether because of age-related decline, medical conditions, or impairments – can significantly affect an individual's ability to drive safely (e.g. AGS & Pomidor, 2016). Many older drivers will eventually reach a point where they must realise, decide, and accept that it is no longer safe for them to drive. However, it is important to emphasise that age alone is not a reliable indicator of driving ability as the rate and presence of these declines vary significantly between individuals. The next section examines changes in psychomotor, sensory, and cognitive functions that commonly affect driving performance in older adults.

Changes in psychomotor functions

Psychomotor functioning broadly refers to an individual's ability to move and coordinate the movement of various parts of their body. While there is considerable individual variability, and not all older drivers are affected in the same way or at the same pace, several key declines are commonly observed that significantly affect the ability to drive safely.

Sufficient motor conditions, such as leg strength and mobility, as well as flexibility in the head, neck, and shoulders, are crucial for driving. There is a significant reduction in muscle strength and mass, which can make it challenging to maintain a firm grip on the steering wheel or affect the ability to reach for controls or operate and apply pressure to pedals (e.g. Huang et al., 2020). Diminished arm strength can make it difficult to turn the steering wheel quickly and effectively. Leg pain or decreased leg strength can hinder the ability to move the foot swiftly between the accelerator and brake pedals (Lang et al., 2013). Studies have found that older drivers who make more pedal errors, such as failing to stop or accelerating inappropriately, face a higher risk of being involved in crashes (Freund et al., 2008). Problems with improper steering and/or braking can also be caused by a loss of tactile sensation, making individuals unaware that their foot is on the pedal or how much pressure they are applying. Impaired proprioception – the body's ability to sense movement, position, and orientation – can lead to difficulty in determining the exact position or angle of the foot (RSF, 2021).

Reduced joint flexibility and a limited range of motion are common, particularly in the neck and trunk. This can make it harder to look over one's shoulder to check blind spots or change lanes, or to check for other traffic and pedestrians when turning left or right at intersections (Lang et al., 2013; Vichitvanichphong et al., 2015). This is particularly problematic for older drivers, who often rely on neck rotation to compensate for restricted visual fields (Eby et al., 2015). Reduced coordination can affect overall vehicle control. A slower physical reaction time, which affects the ability to react quickly and accurately, could indirectly involve motor responses, such as moving the foot to the brake pedal. While modern vehicles are designed to be less physically demanding because of features such as power steering and automatic transmission, basic motor skills are still essential.

Changes in sensory functions

Vision is considered the primary sense used in driving. As people age, their visual and perceptual abilities tend to decline, although the rate and specific changes vary significantly among individuals. These changes can make various driving tasks more challenging and may increase the risk of accidents. Major age-related changes in vision that affect driving are related to visual acuity, visual field, contrast sensitivity, changes in glare sensitivity and night vision, depth perception, and useful field of view.

Visual acuity, the ability to discriminate fine, stationary, high-contrast details, commonly declines with age, particularly after the age of 50. This decline is linked to physiological changes in the eye and an increased incidence of conditions such as cataracts, glaucoma, diabetic retinopathy, and age-related macular degeneration (e.g. Lavallière et al., 2017). Dynamic visual acuity, the ability to perceive a moving target, also declines with age and has been shown to have a stronger relationship to driving errors than static acuity does (Duc-Nghiem et al., 2016; Lang et al., 2013). Reduced visual acuity can make it harder to identify objects on and along the road and to decipher traffic signs. While visual acuity is a standard test for licensing, some research suggests a weak or insignificant correlation with crash risk (e.g. Kobal & Hawlina, 2022; Ortiz-Peregrina et al., 2020). However, some studies indicate that visual screening laws are associated with reduced motor vehicle crash fatality rates (McGwin et al., 2008).

The peripheral vision, or visual field – the area where information can be detected with the head and eyes still – can also decline with age. This can be due to natural ageing changes such as ptosis (drooping eyelid) or narrowed pupils, but more significantly as a consequence of medical conditions such as glaucoma, stroke, or traumatic brain injury. Peripheral vision loss can make it difficult for drivers to notice traffic signs, cars in adjacent lanes, or pedestrians about to cross their path, especially when changing lanes, keeping the car centred, or approaching intersections (e.g. Duc-Nghiem et al., 2016).

Older adults typically experience reduced *contrast sensitivity*, requiring approximately three times more contrast than younger adults to distinguish objects from their background. This deficit is exacerbated by low light levels (Kobal & Hawlina, 2022). Poor contrast sensitivity can make it difficult to discern subtle details, such as deteriorated road markings, signals, and bumps, or to see traffic lights and vehicles at night. It has been identified as a valid predictor of crash risk among older drivers (Ortiz-Peregrina et al., 2020). Severe contrast sensitivity impairment resulting from cataract increases the at-fault crash risk among older drivers, even when present in only one eye. Cataract removal can improve contrast sensitivity and reduce the likelihood of a collision (Mennemeyer et al., 2013; Owsley et al., 2001).

Glare sensitivity and night vision also worsen with age. Older adults often experience increased sensitivity to glare from strong light sources, such as headlights, and it takes them significantly longer to recover their vision after being exposed to glare (Kobal & Hawlina, 2022). The increased sensitivity can affect object detection, particularly at night or on dark, winding roads, which makes night driving particularly challenging for older drivers.

Another age-related visual change is *declines in depth perception*, which make it more difficult to judge gaps in traffic accurately when merging or turning, as well as to estimate the distance and speed of approaching vehicles (e.g. Feldstein, 2019). *Additionally, reductions in visual attention and processing speed*, particularly in the *Useful Field of View* (UFOV), further impair the ability to respond quickly and safely in complex traffic situations. Shrinkage of the UFOV, which measures the ability to divide attention between a central location and the periphery, is correlated with the difficulty older adults experience with driving (Vichitvanichphong et al., 2015) and is a strong predictor of crash rates (Ball et al., 1993).

The reduction of visual functions often develops slowly and gradually, and drivers may not always be aware of the extent of their decline and therefore often overestimate their visual powers (e.g. EC, 2021). While many older drivers employ self-regulation strategies to compensate for these changes (e.g. Huang et al., 2020), some research indicates that these strategies alone may not always fully counterbalance the functional decline to maintain safety (Boot et al., 2013; Ortiz-Peregrina et al., 2020).

While vision plays a key role in driving, *hearing* serves as a valuable perceptual channel. Impaired hearing is a common and natural form of age-related sensory loss. This is particularly true for men and affects higher frequency ranges (Boot et al., 2013). In a study by Brayne et al. (2000), 23% of drivers aged 84 and over reported hearing loss. The ability to process auditory signals in the presence of noise also shows a substantial age-related decline. Hearing provides "all-around information", including warnings and alarms from the traffic environment, and can help calibrate a driver's perception of speed and vehicle performance (Lang et al., 2013). Road noise could potentially mask relevant signals. Additionally, the ability to detect vibration signals, such as those from raised centre lane markers or rumble strips, also declines with age, potentially reducing the value of these multimodal warnings for correcting lane drift (Boot et al., 2013). A diminished ability to discern multiple auditory streams is also a perceptual challenge for older drivers (Huang et al., 2020). However, Dobbs et al. (2005), in a meta-analysis of existing studies, found that even profound deafness does not represent a barrier to driving. While the relationship between deafness and driving performance varied across studies, there was no consistent evidence of an increased risk of crashes.

Changes in cognitive functions

Cognitive functions are crucial for safe driving, as they enable a driver to comprehend complex situations and make informed, timely decisions. Several key cognitive functions, including executive function, complex attention, and dual-tasking, have been associated with driving performance. They tend to decline with age, affecting the driving performance of older drivers (e.g. Depestele et al., 2020).

Divided *attention*, the ability to share attention between various tasks, such as keeping track of speed and other cars simultaneously, shows more pronounced changes than selective attention, which is the ability to separate important from unimportant information (e.g. Cicchino & McCartt, 2014). The ability to switch attention rapidly also degrades (Boot et al., 2013). This can make it harder for older drivers to divide their attention effectively between multiple activities, such as keeping track of road signs, signals, markings, and other traffic and pedestrians. It also affects their ability to scan the visual field for relevant objects and respond appropriately. It can contribute to "failure to look" crashes, particularly at intersections or when yielding the right of way. Older drivers are over-represented in "failure to look" traffic accidents (Reed et al., 2012) and may have a lower state of alertness, particularly in recognising and responding to signs indicating potentially dangerous traffic situations (López-Ramón et al., 2011).

Executive functions are a set of higher-order cognitive processes that are crucial for goal-directed behaviour, problem-solving, planning, sequencing, decision-making, and adapting to novel situations. As individuals age, a natural decline in some executive functions can occur. Failures in judgement, decision-making, and understanding road rules can contribute to crashes. While overlearned driving abilities might allow older drivers with executive function deficits to manage familiar routes, unexpected events (e.g. a child running into the street and road closures) can put them and others at risk (Lang et al., 2013).

Declines in *processing speed* and reaction time in response to demands from the traffic environment are hallmarks of ageing. Older adults typically exhibit slower simple and choice reaction times, taking significantly longer to perform basic information-processing tasks. According to Jastrzembski and Charness (2007), older adults require approximately 1.7–2.0 times more time than younger adults to perform basic information processing tasks, including perceptual, cognitive, and motor functions. This delay can be more pronounced in complex driving situations requiring more time, space, and attention than young drivers need. Reduced reaction time and slower information processing are key factors that contribute to an increased risk for older drivers.

Working memory refers to the ability to hold and manipulate information in one's mind over short periods. In driving, it is essential to remember a sequence of turns, process multiple pieces of information simultaneously (e.g. speed, traffic lights, and other vehicles), and anticipate future events. Memory problems, including forgetfulness or confusion, are red flags for unsafe driving as they can contribute to risky driving behaviour and errors.

Despite the aforementioned various risk factors that can affect driving performance, many older drivers adapt their driving behaviour to their limitations strategically, for example, by avoiding night driving or driving in heavy traffic, or reducing their speed, which can help them maintain safe driving habits (e.g. Boot et al., 2013; Eby et al., 2015; Huang et al., 2020). Additionally, advancements in vehicle technology, such as power-assisted braking and steering, automatic transmissions, and advanced driver assistance systems, have made driving less physically demanding, potentially enabling physically frail older adults to continue driving safely (Emmerson et al., 2013; Lang et al., 2013).

Impact of specific diseases and medications on driving

Medical conditions and medication classes often compound age-related changes in sensory, psychomotor, and cognitive abilities. Some of them, such as arthritis, Parkinson's disease, eye disorders, dementia, stroke, cardiovascular diseases, or diabetes, and the medications used to treat them, are more common among older drivers and relevant to driving safety, often impacting on driving performance and increasing crash risk (e.g. O'Neill, 2015). Evidence-based research emphasises the importance of these medical conditions rather than chronological age when assessing fitness to drive at an older age.

Parkinson's disease can impair driving performance as a result of a combination of motor and cognitive symptoms, including tremors, stiffness, akinesia (slowness of movement), and impaired postural reflexes, all of which can significantly prolong reaction times and interfere with safe driving. Excessive daytime sleepiness and sudden sleep onset – both common in Parkinson's disease – further increase the crash risk, especially during longer drives or when medication dosages are poorly managed or the disease has progressed (Charlton et al., 2010; Holland et al., 2003).

Mild cognitive impairment (MCI) and dementia (including mild Alzheimer's disease pathology) are strongly associated with an increased crash risk and abnormal driving behaviours (e.g. Charlton et al., 2010). This impairs critical cognitive functions necessary for driving, including attention, visuospatial skills, executive functions (such as planning, problem-solving, and decision-making), and memory. A significant percentage of older adults with MCI (up to 60%) and dementia (up to 30%) continue to drive, highlighting the critical need for appropriate assessment tools (Simons-Morton & Ehsani, 2016).

Elderly individuals with *diabetes* may experience a higher rate of motor vehicle crashes as a result of complications caused by advanced disease, such as retinopathy and neuropathy, or from hypoglycaemia, a common side effect of certain anti-diabetic medications. This side effect can lead to cognitive and motor slowing and even loss of consciousness, which may significantly impair driving ability, especially in older adults. Peripheral neuropathy, a complication of Type 2 diabetes, can cause a loss of sensation and proprioception in the feet, contributing to this issue (ODTF, 2021). While some epidemiological studies have reported an increased crash risk among individuals with diabetes, others have found no such association (Hemmelgarn et al., 2006). According to Hostiuc et al. (2016), patients with diabetes do not show a statistically significant increase in risk for adverse traffic events overall. However, older individuals and those who are insulin-dependent tend to face a higher risk. They conclude that, despite progress in the management of diabetes, road safety regulations, and automotive technology, the collision risk for drivers with diabetes has not declined significantly over the past 50 years.

Vision disorders, such as refractive error, cataract, glaucoma, age-related macular degeneration, and diabetic retinopathy, are significantly more prevalent in the older population. These conditions impair visual acuity, contrast sensitivity, and visual field, making it challenging to detect cues and judge distances. Refractive error is the most common and easily corrected vision issue. Cataracts, also common in older adults, are treatable with surgery. The other major vision conditions can often be managed if detected early, but some vision loss may persist, potentially increasing the risk of a car accident for those who continue driving (e.g. Charlton et al., 2010).

A stroke can result in motor deficits, such as paralysis, which may necessitate vehicle adaptations and retraining for driving (EC, 2021). *Musculoskeletal disorders*, such as osteoarthritis, can cause pain and limitations on the range of motion, directly affecting driving ability and physical activity. *Cardiovascular disease*, such as myocardial infarction, can lead to elevated crash rates and sudden incapacitation. Post-cardiac surgery, patients may experience *cognitive decline*, with an impact on driving (e.g. Sagberg, 2006).

Medications

Older adults often have altered sensitivity to medications, which should be carefully considered when they are being prescribed. They tend to experience stronger effects, including side effects and adverse reactions, and medications may stay in their system longer than in younger individuals. Because many older people live with multiple health conditions (comorbidity), they are more likely to be prescribed several different drugs at once. The more medications a person takes, the higher the risk of drug interactions, whether because of how the body processes the drugs or how the drugs affect the body. Additionally, over-the-counter medications can further increase the chance of these interactions (Holland et al., 2003; Ngcobo, 2025). Many medication classes can impair driving ability through side effects such as sedation, drowsiness, reduced motor coordination, hypoglycaemia, blurred vision, hypotension, syncope, and ataxia (Carr et al., 2023; Hill et al., 2020).

Antidepressants, including selective serotonin and norepinephrine reuptake inhibitors (SNRIs) and selective serotonin reuptake inhibitors (SSRIs), are associated with a higher risk of impaired driving performance, which may be affected by side effects such as sleep disturbance, agitation, dizziness, headache, and fatigue (Carr et al., 2023). *Benzodiazepines* are linked to an increased risk of at-fault crashes and can impair vision, attention, information processing, memory, motor coordination, and complex task performance (McGwin et al., 2000). The risk is particularly high with long-acting (long half-life) benzodiazepines and higher doses, especially during the first four weeks of therapy (Charlton et al., 2010). Other types of medication that significantly increase the risk of poor road test performance include *sedatives*, *hypnotics*, and even commonly used analgesics such as *nonsteroidal anti-inflammatory drugs (NSAIDs)* and *acetaminophen (paracetamol)*, which are typically taken to reduce pain, inflammation, and fever. Last but not least, *anticoagulants* and *angiotensin-converting enzyme (ACE) inhibitors*, commonly prescribed to patients with cardiovascular disease, may also compromise driving performance (Carr et al., 2023; McGwin et al., 2000).

It is important to note that the relationships among driving ability, underlying medical conditions, and the medications used to treat them are complex, making it difficult to isolate the effects of the medications from those of the diseases themselves. Nevertheless, the use of certain medications, as well as the overall number of medications, can serve as indicators of increased driving risk and may assist researchers and clinicians in identifying individuals at higher risk who could benefit from targeted interventions (Hill et al., 2020).

Driving transition and cessation

Although older drivers often experience a decline in their fitness to drive caused by the aforementioned decrease in psychomotor, sensory, and cognitive skills, as well as health complications, their driving experience frequently helps them compensate for these limitations (e.g. Andrews & Westerman, 2012). Enhancing driver training programmes or refresher courses and incorporating supportive technologies could assist further in addressing these challenges (Reed et al., 2012). Most older drivers also adopt safer driving habits and avoid challenging driving conditions. For instance, according to Bernstein et al. (2022), older age was associated with driving less per day, fewer trips, and reduced night-time driving. After executive functions had been accounted for, older age was related to increased self-regulation in driving but not dangerous driving behaviours in healthy older adults, as age was not associated with braking hard or accelerating hard. However, self-regulation relies heavily on accurate self-perception, and research indicates that some older adults tend to overestimate their driving abilities. This issue is of particular concern among those with cognitive impairment, who may lack the insight needed to recognise when their driving has become unsafe (e.g. Freund et al., 2005; Huang et al., 2020).

Challenges of driving cessation

Substantial psychological importance is attached to driving in adulthood (Windsor et al., 2007). A driving licence represents independence and autonomy for many seniors, and stopping driving can constitute a major life event with long-term consequences for individuals, their caregivers, and society at large (Piatkowski, 2020). The results of both older and more recent studies indicate a significant decrease in mobility and a substantial deterioration in mental and physical well-being following forced driving cessation (Rosenbloom, 2001; Fonda et al., 2001; Ragland et al., 2005; Liddle et al., 2014). The decision to stop driving stands as a robust predictor of heightened depressive symptoms (Marottoli et al., 2000; Windsor et al., 2007). It leads to a notable decrease in activities undertaken outside the home, perceived dependence on others, and loss of freedom (Davey, 2007) and identity (Ziegler & Schwanen, 2011). Car use and driving cessation affect the person and their friends and family, the way the person is viewed, the way they fit or do not fit into the group, and the person's ability to participate in a community (Goins et al., 2015). For instance, in a study by Pachana et al. (2017), the participants described stopping driving as a significant life event linked to a subjective sense of ageing. Regardless of whether they were still driving, older individuals associated having ceased driving with feeling older than their actual age. Expectations concerning practical challenges and changes in social identity both significantly predicted the level of stress related to giving up driving.

Driving cessation and the unavailability of a personal automobile are closely linked to unmet or unfulfilled mobility needs, which can be defined as those mobility needs that remain unfulfilled because of the impossibility of undertaking necessary or desired trips and activities. On average, at least one-third of older adults experience unmet travel needs, with older women and those aged 75 and above being the most affected (Luiu et al., 2017, 2018). Additionally, seniors living in rural areas with insufficient transportation services are at a greater risk of unmet mobility needs in the event of the unavailability of a private car compared to their urban counterparts facing similar living situations (e.g. Shergold & Parkhurst, 2012). The negative consequences of driving cessation primarily affect optional needs and discretionary activities, the importance of which is often underestimated in travel surveys (Graham et al., 2020; Siren et al., 2015).

It is not surprising that giving up a driver's licence is often an involuntary act and one that impairs quality of life. Many elderly drivers find themselves unprepared for this life-altering change and often do not foresee or accept the potential loss of their driving licence. Several characteristics influence the need and desire to continue driving as age increases. A meta-analysis by Ang et al. (2019) identified age, sex, education, income, race, attitudes and beliefs, and road and transportation systems as significant factors in the self-regulation of driving behaviour in older adults. Physical fitness, mental health, social influence, and support systems received by older adults are important driving forces influencing mobility and adjustments to travel patterns. Gender differences are significant in this area, with women giving up their car keys more frequently and at an earlier age, while men often have a stronger attachment to driving and automobiles. Evidence suggests that older women are more likely to relinquish their driving licence even when they can still drive safely (Siren et al., 2004), whereas for men, giving up driving is often a major life change, potentially because of their frequent role as primary drivers within the household. Additionally, the symbolic and status-related significance associated with cars contributes to this stronger attachment among men (Balkmar & Mellström, 2018).

Planning for driving transition

Driving cessation poses a significant threat to perceived control, which is a significant contributor to mental and physical health (Li et al., 2019). Perceived control is an important factor in the psychological and emotional well-being of older adults as they face changes in their driving abilities. Those who actively prepare for the possibility of stopping driving are more likely to maintain a sense of control and may be better able to manage the practical and emotional difficulties that can result from a sudden and unanticipated end to driving (Musselwhite & Shergold, 2013; Windsor et al., 2007). Nevertheless, most older adults struggle to envision a future without driving and have not made plans for alternative transportation. The lack of proactive planning for driving cessation among older adults has been consistently observed in research (Feng & Meuleners, 2020; Harmon et al., 2018).

Several self-assessment and evaluation tools have been developed over the past 15 years with the goal of helping older adults remain safe drivers for as long as possible while also enabling them to recognise potential current driving challenges and assess their readiness to give up driving. Exploring whether older adults expect to live longer than they will continue driving could prompt valuable conversations with family members or professionals and serve as an important first step in considering alternative mobility options (e.g. Babulal et al., 2019). A recent review by Dickerson et al. (2024) of interventions and strategies to support the transition of older adults from driving to driving cessation highlights the importance of a cooperative approach involving all relevant stakeholders. A move beyond the traditional emphasis on assessing fitness to drive is recommended, advocating instead a more proactive and collaborative approach to planning for future changes in driving status with early engagement and planning for driving cessation rather than a reactive approach. Empowering older adults in the decision-making process regarding retirement from driving, including both the potential loss of licensure and the selection of alternative transportation options, is essential. To maintain outdoor mobility after giving up their car keys, older adults can walk, use powered mobility devices (if cycling is no longer an option), or rely on others for transportation. This includes public transit, paratransit services, accessible taxis, or rides from family, friends, and community programmes. According to the existing literature, the most commonly reported alternative transport source is riding as a passenger with a friend or family member, especially for women and those in rural environments (e.g. Bird et al., 2017; Jones et al., 2018; Zeitler & Buys, 2015). For further information on the types and classifications of innovation to enhance mobility among older adults and ease the transition to non-driving, see, for example, Cirella et al. (2019).

References

American Geriatrics Society, & Pomidor, A. (Ed.). (2016, January). *Clinician's guide to assessing and counseling older drivers* (3rd ed., Report No. DOT HS 812 228). National Highway Traffic Safety Administration.

Andrews, E. C., & Westerman, S. J. (2012). Age differences in simulated driving performance: Compensatory processes. *Accident Analysis & Prevention*, *45*, 660–668.

Ang, B. H., Oxley, J. A., Chen, W. S., Yap, K. K., Song, K. P., & Lee, S. W. H. (2019). To reduce or to cease: A systematic review and meta-analysis of quantitative studies on self-regulation of driving. *Journal of Safety Research*, *70*, 243–251.

Babulal, G. M., Vivoda, J., Harmon, A., Carr, D. B., Roe, C. M., & Zikmund-Fisher, B. (2019). Older adults' expectations about mortality, driving life and years left without driving. *Journal of Gerontological Social Work*, *62*(8), 912–929.

Balkmar, D., & Mellström, U. (2018). Masculinity and autonomous vehicles: A degendered or resegregated future system of automobility? *Transfers*, *8*(1), 44–63.

Ball, K., Owsley, C., Sloane, M. E., Roenker, D. L., & Bruni, J. R. (1993). Visual attention problems as a predictor of vehicle crashes in older drivers. *Investigative Ophthalmology & Visual Science*, *34*(11), 3110–3123.

Bernstein, J. P. K., Mattek, N., Dorociak, K. E., Beattie, Z. T., Kaye, J. A., Ferguson, J. E., & Hughes, A. M. (2022). Age predicts older adults' driving self-regulation but not dangerous driving behaviors after controlling for executive function. *Gerontology*, *68*(1), 98–105. https://doi.org/10.1159/000515497

Bird, D. C., Freund, K., Fortinsky, R. H., Staplin, L., West, B. A., Bergen, G., & Downs, J. (2017). Driving self-regulation and ride service utilization in a multicommunity, multistate sample of U.S. older adults. *Traffic Injury Prevention*, *18*, 267–272.

Boot, W. R., Stothart, C., & Charness, N. (2013). Improving the safety of aging road users: A mini-review. *Gerontology*, *60*(1), 90–96.

Brayne, C., Dufouil, C., Ahmed, A., Dening, T. R., Chi, L. Y., McGee, M., & Huppert, F. A. (2000). Very old drivers: Findings from a population cohort of people aged 84 and over. *International Journal of Epidemiology*, *29*(4), 704–707.

Carr, D. B., Beyene, K., Doherty, J., Murphy, S. A., Johnson, A. M., Domash, H., Riley, N., Walker, A., Sabapathy, A., Morris, J. C., & Babulal, G. M. (2023). Medication and road test performance among cognitively healthy older adults. *JAMA Network Open*, *6*(9), e2332579.

Carson, J., Jost, G., & Meinero, M. (2023, November). *Reducing older people's deaths on European roads. PIN flash report* (No. 45). European Transport Safety Council (ETSC).

Charlton, J., Koppel, S., Odell, M., Devlin, A., Langford, J., O'Hare, M., Kopinathan, C., Andrea, D., Smith, G., Khodr, B., & Scully, M. (2010). Influence of chronic illness on crash involvement of motor vehicle drivers. *Accident Analysis & Prevention*, *42*(2), 436–444.

Cicchino, J. B., & McCartt, A. T. (2014). Trends in older driver crash involvement rates and survivability in the United States: An update. *Accident Analysis & Prevention*, *72*, 44–54.

Cirella, G. T., Bąk, M., Kozlak, A., Pawłowska, B., & Borkowski, P. (2019). Transport innovations for elderly people. *Research in Transportation Business & Management*, *30*, 100381.

Czech Statistical Office (CSU). (2023a). *Senioři v ČR v datech* [Seniors in the Czech Republic in data] (Publication No. 310034-23). Český statistický úřad.

Czech Statistical Office (CSU). (2023b). *Demografická ročenka České republiky – 2022* [Demographic yearbook of the Czech Republic – 2022]. https://csu.gov.cz/produkty/demograficka-rocenka-ceske-republiky-2022

Davey, J. A. (2007). Older people and transport: Coping without a car. *Ageing & Society*, *27*(1), 49–65.

Department for Transport. (2024). *National travel survey: Driving licence holders (NTS02)*. GOV UK. www.gov.uk/government/statistical-data-sets/nts02-driving-licence-holders#driving-licence-holding

Depestele, S., Ross, V., Verstraelen, S., Brijs, K., Brijs, T., van Dun, K., & Meesen, R. (2020). The impact of cognitive functioning on driving performance of older persons in comparison to younger age groups: A systematic review. *Transportation Research Part F: Traffic Psychology and Behaviour*, *73*, 433–450.

Dickerson, A. E., Stapleton, T., Bloss, J., Géinas, I., Harries, P., Choi, M., Margot-Cattin, I., Mazer, B., Patomella, A. H., Swanepoel, L., Van Niekerk, L., Unsworth, C. A., & Vrkljan, B. (2024). A systematic review of effective interventions and strategies to support the transition of older adults from driving to driving retirement/cessation. *Innovation in Ageing*, *8*(6), igae054.

Dobbs, B. M., Wodzin, E. P., & Vegega, M. (2005). *Medical conditions and driving: A review of the literature (1960–2000)*. National Highway Traffic Safety Administration.

Duc-Nghiem, N., Hoang-Tung, N., Kojima, A., & Kubota, H. (2016). Traffic safety of elderly road user: The global trend and the Japanese case. 土木学会論文集 *D3 (土木計画学)*, *72*(5), I_1249–I_1264.

Eby, D. W., Molnar, L. J., Zhang, L., St. Louis, R. M., Zanier, N., & Kostyniuk, L. P. (2015). Keeping older adults driving safely: A research synthesis of advanced in-vehicle technologies, a LongROAD study. *Transportation Research Part F: Traffic Psychology and Behaviour*, *32*, 117–129.

Emmerson, C., Guo, W., Blythe, P., Namdeo, A., & Edwards, S. (2013). Fork in the road: In-vehicle navigation systems and older drivers. *Transportation Research Part F: Traffic Psychology and Behaviour*, *21*, 173–180.

European Commission. (2021). *Facts and figures seniors*. European Road Safety Observatory, Directorate-General for Transport. https://ec.europa.eu/transport/road_safety/specialist/observatory_en

European Commission. (2024). *Road safety thematic report – seniors*. European Road Safety Observatory. European Commission, Directorate General for Transport.

Federal Highway Administration. (2023). *Highway statistics series. Distribution of licensed drivers – 2023 by sex and percentage in each age group and relation to population.* www.fhwa.dot.gov/policyinformation/statistics/2023/dl20.cfm

Feldstein, I. T. (2019). Impending collision judgment from an egocentric perspective in real and virtual environments: A review. *Perception*, *48*(9), 769–795.

Feng, Y. R., & Meuleners, L. (2020). Planning for driving cessation in older drivers. *Transportation Research Part F: Traffic Psychology and Behaviour*, *72*, 62–70.

Fonda, S. J., Wallace, R. B., & Herzog, A. R. (2001). Changes in driving patterns and worsening depressive symptoms among older adults. *Journals of Gerontology. Series B: Psychological Sciences and Social Sciences*, *56B*(5), 343–351.

Freund, B., Colgrove, L. A., Burke, B. L., & McLeod, R. (2005). Self-rated driving performance among elderly drivers referred for driving evaluation. *Accident Analysis & Prevention*, *37*(4), 613–618.

Freund, B., Colgrove, L. A., Petrakos, D., & McLeod, R. (2008). In my car the brake is on the right: Pedal errors among older drivers. *Accident Analysis and Prevention*, *40*(1), 403–409.

Goins, R. T., Jones, J., Schure, M., Rosenberg, D. E., Phelan, E. A., Dodson, S., & Jones, D. L. (2015). Older adults' perceptions of mobility: A metasynthesis of qualitative studies. *The Gerontologist*, *55*(6), 929–942.

Graham, H., De Bell, S., Flemming, K., Sowden, A., White, P., & Wright, K. (2020). Older people's experiences of everyday travel in the urban environment: A thematic synthesis of qualitative studies in the United Kingdom. *Ageing & Society*, *40*(4), 842–868.

Harmon, A. C., Babulal, G. M., Vivoda, J. M., Zikmund-Fisher, B. J., & Carr, D. B. (2018). Planning for a nondriving future: Behaviors and beliefs among middle-aged and older drivers. *Geriatrics*, *3*(2), 19.

Hemmelgarn, B., Lévesque, L. E., & Suissa, S. (2006). Anti-diabetic drug use and the risk of motor vehicle crash in the elderly. *The Canadian Journal of Clinical Pharmacology*, *13*(1), e112–e120.

Hill, L. L., Andrews, H., Li, G., DiGuiseppi, C. G., Betz, M. E., Strogatz, D., Pepa, P., Eby, D.W., Merle, D., Kelley-Baker, T., Jones, V., & Pitts, S. (2020). Medication use and driving patterns in older drivers: Preliminary findings from the LongROAD study. *Injury Epidemiology*, *7*(1), 1–11.

Holland, C., Handley, S., & Feetam, C. (2003). *Older drivers, illness and medications. Road safety research report.* Department of Transport. ISSN: 468–9138.

Hostiuc, S., Negoi, I., & Hostiuc, M. (2016). Diabetes and collision risk. A meta-analysis and meta-regression. *International Journal of Clinical Practice*, *70*(7), 554–568.

Huang, G., Luster, M., Karagol, I., Park, J. W., & Pitts, B. J. (2020). Self-perception of driving abilities in older age: A systematic review. *Transportation Research Part F: Traffic Psychology and Behaviour*, *74*, 307–321.

Jastrzembski, T. S., & Charness, N. (2007). The model human processor and the older adult: Parameter estimation and validation within a mobile phone task. *Journal of Experimental Psychology: Applied*, *13*(4), 224.

Jones, V. C., Johnson, R. M., Rebok, G. W., Roth, K. B., Gielen, A., Molnar, L. J., Pitts, S., DiGuiseppi, C., Hill, L., Strogatz, D., & Li, G. (2018). Use of alternative sources of transportation among older adult drivers. *Journal of Transport & Health*, *10*, 284–289.

Kobal, N., & Hawlina, M. (2022). Comparison of visual requirements and regulations for obtaining a driving license in different European countries and some open questions on their adequacy. *Frontiers in Human Neuroscience*, *16*, 927712.

Lang, B., Parkes, A., & Fernandez-Medina, K. (2013). *Driving choices for the older motorist: The role of self-assessment tools.* RAC Foundation.

Lavallière, M., D'Ambrosio, L., Gennis, A., Burstein, A., Godfrey, K. M., Waerstad, H., Puleo, R. M., Lauenroth, A., & Coughlin, J. F. (2017). Walking a mile in another's shoes: The impact of wearing an age suit. *Gerontology & Geriatrics Education*, *38*(2), 171–187.

Li, G., Andrews, H. F., Chihuri, S., Lang, B. H., Leu, C. S., Merle, D. P., Gordon, A., Mielenz, T. J., Strogatz, D., Eby, D. W., & LongROAD Research Team David LeBlanc Lindsay Ryan. (2019). Prevalence of potentially inappropriate medication use in older drivers. *BMC Geriatrics*, *19*, 1–8.

Liddle, J., Reaston, T., Pachana, N., Mitchell, G., & Gustafsson, L. (2014). Is planning for driving cessation critical for the well-being and lifestyle of older drivers? *Internal Psychogeriatrics*, *26*(7), 1111–1120.

López-Ramón, M. F., Castro, C., Roca, J., Ledesma, R., & Lupianez, J. (2011). Attentional networks functioning, age, and attentional lapses while driving. *Traffic Injury Prevention*, *12*(5), 518–528.

Luiu, C., Tight, M., & Burrow M. (2017). The unmet travel needs of the older population: A review of the literature. *Transport Reviews*, *37*(4), 488–506.

Luiu, C., Tight, M., & Burrow, M. (2018). An investigation into the factors influencing travel needs during later life. *Journal of Transport & Health*, *11*, 86–99.

Marottoli, R. A., Mendes, C. F. Glass, T. A., & Williams, C. S. (2000). Consequences of driving cessation: Decreased out-of-home activity levels. *Journals of Gerontology: Social Sciences*, *55*(6), 334–340.

McGwin, G., Sarrels, S. A., Griffin, R., Owsley, C., & Rue, L. W. (2008). The impact of a vision screening law on older driver fatality rates. *Archives of Ophthalmology*, *126*(11), 1544–1547.

McGwin, G., Jr., Sims, R. V., Pulley, L., & Roseman, J. M. (2000). Relations among chronic medical conditions, medications, and automobile crashes in the elderly: A population-based case-control study. *American Journal of Epidemiology*, *152*(5), 424–431.

Mennemeyer, S. T., Owsley, C., & McGwin, G., Jr. (2013). Reducing older driver motor vehicle collisions via earlier cataract surgery. *Accident Analysis & Prevention*, *61*, 203–211.

Musselwhite, C. B., & Shergold, I. (2013). Examining the process of driving cessation in later life. *European Journal of Ageing*, *10*, 89–100.

National Highway Traffic Safety Administration. (2023). *Overview of motor vehicle traffic crashes in 2023* (Research Note. DOT HS 813 705). U.S. Department of Transportation, National Highway Traffic Safety Administration. https://crashstats.nhtsa.dot.gov/Api/Public/ViewPublication/813705

Ngcobo, N. N. (2025). Influence of ageing on the pharmacodynamics and pharmacokinetics of chronically administered medicines in geriatric patients: A review. *Clinical Pharmacokinetics*, 1–33.

Older Drivers Task Force (ODTF). (2021, November). *Supporting safe driving into old age*. Technical Report. Road Safety Foundation. https://roadsafetyfoundation.org/project/safe-driving-into-old-age

O'Neill, D. (2015). Transport, driving and ageing. *Reviews in Clinical Gerontology*, *25*(2), 147–158.

Ortiz-Peregrina, S., Ortiz, C., Casares-López, M., Castro-Torres, J. J., Jimenez del Barco, L., & Anera, R. G. (2020). Impact of age-related vision changes on driving. *International Journal of Environmental Research and Public Health*, *17*(20), 7416.

Owsley, C., Stalvey, B. T., Wells, J., Sloane, M. E., & McGwin, G. (2001). Visual risk factors for crash involvement in older drivers with cataract. *Archives of Ophthalmology*, *119*(6), 881–887.

Pachana, N. A., Jetten, J., Gustafsson, L., & Liddle, J. (2017). To be or not to be (an older driver): Social identity theory and driving cessation in later life. *Ageing & Society*, *37*(8), 1597–1608.

Piatkowski, D. P. (2020). Exploring support for and solutions to family CABs (Chauffeur-Associated Burdens). *Transportation Research Record*, *2674*(10), 874–885.

Ragland, D. R., Satariano, W. A., & MacLeod, K. E. (2005). Driving cessation and increased depressive symptoms. *The Journals of Gerontology. Series A: Biological Sciences and Medical Sciences*, *60*(3), 399–403.

Reed, N., Kinnear, N., & Weaver, L. (2012). *Why do older drivers have more 'failed to look' crashes? A simulator based study* (No. PPR635). Transport Research Laboratory.

Road Safety Foundation. (2021, November). *Supporting safe driving into old age. Technical report*. https://roadsafetyfoundation.org/project/safe-driving-into-old-age

Rosenbloom, S. (2001). Driving cessation among older people: When does it happen and what impact does it have? *Transportation Research Record: Journal of the Transportation Research Board, 1779*(1), 93–99.

Sagberg, F. (2006). Driver health and crash involvement: A case-control study. *Accident Analysis & Prevention*, *38*(1), 28–34.

Shergold, I., & Parkhurst, G. (2012). Transport-related social exclusion amongst older people in rural Southwest England and Wales. *Journal of Rural Studies 28*, 412–421.

Simons-Morton, B., & Ehsani, J. P. (2016). Learning to drive safely: Reasonable expectations and future directions for the learner period. *Safety*, *2*(4), 20.

Siren, A., Hakamies-Blomqvist, L., & Lindeman, M. (2004). Driving cessation and health in older women. *Journal of Applied Gerontology*, *23*(1), 58–69.

Siren, A., Hjorthol, R., & Levin, L. (2015). Different types of out-of-home activities and well-being amongst urban residing old persons with mobility impediments. *Journal of Transport & Health*, *2*(1), 14–21.

Vichitvanichphong, S., Talaei-Khoei, A., Kerr, D., & Ghapanchi, A. H. (2015). What does happen to our driving when we get older? *Transport Reviews*, *35*(1), 56–81.

Windsor, T. D., Anstey, K. J., Butterworth, P., Luszcz, M. A., & Andrews, G. R. (2007). The role of perceived control in explaining depressive symptoms associated with driving cessation in a longitudinal study. *The Gerontologist*, *47*(2), 215–223.

Zeitler, E., & Buys, L. (2015). Mobility and out-of-home activities of older people living in suburban environments: 'Because I'm a driver, I don't have a problem'. *Ageing & Society*, *35*, 785–808.

Ziegler, F., & Schwanen, T. (2011). 'I like to go out to be energised by different people': An exploratory analysis of mobility and wellbeing in later life. *Ageing & Society*, *31*(5), 758–781.

9 Assessment of fitness to drive

Dana Černochová and Vlasta Rehnová

Introduction

Obtaining a driving licence is conditional upon demonstrating professional, medical, and psychological fitness to drive. The requirements for each type of fitness, as well as the methods used for their assessment, are governed by legal regulations. In addition to demonstrating knowledge of traffic rules and vehicle operation, a driver must also possess the psychological characteristics and state of health necessary for safe participation in road traffic.

Professional fitness to drive

A person is deemed *professionally fit* if they possess the skills and knowledge required to operate a motor vehicle. The competent authority issues a driving licence, which serves as proof of the holder's qualification for driving, as well as certifying their authorisation for a specific category of vehicles. The authority also has the power to make the driving licence conditional (in cases of defined medical requirements), to restrict it (if the holder has lost their professional qualification for a certain category of vehicles or decided to stop driving), to revoke it (in cases of medical or professional unfitness), or to suspend it (as a provisional measure during proceedings concerning conditional, restricted, or revoked entitlements).

The examination to assess *professional fitness* involves testing on traffic regulations, basic first aid, knowledge of vehicle control and maintenance, and a practical driving test. The law also sets out the curriculum for training and instruction, as well as the qualifications required of those authorised to provide such training. One of the conditions is undergoing a psychological assessment, during which the applicant must demonstrate psychological fitness to drive as well as teaching competence.

Certificate of professional competence

The law requires drivers of vehicles falling within the licence categories D1, D1+E, D, or D+E to hold a certificate of professional competence. The content,

DOI: 10.4324/9781003608394-9

scope, and method of the examination, along with the provisions regarding initial and periodic training, are stipulated in a regulation which also specifies the number of hours of training and the validity of the certificate (five years). No further examination is required if, within this five-year period, the driver completes regular training provided by an accredited institution. The obligation to hold a certificate of professional competence does not apply to drivers engaged in private road transport not associated with remuneration.

Medical fitness to drive

The link between medicine and psychology in assessing fitness to drive varies from country to country.

In Germany, fitness-to-drive evaluation incorporates close collaboration between medicine and psychology. Drivers undergo what is known as a medical-psychological assessment, where both a specialist physician and a psychologist must jointly decide on the individual's fitness. The assessment guidelines are developed and published jointly for both disciplines (BASt, 2019), and so is their commented version (Schubert et al., 2005).

In Austria, the cooperation between medicine and traffic psychology is conceived differently. When doubts arise about a person's fitness to drive, an official medical examiner issues a report, which must, in specific cases, include the result of a psychological assessment. Thus, the opinion of the traffic psychologist is not issued independently but is integrated into the official medical report. Specialist physicians and psychologists work in institutionally separate and largely independent frameworks but under the overall responsibility of official medical examiners.

In Switzerland, traffic psychologists and medical doctors generally operate independently, although in some cases, traffic medicine institutions have expanded their remit to include psychological assessments by employing traffic psychologists. It is up to the competent authority – or more specifically the designated medical officer – to decide whether the driver should undergo a medical and/or psychological assessment for their fitness to drive.

In Slovakia, special rules govern the review of driving fitness in relation to alcohol and other psychoactive substances, including medication. If a driver is reasonably suspected of violating road traffic rules by driving under the influence of alcohol, drugs, or medication – or refuses to undergo testing for such substances – the law enforcement authority may order, without initiating further proceedings, the person to undergo a psychiatric assessment for substance use disorders (Slovak Republic, 2009a). This applies even if the offence was committed abroad. If no substance use disorder is established, the driver is required to participate in a traffic-specific psychological counselling programme.

In the Czech Republic, an initial medical examination is mandatory for applicants for a driving licence, and periodic medical check-ups are required for drivers of emergency vehicles, employees who drive as part of their work, drivers of vehicles in categories C, D, and E, and driving school instructors. Medical fitness assessments are also mandatory for drivers seeking to have their licences reinstated following a driving ban or the accumulation of maximum penalty points under the demerit point system.

Medical fitness to drive refers to the physical and mental ability to operate a motor vehicle. A person holding or applying for a driving licence may be declared medically fit, medically unfit, or medically fit with conditions.

Legislation specifies diseases, impairments, and conditions that either exclude or condition medical fitness to drive. These include issues related to vision, hearing, the musculoskeletal and connective tissue systems, the cardiovascular system, diabetes mellitus, the nervous system, mental disorders, substance use disorders (including those caused by polydrug use), renal insufficiency, organ transplants, and artificial implants. A physician may impose conditions on fitness by means of harmonised codes or time limits on the validity of the medical report. If the physician ascertains that the examinee is medically unfit to drive, this must be reported to the authorities without delay. Medical fitness is assessed and the medical certificate issued by the examining physician on the basis of the applicant's or licence holder's declaration, the medical examination, and any additional specialist tests deemed necessary. The physician may request further expert assessments where appropriate.

Psychological fitness to drive

Determining psychological fitness to drive a motor vehicle should indicate that the individual does not pose a foreseeable risk to road safety in terms of their psychological characteristics. In other words, the public interest takes precedence over the interests of the individual in this respect. Different jurisdictions approach this by enacting legislation that defines the key criteria for assessing such fitness.

Psychological fitness to drive refers to the state of psychological traits relevant to this activity. In the Czech Republic, this is assessed through a specialised psychological evaluation.

Assessment of psychological fitness to drive

The purpose of this *traffic-specific psychological assessment* is to evaluate whether an individual is fit or unfit to drive a motor vehicle. In rare cases, this assessment may be requested by drivers themselves if in doubt about their own fitness to drive. Such cases typically involve older individuals who are self-critical in viewing their own capabilities. However, the majority of people who undergo this evaluation are required to do so by law. The role of the traffic psychologist is to weigh the individual's interest in participating in road traffic against the public interest in maintaining road safety.

The legislation defines which individuals are required to undergo a *fitness-to-drive assessment* and sets out the qualifications a psychologist must meet in order to be accredited to carry out such examinations.

Mandatory evaluations apply to holders of a category C driving licence if they drive vehicles exceeding 7,500 kg in weight, holders of category D and E licences, drivers who have reached 12 points in the demerit point system, and those who have been banned from driving for at least six months either by an administrative authority or by a court. In addition, a medical examiner may request a supplementary expert assessment, including a psychological evaluation, as part of the assessment of medical fitness to drive. In Germany and Austria, drivers may also be referred for psychological examinations for other reasons (Bukasa & Utzelmann, 2009), including obtaining a driving licence before the age of 18 (to assess maturity in terms of awareness of responsibility), difficulties at a driving school, physical disabilities, mental and physical illnesses, and advanced age.

The requirements which psychologists must meet to be qualified to assess psychological fitness to drive are defined as follows: completion of an accredited master's degree in psychology, postgraduate training in traffic psychology, three years of professional experience in the field of traffic psychology, and access to the equipment necessary to conduct such evaluations.

The legislation also lays down rules for maintaining documentation pertaining to the assessment of fitness to drive, issuing and submitting an expert report, the time frame for a possible follow-up examination (no sooner than three months later), procedures for appeals, the appointment of an appeals board, and the potential conducting of an appellate assessment. The prescribed structure of the traffic-specific psychological evaluation consists of an introductory interview, assessment of cognitive performance in terms of quantity and quality, evaluation of personality traits, a one-on-one interview, communication of the results to the examinee, and the issuance of a written report. The assessment includes the collection of background information relevant to the driver's fitness to drive, including records from the driver's licensing history. The final interview focuses on reviewing the findings and may include a brief counselling intervention addressing driving behaviour. Psychologists are required to follow this structured process during the assessment.

The *history taking* also includes information on both at-fault and not-at-fault accidents, traffic violations, the number of points accumulated under the demerit point system, any previous licence suspensions, and the causes and circumstances of all instances of traffic law violations or behaviour that endangered road safety.

Relatively stable personality traits and performance parameters are primarily assessed using standardised personality questionnaires and performance tests. The interview and background data provide information about life events and conditions that may have led to problematic driving behaviour. Any additional input available to the psychologist, particularly data obtained through the interview, history taking, and observations made during the evaluation, must be taken into account for a comprehensive assessment and final decision.

Chaloupka-Risser et al. (2011) point out the distinction between a *psychological assessment of fitness to drive* – the purpose of which is typically to evaluate the performance and behaviour of healthy individuals and to protect the public from high-risk drivers – and a clinical psychological evaluation, which addresses pathological performance and personality changes and is generally aimed at assisting and supporting the individual. They also note that effective assessment of fitness to drive requires a good understanding of the links between psychological methods and driving behaviour (e.g. the influence of the traffic system on driver behaviour, perception processes while driving, compensation for lower performance, risky driving behaviours, and formal and informal social norms).

In addition, Chaloupka-Risser et al. (2011) summarise the requirements for *assessing psychological fitness to drive* in the following points: analysis of the accident event (considering all conditions of the traffic situation), performance parameters (psychological functions needed for safe driving), personality traits (characteristics that may pose a risk to safe driving), attitudes (safety perspectives and work motivation), and usual behaviour patterns (alcohol consumption, mental illness, criminal offences, misinterpretation of social aspects, behavioural stability, or capacity for change).

A psychological assessment of fitness to drive also provides an opportunity for an individual counselling intervention, analysis of the causes of inappropriate driving behaviour, the formulation of remedial measures, and the development of compensatory mechanisms to reduce the risk of future failure in driving practice.

Mental functions and personality traits

Mental performance is assessed in terms of both quantity and quality through the following *performance attributes*: intellect, attention (especially concentration, distribution, and range), sensorimotor reactivity and coordination (particularly the speed and accuracy of sensorimotor responses to a series of visual, and where applicable, auditory stimuli under time pressure), decision-making (especially the speed and reliability under time constraints), memory (mainly visual), and other mental functions as needed. Regarding *personality traits*, it is necessary to examine emotional stability, lability and impulsivity, the capacity for adaptive behaviour, risk-taking tendencies, stress resilience, psychopathological symptoms, and other personality traits where relevant.

The *personality characteristics* typically associated with "risky drivers", that is, individuals with diminished psychological fitness to drive, include insufficient self-control, irresponsibility, recklessness, a strong need for self-assertion, hostility, and a low capacity for anticipation and discipline. Risky drivers tend to exhibit high scores for sensation-seeking, a trait marked by a pursuit of excitement, adventure, and novel experiences, a lack of inhibition, and low tolerance for monotony. In comparison to the general driving population, risky drivers are also likely to show lower levels of altruistic behaviour, a greater willingness to take risks, lower frustration tolerance, and features consistent with Type A behaviour patterns. A serious indicator of reduced psychological fitness to drive is aggressive traffic behaviour, which is often associated with personality traits such as competitiveness, impulsivity, or an excessively high level of self-confidence. Long-term psychological fitness is determined on the basis of permanent personality traits, the performance of cognitive functions, attitudes, and lifestyle. Current psychological fitness may be affected by fatigue, illness, stress, emotions, or current psychological strain (Šucha et al., 2016).

A specific area in the assessment of psychological fitness concerns *personality disorders*. Wagner et al. (2020) highlight particularly those personality disorders marked by increased impulsivity or reduced self-control. Prominent among these is antisocial personality disorder, which is typically characterised by irresponsibility, an inability to experience guilt, and a disregard for social norms and rules. Diagnostic criteria suggest that individuals with this disorder can be expected to be frequent perpetrators of infractions and criminal offences. In terms of their participation in road traffic, a risk of aggressive behaviour and a reduced fear of danger also need to be taken into account. Another personality disorder that requires attention when assessing fitness to drive is emotionally unstable personality disorder, which is primarily manifested in impulsive behaviour, mood swings without apparent cause, and a tendency to ignore the consequences of one's actions. Despite the limited number of studies on implications for traffic behaviour, the symptoms suggest a risk of unpredictable and aggressive behaviour. Kranich and Reschke (2021) also include schizoid personality disorder among the conditions that can present a risk to safe driving, particularly when the individual displays a poor capacity to recognise and adhere to social rules. These features are also identified as road safety risks by Madea et al. (2007). While the Czech regulations covering the assessment of fitness to drive do not mention personality disorders, the German directive states that persons who display a high level of aggressive potential combined with a strong tendency to assert their own interests and act impulsively – that is, symptoms which to some extent correspond to those of personality disorders – are considered unfit to drive (Schubert et al., 2005).

Legislative frameworks governing the psychological assessment of drivers vary across countries. These differences may concern the requirements for traffic psychologists' qualifications and practice, as well as specific criteria for conducting psychological assessments. For example, in Poland (according to *Metodyka przeprowadzania badań psychologicznych w zakresie psychologii transportu*), such traffic-specific psychological evaluations include the examination of cognitive functions and personality traits, including comprehension, perception, attention, psychomotor performance, reaction times, stress resilience, emotional stability, and temperament in relation to situational coping.

According to the German traffic regulations (Germany, 2010, 2024), the assessment of fitness to drive focuses on physical and psychological performance, the risk of sudden failure, and other relevant capacities, insofar as these may contribute to risky driving behaviour. Drivers undergo what is known as a medical-psychological assessment (MPU), which addresses questions posed by the authority on the basis of specifically formulated hypotheses. Under the German rules (BASt, 2019), reduced fitness to drive may be associated with deficits in the following functions:

- insufficiently fast and reliable perception of visual information
- uncertain orientation in the visual field
- temporary or permanent concentration disorders
- poor distribution of attention
- reduced attention span
- slowed motor responses
- uncertain, inaccurate, and hasty reactions
- lack of balance between speed and accuracy.

In Austria, the regulations define the performance dimensions that must be assessed: situational awareness, reaction capability, concentration, memory, intelligence, and sensorimotor skills. Unlike Germany, where personality traits are typically assessed through interviews and case history, Austria uses personality questionnaires. These assess social responsibility, self-control, psychological stability, risk-taking tendencies, predisposition to aggression, and willingness to respect traffic rules. The methods used must be officially approved, with consideration given to validity, reliability, and objectivity criteria, norm development, and relevance to the purpose of the assessment. As in Germany, Austria applies the same percentile thresholds to determine cut-off scores in performance-based assessments:

- Group 1 (licences A, A1, B, B1): at least the 16th percentile must be reached or exceeded in most tests that are administered.
- Group 2 (licences C, C1, D, C+E, D+E, and vehicles used for passenger transport): at least the 33rd percentile must be reached or exceeded in most tests that are administered.

In Slovakia, no specific psychological assessment methods are prescribed for evaluating fitness to drive, but psychologists are allowed to use only those methods that are standardised for a driver population and such standardisations are not more than 15 years old. Reassessment is possible after 12 months. The legislation (Slovak Republic, 2009a, 2009b) stipulates that the following mental abilities must be examined:

- general intellectual abilities at least at the below-average level of the population
- literacy (alphabetism)

- performance indicators at or above the 33rd percentile for each criterion: that is, attention under strain, psychomotor pace, sensorimotor reactivity and coordination, memory, visual perception, and perceptual speed
- for ADR drivers and drivers of emergency vehicles, peripheral vision must be at least 160° of the visual field.

Contraindications for determining psychological eligibility include the following personality traits:

- uncritical self-image, unreasonable self-esteem
- emotional instability
- low capacity for rational self-regulation and self-control and poor tolerance for negative affect
- maladaptive strategies for coping with life challenges, stress, and assessing traffic situations, including:
- irresponsible, reckless, inconsiderate, impulsive, and aggressive behaviour or driving style
- poor ability to estimate and assess risk and danger, strong risk-taking tendencies
- socially unacceptable behaviour towards other road users, a competitive and uncooperative orientation, intolerance, and inconsideration.

Psychological assessment methods

The selection of psychological methods for assessing fitness for specific tasks is based on occupational profiling data related to that activity. Such assessments require the use of standardised psychodiagnostic tools. The relevant Czech regulation does not specify cut-off scores or contraindications for declaring a person unfit for the activity under consideration. Nor does it define criteria for selecting particular psychodiagnostic methods suitable for psychological assessment in the field of transport. However, it does require that the properties and performance parameters stipulated by the decree be evaluated. Therefore, methods specifically designed to assess these characteristics must be used. A detailed overview of applicable methods for assessing fitness to drive is provided by Šucha et al. (2013).

Given that driving a motor vehicle involves the interaction of various performance functions, psychological assessments must include methods addressing multiple performance attributes. As individual methods alone explain only a small proportion of the possible indicators of risky driving behaviour, a wider range of performance and personality assessment tools should be applied.

When assessing the fitness to drive of a person with low performance scores, it is necessary to consider possible compensatory factors such as personality traits or driving experience. Schubert et al. (2005) argue that driving fitness deficits can only be compensated for to a limited extent. The more extreme the deficit, or the more deficits occur across multiple areas, the less effective such compensation is likely to be. Their commented guidelines identify the following potential compensatory factors: adequate intellectual capacity enabling anticipatory behaviour and the timely recognition of hazardous situations, driving experience and skills, intact sensory perception, and a sufficient level of responsibility, on the basis of which the driver can reasonably be expected to reflect self-critically on their deficits and adjust their driving behaviour accordingly. A person is considered psychologically fit to drive if their mental characteristics do not present a safety risk on the road, that is, if there is no evidence-based likelihood that they will cause a crash event (Schubert et al., 2005).

In Germany, psychodiagnostic methods used in traffic psychology must be standardised and meet scientific criteria. Only those methods that have been validated in terms of traffic safety, that is, those that show relevance to driving behaviour, are considered to meet the validity criteria. Specific tests and test batteries must be formally recognised. The methods focus on assessing performance parameters specified in legislation: orientation, reaction ability, concentration, attention, and resilience to stress. The following cut-off values have been set to standardise assessment criteria:

- Group 1 (licences A, A1, B, B1): the 16th percentile must be reached or exceeded in the majority of tests that are administered.
- Group 2 (licences C, C1, D, C+E, D+E, and vehicles used for passenger transport): because of the higher psychological performance demands, the required threshold is the 33rd percentile.

The same cut-off scores apply in Austria. Chaloupka-Risser et al. (2011) note that these thresholds are based on the normal distribution of the population, with the 16th to 84th percentile range representing the "average". However, drivers of more demanding vehicle types (holders of licences in groups C and D) must demonstrate higher-level performance, which is why the lower threshold for performance test results is set at the 33rd percentile. For these driving licence groups, greater demands are placed on attention and stress resilience for reasons such as longer driving periods, stress related to adherence to schedule, distractions caused by passengers, or driving in adverse weather conditions. Drivers of vehicles used for passenger transport (such as buses and taxis) may have to deal with problematic passenger behaviour. In Austria, psychological assessments of applicants for these professional driving roles include an evaluation of their motivation for the position.

The interview, as part of the psychological assessment of fitness to drive (BASt, 2019), should explore all relevant factors related to driving behaviour, particularly whether the driver:

- responds to questions openly and coherently,
- provides detailed and credible information about the traffic offence in question,

- demonstrates a self-critical insight into their problematic driving behaviour,
- seeks to recognise and change their life circumstances,
- has achieved a stable behavioural change, making a recurrence of their problematic conduct unlikely,
- has developed strategies to cope with maladaptive behaviours that have proved effective over time,
- is aware of the causes of their problematic behaviour, and
- has addressed serious issues (e.g. alcohol dependence) with professional help (e.g. receiving psychotherapy and treatment).

The psychological interview should also help clarify the possible causes of poor test performance and propose appropriate compensatory strategies.

As an example of how specific psychological assessment methods may be applied, Chaloupka-Risser et al. (2011) refer to *Michon's model*, which describes three levels of decision-making in driving. On a strategic level, the interview is used to assess planning abilities, such as selecting appropriate times and routes (looking into the driver's capacity to reflect on their own abilities). On a tactical level, personality assessment methods are employed in addition to the interview. This involves estimating driving behaviours and manoeuvres that allow ample time for decision-making. The operational level relates to managing driving tasks in potentially hazardous situations, where performance-based tests are especially informative.

Chaloupka-Risser et al. (2011) recommend that, particularly for older drivers, psychological assessments be supplemented with observation during a practical driving session to determine whether performance deficits in psychological tests may be compensated for by actual driving behaviour (such as driving skills, interaction and communication abilities, anticipation, and conduct in conflict situations).

Psychological assessment of fitness to drive and prognosis of driving behaviour

Psychological fitness-to-drive assessment serves to predict individuals who may, with a certain degree of probability, engage in problematic driving behaviour. Bukasa and Utzelmann (2009) describe a prognosis of the potential development of behavioural patterns based on such a psychological assessment. Their approach is underpinned by a multimodal three-axis diagnostic framework, which conceptualises driving behaviour as a joint action of personality traits, the environment, and time. The "personality" axis comprises three domains important for assessing fitness to drive: personality traits, attitudes, and somatic aspects (e.g. illnesses, addictions, and age). The "environment" axis covers areas that may be of diagnostic relevance in individual cases (e.g. family or work situation). The "time" axis highlights the importance of learning as the core element in individual development (to what extent the personality and life circumstances have changed since the offence and how stable these changes are).

The purpose of the *psychological assessment* is to evaluate an individual's mental fitness to drive, that is, to predict their potential for safe driving behaviour. The methods used for this purpose need to be verified in terms of their predictive validity. This requires choosing an appropriate criterion that makes it possible to identify a relationship between that criterion and the outcomes of psychodiagnostic methods. The essence of criterion validity lies in determining how closely test results match the selected criterion.

According to Chaloupka-Risser et al. (2011), an external criterion may be how a driver "performs" in road traffic, in other words, to what extent it is possible to predict whether the driver will commit traffic offences or cause crashes. However, it is also pointed out that this approach is not entirely satisfactory, as traffic crashes are partly random events, and it cannot be conclusively argued that a person who has not caused any crashes is necessarily a safer driver. Traffic crashes are a multifactorial phenomenon, and so objections are often raised against their use as a validation criterion.

Chaloupka-Risser et al. (2011) further argue that even if extremely high correlations were found between test variables and the external criterion of a "traffic crash", it would still not be possible accurately to predict the likelihood of a crash occurring. This is because test results are mathematically only an indication of the probability of an event happening. The problem is compounded by the fact that even individuals with excellent test results may still make poor judgements in traffic, as is sometimes seen with young drivers. Currently, no standardised procedure exists that clearly measures motivation for responsible behaviour with respect to safety. Berg et al. (2008) explain low correlations by noting that, among drivers who have caused a crash, the underlying problem may differ for each person or may result from a combination of various performance or personality deficits.

Another validation criterion in traffic psychology is the *assessment of driver behaviour during the actual driving task*. This approach has been known for some time. More recently, it was used by Bukasa et al. (2003) and Sommer and Häusler (2006). However, Berg et al. (2008) and Schubert et al. (2005) point out the difficulty of maintaining consistent conditions for all drivers participating in a validation study. Ideally, the test drive should always occur under the same lighting conditions, in the same vehicle, on the same route, and with the same traffic density. Nevertheless, different traffic situations may arise, placing different and non-comparable demands on attention or the ability to react quickly. Validation studies conducted in Austria involved a driving test, *Wiener Fahrprobe* (Chaloupka & Risser, 1995), which is governed by standardised rules.

In addition to the above, validation criteria may include the withdrawal of a driving licence, the current number of penalty points, or the number of traffic violations. These categories are less dependent on situational factors or other road users and more on the driver's own decision to engage in behaviour that violates traffic safety rules (Šucha et al., 2015).

The results of test methods from a sample of individuals provide data for establishing norms. Since these norms are used to compare the test performance of drivers, Bukasa and Utzelmann (2009) propose two criteria for the normative group: its members must be active drivers, and their motivation for undergoing the assessment must match that of the drivers being evaluated.

While comparing results in individual diagnostic methods against norms is essential, the final verdict on whether a person is psychologically fit or unfit to drive is rendered by the psychologist on the basis of all the information and findings obtained during the assessment process: objective data, interview results, observations of the driver's behaviour during the assessment, their tendency to downplay violations, and the specific driving licence category in question. The psychologist also draws on their knowledge of the traffic environment, the demands of different transport types, and previous assessment practice.

References

Berg, M., Kieschke, U., & Schubert, W. (2008). Artifizielle Validitatsnachweise – Anmerkungen zur verkehrspsychologischen Validierung von Messapparaturen. *Zeitschrift fur Verkehrssicherheit*, *54*(4), 183–187.

Bukasa, A., Strauch, H., & Koch, R. (2003). Validitätsüberprüfung verkehrspsychologischer Leistungstests für Fahreignungsbegutachtung. *Zeitschrift für Verkehrssicherheit*, *45*, 191–197.

Bukasa, B., & Utzelmann, H. (2009). Psychologische Diagnostik der Fahreignung. In *Anwendungsfelder der Verkehrspsychologie* (pp. 237–275). Hogrefe Austria.

Bundesanstalt für Strassenverkehr (BASt) (Hrsg.). (2019). *Begutachtungsleitlinien zur Kraffahreignung*. Berichte der Bundesanstalt für Strassenverkehr.

Chaloupka, C., & Risser, R. (1995). Don't wait for accidents – possibilities to assess risk in traffic by applying the 'Wiener Fahrprobe'. *Safety Science*, *19*(2–3), 137–147.

Chaloupka-Risser, C., Risser, R., & Zuzan, W.-D. (2011). *Verkehrspsychologie. Grundlagen und Anwendungen*. Facultas Verlags- und Buchhandels AG.

Germany. (1997, September 15). *Führerscheingesetz-Gesundheitsverordnung (FSG-GV). Bundesgesetzblatt II Nr. 322/1997*. www.ris.bka.gv.at/geltendefassung.wxe?ShowPrintPreview=True&abfrage=bundesnormen&gesetzesnummer=10012726

Germany. (2010, December 13). *Verordnung über die Zulassung von Personen zum Straßenverkehr (Fahrerlaubnis-Verordnung – FeV). Bundesgesetzblatt I*. www.gesetze-im-internet.de/fev_2010/

Germany. (2013, March 6). *Straßenverkehrs-Ordnung [StVO] (Fassung vom 6.3.2013, zuletzt geändert)*. Gesetze im Internet. www.gesetze-im-internet.de/stvo_2013/BJNR036710013.html

Germany. (2024, Oktober 23). *Straßenverkehrsgesetz [StVG]* (Dern. geänd. durch Art. 70 G vom 23.10.2024, BGBl. I S. 323). Gesetze im Internet. www.gesetze-im-internet.de/stvg/BJNR004370909.html

Kranich, U., & Reschke, K. (2021). *Verkehrspsychologie im automobilen Straßenverkehr: Kompakt* (Theorien und Methoden der Psychologie, Band 7). Verlag Dr. Kovač.

Madea, B., Musshoff, F., & Berghaus, G. (Hrsg.). (2007). *Verkehrsmedizin, Fahreignung, Fahrsicherheit, Unfallrekonstruktion*. Deutscher Ärzte-Verlag.

Schubert, W., Schneider, W., Eisenmenger, W., & Stephan, E. (2005). *BegutachtungsLeitlinien zur Kraftfahrereignung. Kommentar*. Kirschbaum Verlag.

Slovak Republic. (2009a). Act No. 8/2009 Coll., road traffic act. *Collection of Laws of the Slovak Republic*. www.slov-lex.sk/pravne-predpisy/SK/ZZ/2009/8

Slovak Republic. (2009b). Decree of the Ministry of Interior No. 9/2009 Coll., implementing the act on road traffic and on amendment of certain acts. *Collection of Laws of the Slovak Republic*. www.slov-lex.sk/pravne-predpisy/SK/ZZ/2009/9/20220101

Sommer, M., & Häusler, J. (2006). Kriteriumsvaliditaet des Expertensystems Verkehr [Criterion validity of the expert traffic system]. *Zeitschrift für Verkehrssicherheit ZVS*, *52*(2), 83–89.

Šucha, M., Černochová, D., Rehnová, V., Šrámková, L., & Zámečník, P. (2015). *Metody pro dopravněpsychologické vyšetření. Posuzování psychické způsobilosti k řízení silničních motorových vozidel* [Methods for assessing psychological fitness to drive]. Hogrefe-Testcentrum.

Šucha, M., Rehnová, V., Kořán, M., & Černochová, D. (2013). *Dopravní psychologie pro praxi. Výběr, výcvik a rehabilitace řidičů* [Traffic psychology in practice. Selection, training, and rehabilitation of drivers]. Grada.

Šucha, M., Vlčková, I., Černochová, D., Zámečník, P., & Rehnová, V. (Ed.). (2016). *Terminologický a výkladový slovník dopravní psychologie* [Explanatory dictionary of traffic psychology]. Palacky University.

Wagner, T., Müller, D., Koehl, F., & Rebler, A. (2020) *Fahreignungzweifel bei Verkehrsdeliquenz, Aggressionspotenzial und Straftaten, Rechtsgrundlagen und evidenzbasierte Profilbildung der Risikogruppen*. Kirschbaum Verlag.

10 Communication in traffic

Ralf Risser and Petr Zámečník

Introduction

The core topic of this chapter is the understanding of communication between road users, with communication being used as a synonym for interaction or social interaction throughout. It is important to realise that most of the time all road users occupy the same time and space. Therefore, they may, or in fact do, see each other and if they see each other, they may, and in fact most often do, adjust their behaviour accordingly (Risser, 1985). The difficulty of understanding that interaction is the same as communication stems from the fact that we do not usually think of car movements, speed changes or speed characteristics, choice of trajectories, and similar phenomena as acts of communication. However, if a car driver approaches a pedestrian crossing where a pedestrian is waiting without changing their speed, this is certainly a communicative signal given by one person to another. In addition to these (hundreds of?) interactions that could effectively be framed as communicative acts, many more explicit communication processes occur: using indicators, sounding the horn, directing facial expressions at other road users, signalling with one's hands, body parts, or whole body, or even exchanging words when passing each other at very low speeds or when standing in close proximity. No one has ever compiled a dictionary of the "language" used in the public space by individuals occupying different roles and functioning as various types of road users. Nevertheless, this language exists, is in use, and is clearly understood by most people, despite occasional misunderstandings. In what follows, several aspects of the role that communication plays in traffic and mobility will be outlined and discussed.

Another aspect of communication will also be addressed – namely, one-way communication within the framework of social marketing (Kotler et al., 2015). In this context, a "sender" attempts to persuade the "receiver" to adopt a specific behaviour, such as driving at a lower speed. While the feedback component is absent – or rather, there is no immediate feedback to the sender – this process still constitutes a form of communication.

DOI: 10.4324/9781003608394-10

The importance of communication

Effective communication is essential for the smooth, safe, and comfortable cooperation of all those interacting in road traffic, or more generally in public space (Risser & Persson, 1996). *Communication in traffic* is analogous to communication in everyday life, and thus Paul Watzlawick's first Axiom of Communication holds true: "We cannot not communicate." Everything we do in traffic is, to some extent, a form of communication. As soon as two individuals perceive each other, they begin to communicate (usually with the intention of ensuring their own safety or comfort). Any perceivable behaviour, including the absence of (inter)action, is typically interpreted by other road users as having some meaning. In other words, we communicate through our behaviour even when we do not necessarily intend to. Human communication thus involves both social behaviour and communicative acts.

Road accidents, or more accurately, crashes – as they often do not occur "by accident" – are, for the most part, the result of problematic communication between different road users. Only in rare cases are accidents caused by technical defects (Risser et al., 1991). As early as the 1980s, Hydén (1987) highlighted the aspect of interpersonal communication in road traffic, focusing on road user behaviour as social behaviour. Road users are viewed as members of a society, behaving within a complex social context, rather than as individuals acting independently of their social environment. "Communication" does not simply mean the transmission (or omission) of information and responding to it. It also includes infringing upon others' rights and/or feelings by the deliberate neglect of rules, which may lead to dangerous situations and ultimately result in relinquishing one's own rights with the aim to be cooperative and/or polite (Chaloupka & Risser, 2020).

Communication in road traffic can serve multiple functions: it allows individuals to see the consequences of their own behaviour being reflected in the behaviour of others, makes it possible to convey mutual understanding among road users, and, through signalling, improves both safety and traffic flow. However, conflicts and other problems may arise because many of these signals are not explicitly defined, which may lead to issues resulting from communication that is misleading or misunderstood.

This topic has enjoyed significant research interest. For example, the intricate social cues that underlie non-verbal communication involved in pedestrian crossing decisions were explored by Fuest et al. (2019). They showed that pedestrians use vehicle kinematics to deduce social intentions rather than perceiving it as a state of a moving object. They managed to expand the originally accepted basic interpretation proposed by Beggiato et al. (2018) by demonstrating that vehicles following intentionally atypical trajectories led to pedestrian confusion, as pedestrians seemed to evaluate vehicle kinematics in a way that went beyond mere time-to-arrival considerations.

Communication at the society level

At the societal level, communication occurs between various groups of road users or between road users and traffic authorities (such as the police and government agencies). This type of communication encompasses laws, norms, measures, and policies. When shifting from interpersonal communication to that between public authorities and society (e.g. through campaigning), desirable communication is that which is "good", that is, communication which provides information or advice leading to the adoption of desired behaviours or positive changes in behaviours which were previously "bad" (ECMT, 1999).

In principle, this is clear, but is it so straightforward in practice? When one looks into the real effectiveness of communication campaigns, it has rarely been demonstrated that a communication campaign alone can significantly improve road safety (see, e.g. Fujimoto et al., 2022). The European SafetyCube project (n.d.) summarised that well-designed campaigns, when combined with other measures, contribute to better safety, but the average effect of stand-alone campaigns is negligible (at most, a reduction of a few percentage points in numbers of accidents). A small positive influence of campaigns (in the order of a few percentage points) under certain circumstances was also demonstrated by Phillips et al. (2011). Naturally, not every campaign has the same added value, but generally, they are effective when accompanying or amplifying other measures rather than when they are run independently (ECMT, 1999).

Communication with the target group/the public

The theoretical foundation of communication with the public is related to the marketing model, that is, the communication model derived from social psychology, with Philip Kotler (e.g. Kotler et al., 2015) being among the most influential sources. At the heart of this model is the assumption that all measures targeting individuals with the goal of encouraging them to change their behaviour stem from a sound understanding of these individuals. All work related to the marketing model begins with engaging with the target groups in order to learn about them and their motivations. All subsequent marketing steps – including product design and improvement, communication, and incentivising and distribution activities – must be based on this information. The rationale is that under this condition, there is a better chance that the measures will align with the perspectives of the target groups. The stimulus (in our case, a message to the public or segments of it) should correspond to the viewpoints of the individuals being addressed if it is to be noticed by them. A campaign achieves optimal effectiveness when it is used in harmony with a variety of other measures, working as an orchestra. Kotler's model is built upon five fundamental steps.

Marketing steps according to Kotler's social marketing model

- Information (citizens' motivational status): understand road users' preconditions for meeting their needs and avoiding barriers. It is important to gather comprehensive information about the existing and potential customers for the "social product", as well as systematically collecting data about their attitudes and motivations (including their needs, interests, and moral concepts).
- Product: define what is to be marketed and then set clear objectives, including the precise determination of the behavioural change that needs to be achieved. In principle, it is essential to offer attractive products placed in an agreeable context. The traffic "product" then needs to be designed in accordance with the needs and wishes of the citizens, specifically the users of the road/public space.
- Communication: engage citizens and road users with arguments that are acceptable and avoid concealing any downsides. Information, instructions, public relations measures, and advertisements must be provided in such a way as to align with the motivational settings of the audience. Appropriate communication will reach target population groups and help in persuading them to adopt the desired behaviours central to any social marketing activity. Such communication goes beyond merely presenting facts; it gives them a spin and reflects the reasons why individuals should care about these facts and adhere to the ideas they represent. According to the European CAST manual (Campaigns and Awareness-raising Strategies in Traffic Safety; European Commission, 2010), messages are more likely to influence behaviour if the recipient feels motivated to process them, thinks actively about them, and elaborates on their content.
- Incentives: provide the target group with opportunity to test the product and allow them to experience its convenience and utility. Such measures offer direct (contingent) reinforcement for citizens when they engage in wished-for behaviours, such as using public transport.
- Distribution: place communication measures in locations where there is a high probability that citizens will become aware of them.
- An effective message must be plausible to the audience (*credible*), convey a behaviour that is achievable (*realistic*), be heard repeatedly (*consistent*), be easy to understand (*clear*), be capable of generating change (*persuasive*), be *relevant* to the individual, and be appealing (*attractive*).

In terms of content, communication within the framework of social marketing also plays the role of providing a knowledge base for citizens. For instance, the health benefits of walking or the detrimental effects of insufficient exercise as a result of excessive car use should be communicated to the public to encourage them to reconsider certain options – and perhaps change their mode choice behaviour (e.g. Ogilvie et al., 2007; Risser & Šucha, 2020b).

Understanding the citizens

Optimum functioning of a communication measure can be achieved if the target population is properly segmented. Different segments of the population will respond to different types of factual information and arguments. Risser and Šucha (2020b) proposed a segmentation based on the Trans-Theoretical Model (TTM) developed by Prochaska et al. (1992). The model features the following stages: (1) *Pre-contemplation* – where the individual has no intention of changing their behaviour and may even resist change; (2) *Contemplation* – where the individual begins to recognise the drawbacks of their behaviour, but without any concrete intentions yet; (3) *Preparation* – where the individual begins to make preparations for change; (4) *Action* – where change has occurred, although there is still the risk that the individual may revert to their previous behaviour patterns; (5) *Maintenance* – where the new behaviour becomes a habit; and (6) *Termination* – where the new behaviour is fully established, and the individual is unlikely to revert to their old behaviour.

The methods to be used for segmenting the population are not discussed in this chapter. Information on how to proceed in this regard can be found in Creswell and Clark (2017), for example. However, once the target population has been divided into subgroups according to the procedure outlined earlier, a systematic dialogue can begin, aimed at understanding how the different groups perceive the behaviour that is being promoted, seeking the answers to questions such as what makes it attractive in their eyes and what, conversely, makes it difficult to adopt, and what obstacles prevent them from adhering to it. Such knowledge can then serve as the basis for developing suitable communication strategies tailored to each group.

Communication models and theories

As indicated earlier, in order to improve mobility efficiency and increase traffic safety, road users need to understand the intentions of others so that they can adjust their behaviour accordingly. When we want to convey our intentions, we must interact and communicate with each other. In addition to physical forms of communication, we can use facial expressions, gestures, eye contact, visual signs, acoustic signals (Risser, 1985), or movement energy (speed, acceleration/deceleration, and direction) (Lee et al., 2021) as communicative signals. Given that traffic is full of impersonal and short-term interactions, it is not possible to use "rich" verbal communication. Instead, such exchange relies more on signals or cues, as on the road, you will need to use signals to inform other road users about what you are planning to do. It is key that communication occurs both "between participants" and "about the subject". The former is especially important for safety and comfort, while the latter affects the quality of life. This type of communication can be described using various models: one-way communication models, interactive models (two-way communication), and transactional models, involving two-way communication, where the message becomes more complex as the communication event progresses (Risser & Šucha, 2020a).

One of the classic approaches to one-way communication which remains widely cited is the model designed by Shannon and Weaver (1949). The model consists of six elements – the information source, the transmitter, the receiver, the information sink, the information channel, and the noise source – which determine whether communication can proceed in accordance with the original intention and content. In addition to these six elements, emotions also play an important role in the exchange of information: the expectations one has of others, stress, anger, or fear can significantly influence the quality of information exchange, and can thus also act as sources of noise in the communication process (McQuail, 2008). The key message of this theory for practice is that "what someone says is not the same as what someone else hears". The amber phase at a traffic light-controlled junction can serve as an example. The message that road designers are communicating is that you should stop (brake or remain stationary). However, drivers decode this information differently. Empirical evidence shows that there is considerable variation in the way drivers react (while driving schools, for instance, always stop, the majority of experienced drivers accelerate to "catch the amber").

Regarding interactive models, *Schulz von Thun's approach, also known as the "four ears model"*, can be applied. This model proposes that what we say has a fourfold effect (Schulz von Thun, 2005): (1) what we inform about, (2) what we reveal about ourselves, (3) what we think of the receiver and how we relate to them, and (4) what we want to make the receiver do. In addition to these four layers, both the sender and the receiver are responsible for the quality of communication, with clear communication being more of an ideal than a rule. Extending the Shannon–Weaver model, this model asserts that the four levels on both sides of the communication process must be correctly aligned for communication to be successful. The "four ears model" can be used in campaigns promoting road safety or sustainable modes of transport because it helps convey the intended message to the audience (target group) as it was meant to be.

Another *four-layer model, the "integrative sustainable development communication model"*, was developed by Putri et al. (2020). It begins with assessment, which consists of analysing the situation before the policy is created and after the policy has been implemented. This stage is followed by planning, which involves developing communication strategies (interests, mass media involvement, and the changing interests of the local community), and production, comprising the preparation of messages and media production (integrating perspectives from the breadth of the horizon, knowledge, and public education). Each actor is responsible for the message or information communicated to the public. Finally, the action and reflection stage focuses on whether the message can influence the target group and have an impact on them. These impacts can be both positive and negative.

Communication in the automation era

In addition to communication between road users themselves and between them and authorities or transport system designers, the issue of interaction between human road users and automated vehicles (AVs) is becoming increasingly important. Progress in automation poses new challenges regarding how to ensure comprehensible and effective human–machine communication.

The taxonomy introduced by the Society of Automotive Engineers (SAE) distinguishes six levels of automation, with Layers 4 and 5 representing highly and fully automated driving, where the entire dynamic driving task is handed over to an artificial intelligence system without the need for human driver intervention (SAE, 2021). Although fully automated vehicles are not yet in widespread use, real-world testing – including that conducted in mixed traffic environments with pedestrians, cyclists, and conventional vehicles – is on a constant rise. However, such interaction places significant demands on the social capabilities of autonomous vehicles. Given the complexity of urban environments and human needs, there will always be a degree of mixing between automated and non-automated vehicles, as well as with other road users.

Communication between humans and vehicles controlled by artificial intelligence has therefore become a highly appealing subject (e.g. Chaloupka & Risser, 2020; Markkula et al., 2020; Rasouli & Tsotsos, 2019; Rothenbucher et al., 2016). However, research into communication between AVs and other road users remains in its very early stages, as it is difficult to study something that has not yet been fully implemented. AVs will face challenges in navigating urban settings, where the diversity of travel modes creates complex interactions. Human road users rely on experience, instinct, or cultural knowledge, and are accustomed to analysing many different forms of signals and social cues in order to anticipate the intentions of others and support their own decision-making. AVs will therefore need not only to detect other road users but also to analyse and predict their behaviour in order to communicate the vehicle's intentions effectively.

A key factor will be the ability of AVs to communicate with human road users in a way that is comprehensible, predictable, and safe. Research shows that people, particularly pedestrians, often base their decisions on non-verbal and social cues, such as eye contact, gestures, or the movement patterns of vehicles (Šucha, 2014; Lundgren et al., 2017). However, automated vehicles lack these means of communication and therefore require new forms of interaction, such as external human–machine interfaces (eHMIs), including light displays, projections, or other visual signals (Mariani et al., 2020; Schroeder & Demetre, 2021).

However, existing studies show mixed results: while some forms of eHMI increase perceived certainty and understanding, others, especially if not sufficiently intuitive or familiar, may confuse people (Mariani et al., 2020). At the same time, people maintain their tendency to seek eye contact even with

AVs, which suggests a deeply rooted expectation of this form of interaction, despite the driver's seat being vacant (Schroeder & Demetre, 2021).

In addition to technical and perceptual challenges, questions of behaviour and ethics also arise. Some studies warn, for example, that the excessive predictability of AVs, such as their willingness to stop for pedestrians at all times, could lead to people taking advantage of this and engaging in risky crossing practices, knowing that the AV will always stop to avoid a collision (Fujimoto et al., 2022). This creates a new type of asymmetry in road traffic that will need to be considered when developing rules and communication technologies.

It appears that the successful integration of AVs into regular traffic depends not only on technological progress but also on the ability to create comprehensible communication mechanisms. Research evidence accumulated since the 1960s has shown that road users do not move irrespective of others, but instead coordinate and adjust their movements to avoid collisions or mutual obstruction (see Lee et al., 2021). This topic was also addressed in the work of Erving Goffman (for more recent studies, see Haddington & Rauniomaa, 2014 or Markkula et al., 2020). In traffic, interpretative communication and interactions are often influenced by multiple factors, including road infrastructure, traffic rules, and cultural expectations (Markkula et al., 2020). The interpretational context involves various forms of non-verbal interaction, including those that are officially recommended (e.g. ADT, 2018). These will be absent in contact with *automated vehicles*. Therefore, autonomous cars must ensure alternative forms of communication to replace the human non-verbal component.

References

Arizona Department of Transportation (ADT). (2018). *Arizona driver license manual and customer service guide* (Publication No. 99-0117). https://apps.azdot.gov/files/mvd/mvd-forms-lib/99-0117.pdf

Beggiato, M., Witzlack, C., Springer, S., & Krems, J. (2018). The right moment for braking as informal communication signal between automated vehicles and pedestrians in crossing situations. In *Advances in human aspects of transportation* (pp. 1072–1081). Springer.

Chaloupka, C., & Risser, R. (2020). Communication between road users and the influence of increased car automation. *Transactions on Transport Sciences*, *10*(2), 5–17.

Creswell, J. W., & Clark, V. L. P. (2017). *Designing and conducting mixed methods research*. Sage Publications.

European Commission. (2010). *Road safety communication campaigns: A final deliverable from the project CAST – campaigns and awareness-raising strategies in traffic safety*. Publications Office of the European Union. https://road-safety-charter.ec.europa.eu/sites/default/files/castroadsafetycommunicationcampaigns.pdf

European Conference of Ministers of Transport (EMCT). (1999). *Communication in road safety: International seminar – Warsaw, 2–3 October 1997*. OECD Publishing. https://doi.org/10.1787/9789264173057-en

Fuest, T., Michalowski, L., Schmidt, E., & Bengler, K. (2019, October). Reproducibility of driving profiles – application of the Wizard of Oz Method for vehicle pedestrian interaction. In *2019 IEEE intelligent transportation systems conference (ITSC)* (pp. 3954–3959). IEEE.

Fujimoto, S., Kiyohara, K., Matsuyama, T., Kitamura, T., & Kiguchi, T. (2022). Assessing the impact of the national traffic safety campaign: A nationwide cohort study in Japan. *BMJ Open*, *12*(1), e054295. https://doi.org/10.1136/bmjopen-2021-054295

Haddington, P., & Rauniomaa, M. (2014). Interaction between road users: Offering space in traffic. *Space and Culture*, *17*(2), 176–190.

Hydén, C. (1987). The development of a method for traffic safety evaluation: The Swedish traffic conflicts technique. *Bulletin Lund Institute of Technology Nr. 70*. Lund Technical University.

Kotler, P., Burton, S., Deans, K., Brown, L., & Armstrong, G. (2015). *Marketing*. Pearson Higher Education AU.

Lee, Y. M., Madigan, R., Giles, O., Garach-Morcillo, L., Markkula, G., Fox, C., Camara, F., Rothmueller, M., Vendelbo-Larsen, S. A., Rasmussen, P. H., & Merat, N. (2021). Road users rarely use explicit communication when interacting in today's traffic: Implications for automated vehicles. *Cognition, Technology & Work*, *23*, 367–380. https://doi.org/10.1007/s10111-020-00635-y

Lundgren, V. M., Habibovic, A., Andersson, J., Lagström, T., Nilsson, M., Sirkka, A., Fagerlönn, J., Fredriksson, R., Edgren, C., Krupenia, S., & Saluäär, D. (2017). Will there be new communication needs when introducing automated vehicles to the urban context? In *Advances in human aspects of transportation* (pp. 485–497). Springer.

Mariani, S., Cabri, G., & Zambonelli, F. (2020). Coordination of autonomous vehicles: Taxonomy and survey. *arXiv preprint* arXiv. https://arxiv.org/abs/2001.02443

Markkula, G., Madigan, R., Nathanael, D., Portouli, E., Lee, Y. M., Dietrich, A., Billington, J., Schieben, A., & Merat, N. (2020). Defining interactions: A conceptual framework for understanding interactive behaviour in human and automated road traffic. *Theoretical Issues in Ergonomics Science*, *21*(6), 728–752. https://doi.org/10.1080/1463922X.2020.1736686

McQuail, D. (2008). Models of communication. In *The International Encyclopedia of communication*. Wiley Online Library.

Ogilvie, D., Foster, C. E., Rothnie, H., Cavill, N., Hamilton, V., Fitzsimons, C. F., & Mutrie, N. (2007). Interventions to promote walking: Systematic review. *BMJ*, *334*(7605), 1204.

Phillips, R. O., Ulleberg, P., & Vaa, T. (2011). Meta-analysis of the effect of road safety campaigns on accidents. *Accident Analysis & Prevention*, *43*(3), 1204–1218. https://doi.org/10.1016/j.aap.2011.01.002

Prochaska, J. O., DiClemente, C. C., & Norcross, J. C. (1992). In search of how people change: Applications to addictive behaviors. *American Psychologist*, *47*(9), 1102–1114.

Putri, N. E., Frinaldi, A., & Mubarak, A. (2020, December). Communication model for sustainable development of public transportation infrastructure in West Sumatra. In *6th International conference on social and political sciences (ICOSAPS 2020)* (pp. 103–109). Atlantis Press. https://doi.org/10.2991/assehr.k.201219.016

Rasouli, A., & Tsotsos, J. K. (2019). Autonomous vehicles that interact with pedestrians: A survey of theory and practice. *IEEE Transactions on Intelligent Transportation Systems*, *21*(3), 900–918.

Risser, R. (1985). Behavior in traffic conflict situations. *Accident Analysis & Prevention*, *17*(2), 179–197.

Risser, R., & Persson, H. (1996). *Kommunikation im Straßenverkehr – Ein Grundlagenpapier*. Lund Institute of Technology and Factum OG.

Risser, R., & Šucha, M. (2020a). Communication between road users and the influence of increased car automation. *Transportation on Transport Sciences*, *10*(2), 5–17. https://doi.org/10.5507/tots.2019.014

Risser, R., & Šucha, M. (2020b). Start walking! How to boost sustainable mode choice – psychological measures to support a shift from individual car use to more sustainable traffic modes. *Sustainability*, *12*(2), 554.

Risser, R., Tamme, W., Steinbauer, J., & Kaba, A. (1991). *Handbuch zur Erhebung von Verkehrskonflikten mit Anleitungen zur Beobachterschulung*. Kuratorium für Verkehrssicherheit.

Rothenbucher, M., Habibovic, M., & Bengler, K. (2016). The role of external human-machine interfaces in road safety: A study on pedestrian perception of automated vehicles. *Transportation Research Part F: Traffic Psychology and Behaviour*, *41*, 1–14. https://doi.org/10.1016/j.trf.2016.05.004

SAE International. (2021). *Taxonomy and definitions for terms related to driving automation systems for on-road motor vehicles* (J3016_202104). https://doi.org/10.4271/J3016_202104

SafetyCube Project. (n.d.). *SafetyCube: Safety CaUsation, benefits and efficiency*. Retrieved March 31, 2025, from www.safetycube-project.eu/

Schroeder, B. J., & Demetre, J. D. (2021). Pedestrian interaction with autonomous vehicles: The effects of vehicle movement and visual cues on crossing decisions. *Accident Analysis & Prevention*, *153*, 106038. https://doi.org/10.1016/j.aap.2021.106038

Schulz von Thun, F. (2005). *Jak spolu komunikujeme: Překonávání nesnází při dorozumívání*. Grada Publishing.

Shannon, C. E., & Weaver, W. (1949). *The mathematical theory of communication*. University of Illinois Press.

Šucha, M. (2014). Road users' strategies and communication: Driver-pedestrian interaction. *Transport Research Arena (TRA)*, *1*.

11 Traffic system

Culture and norms

Anton Pashkevich, Petr Zámečník, and Václav Linkov

Introduction

As a broadening of perspectives on accident phenomena, Parker (2007) described three eras of safety, highlighting their complementary characteristics. The first period was primarily associated with technology and technical measures, the second focused on systems, and the third was identified as the era of safety culture. All three periods are illustrated in Figure 11.1. It can be argued that the era of *safety culture introduced a "new paradigm"* which can be used to explain the main reasons behind the dramatic differences in road traffic safety outcomes across countries and promote global reductions in road accidents. This also applies to countries that have already achieved a high level of road safety, where further improvements may be more difficult to attain. The concept is useful "to explain observed differences in international, regional, and demographic crash risk, as well as the propensity to commit high-risk behaviours" (Ward et al., 2010).

Although the concept of *safety culture* emerged over 25 years ago (Guldenmund, 2000), and its practical application to traffic safety began a few years later (Wiegmann et al., 2007), it remains a relatively new topic within the traffic literature. This is especially true in relation to explaining the main causes of the striking differences in the number of road crashes, injuries, and fatalities both between countries and within national contexts. Surprisingly, even the second era, associated with systems theory, has not been fully utilised to develop existing road safety strategies. This issue was addressed by Hughes and co-authors (2015), who evaluated Sweden's *Vision Zero*, the United Kingdom's *Tomorrow's Roads – Safer for Everyone*, the Netherlands' *Sustainable Safety*, and Australia's *National Road Safety Strategy: 2011–2020.* This assessment, grounded in fundamental elements of systems theory (i.e. key components, relationships, and goals) and their interdependencies, led the researchers to conclude that *the analysis did not provide evidence that road safety strategies have a solid foundation in systems theory*. They pointed out that the strategies in question did not involve essential aspects of systems theory, which would make it possible to describe the interdependence and interaction between the individual key components. The team led by Brett P.

DOI: 10.4324/9781003608394-11

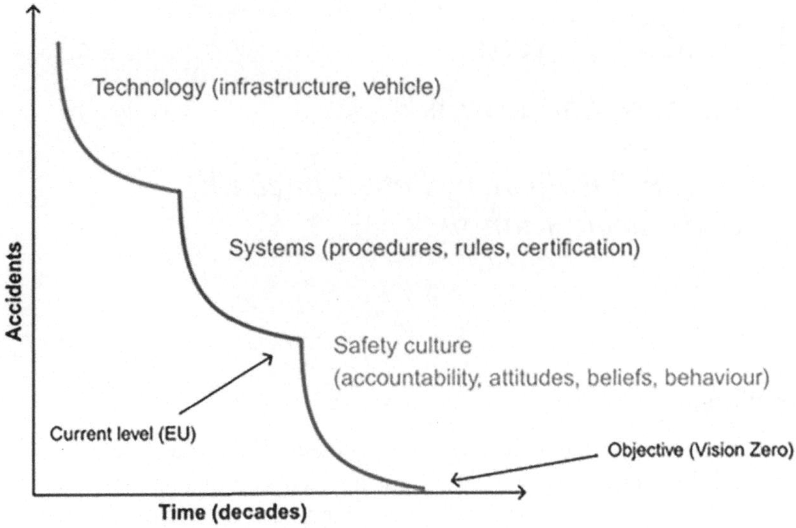

Figure 11.1 Traffic culture as a new paradigm.

Hughes also proposed that, in order to enhance the value of road safety strategies, social factors (such as culture) and economic factors should be taken into account. This served as a strong motivation for the team behind the project *Traffic Safety Cultures and the Safe Systems Approach: Towards a Cultural Change Research and Innovation Agenda for Road Safety (TraSaCu)* to develop a comprehensive model aimed at presenting a system of traffic safety culture (outlined next), in which transport is viewed as an open system, with the cultural component occupying a central position.

Background

In a broader sense, culture can be defined as the *shared products of society* (Flora et al., 1992), with an emphasis on ideas, norms, and objects (Giddens et al., 2011). One of the challenges in incorporating culture into transport research lies in the difficulty of quantifying or operationalising it. Social and cultural norms and ideas represent rules or expectations that are applied to behaviour and thinking on the basis of our shared beliefs within a specific cultural or social group. Although not formally visible, norms offer social standards for appropriate and inappropriate behaviour that establish what is (and is not) acceptable in interpersonal interactions. The focus on the cultural aspect in traffic (the concept of traffic safety culture) has a long history dating back to the 1990s. For example, Zaidel (1992) noted that drivers want to drive like others and are more likely to conform to majority norms than to official rules. *Informal rules* based on behaviours learned by observation of other community members, in turn, influence the behavioural choices of individual group members. Social and cultural norms have a significant impact on individual behaviour across a variety of contexts, including unsafe driving and its prevention, as informal norms can create an environment that may either encourage or discourage unsafe driving. In their *social accident* model, Factor and co-authors (2007), for example, argue that drivers from different social groups interpret a given situation differently, and that such varied interpretations can lead to crashes. If a social norm suggests that certain risky driving behaviour is normal and commonplace among members of a specific (sub)culture, each individual driver may then rationalise such risky behaviour (Ward et al., 2010). Traffic safety culture is thus regarded as a contextual concept that can be used to explain or predict traffic safety issues across different cultures (Ward et al., 2015).

When referring to culture or informal norms in the context of traffic, we usually mean traffic safety culture. However, a clear definition of traffic safety culture is complicated by a lack of knowledge and a range of ambiguities, as the concept is highly context-dependent and is not based on a rigorous operational model (Timmermans et al., 2020). One of the earliest definitions was proposed by Leviäkangs (1998). He defines traffic culture as the sum of all factors that affect drivers' skills, attitudes, and behaviour, a notion also endorsed by Lajunen and co-authors (2017). The Safety Council of the United States Department of Transportation (Morrow & Coplen, 2017) defines traffic safety culture as *the shared values, actions, and behaviours that demonstrate a commitment to safety over competing goals and demands*. In similar terms, Otto and co-authors (2016) described *traffic safety culture* as the values and beliefs shared by groups of road users and other stakeholders which influence their decisions to behave or act in ways that affect traffic safety.

While theoretical definitions refer to skills, attitudes, actions, and behaviours, empirical studies work with aspects at varying levels of generality. An emphasis may be placed on cultures, norms, subcultures, nations, and similar categories. For example, Timmermans with co-authors (2020) identified a total of 11 subculturally conditioned driving themes: personal acceptance of speeding, personal acceptance of driver distraction and negligence, personal acceptance of extreme risk-taking, social norms accepting speeding, social norms accepting hands-free phone use, social norms accepting driver distraction and negligence, social norms accepting extreme risk-taking, legislation against speeding, legislation against distraction, legislation against driving under the influence of psychoactive substances, and legislation against running a red light. While many factors are known to be interrelated, it is often unclear *why* or *how* exactly safety is affected. For example, a number of studies have reported higher levels of anger among female drivers with no apparent rationale (e.g. Willemsen et al., 2008). There are also many other aspects within the broader concept of traffic culture, including the total number of registered vehicles per capita, the proportion of two- and three-wheeled vehicles and bicycles, and the length of paved roads. All of these factors further influence traffic safety measures, which in turn have a direct impact on accident rates. Arguably, the aim of traffic safety culture is to adjust shared cultural cognitions, beliefs, and normative perceptions in order to increase overall compliance with traffic safety regulations (Ward & Özkan, 2014).

However, all these aspects vary in relevance. There are universally applicable aspects where the cultural dimension plays only a minimal role – these include issues concerning young drivers (mainly attributed to the immaturity of the brain and only marginally influenced by culture) or the poor enforcement of traffic laws as a reason for more frequent speeding and traffic violations (Stanojević et al., 2013). The majority of the aspects, however, are culturally conditioned, as behavioural differences exist even between individual regions (Sharda et al., 2017), with cultural diversity being apparent not only between countries but also within a single nation or across different subpopulations. For instance, in the Global North, religion is seldom associated with driving in either research or planning practice (Newmark & Rearick, 2021), whereas in the Global South, religious institutions are among the most powerful transmitters or "agents" of culture (Flora et al., 1992). Nonetheless, the connection between religion and driving behaviour has its limits (Newmark & Rearick, 2021). According to Van den Berghe and co-authors (2020), too, the high rate of road fatalities in low-income countries in comparison with wealthier ones may, to a greater extent, be attributed to cultural factors. Therefore, the study of traffic safety culture must adopt a holistic approach involving the exploration of cultural influences at national, subcultural, and regional levels by examining smaller cultural groups within countries (Edwards et al., 2014).

Some aspects are much easier to measure, and some have greater importance than others. For example, educational, economic, or engineering aspects are easier to model and address, as there is ample data available for them. Education levels generally serve as predictive factors for traffic accidents (Whitley et al., 2010). Nevertheless, a more culturally oriented perspective reveals that, for example, highly educated Japanese drivers tended to evaluate other road users as safe, while highly educated Chinese drivers viewed others as aggressive (Hussain et al., 2023). Presumably, the level of road safety is based on traffic accident data, self-reported data, and the interpretation of these. However, such data depends on several cultural factors, including national road traffic legislation, cultural values, and cultural specifics. This means that the data is biased, but, more importantly, that the cultural dimension has a strong influence on road safety.

One attempt to integrate a wide range of cultural factors into a single model was made within the *TraSaCu* project. It gave rise to a general traffic safety culture system (G-TraSaCu-S), which currently represents the most comprehensive and integrative model of the traffic system that also incorporates a cultural component (Özkan & Lajunen, 2015). The traffic system is underpinned by its internal components: the human factor, vehicles, and the environmental/behavioural context. In addition, the main outcomes/outputs and consequences of the system are manifested at different levels. "Controllers" and/or "decision-makers" (e.g. individual road users, company managers, mayors, and national ministers) operate at these levels. These key individuals at different levels can play critical roles in other components of the traffic safety system (e.g. scientists in relation to outside influences, family members and teachers in the transfer of values and norms, and instructors responsible for driver training). Importantly, the main goals/aims of the traffic system can influence the system's internal factors and have the potential to provide feedback to other components. There is also a feedforward mechanism. While the feedback mechanism is based on consequences, feedforward can be regarded as being based on pre-consequences and inputs, primarily derived from the internal factors (e.g. unsafe acts) of the traffic system. In their model, Özkan (2006) and Özkan and Lajunen (2011) adapted the system environment, that is, the *factors of the socio-ecological theory*, to the traffic domain. Cultural components thus become the primary predictors of the development and maintenance of behavioural patterns and play a central and significant role in the entire system, functioning as the "software".

Global perspectives and cultural variability

Educational aspects

Education is one of the most important tools for improving traffic safety. In almost every country worldwide, traffic safety education forms an integral part of the process of obtaining a driving licence. It also features in both formal and informal general education. However, there is often a considerable gap between theoretical approaches or teaching methods and their practical application which affects overall effectiveness. Thanks to evaluation studies, it is possible to compare the outcomes of various measures and identify those which significantly reduce road accidents (Mayhew et al., 2017; Shell et al., 2015). These studies show that teenagers who completed a dedicated driver education programme were less likely to be involved in accidents during their first two years of driving in comparison to those who had not undergone such training (Topolšek et al., 2019). On the basis of an analysis of the effectiveness of road safety education, Shell and co-authors (2015) found that individuals without such education were 1.22 times more likely to be involved in a road accident than those who had completed a programme. There are notable differences in both the length and content of education across countries. In the Czech Republic, for example, formal driver education lasts only a few months, whereas in the Netherlands, it can extend over several years. Moreover, risk perception tests are part of the education systems in countries with the highest levels of road safety, such as the United Kingdom, Sweden, and the Netherlands, but are often absent in low- and middle-income countries (LMICs). Cultural differences also extend beyond education into how traffic safety knowledge is assessed. In China, for example, written tests with challenging questions are emphasised, but the practical application of learned knowledge is less developed. Similarly, in Japan, accuracy is highly prioritised, and the testing process is twice as long as in EU countries.

Left-hand and right-hand culture

At present, around two-thirds of the world's population drive on the right-hand side of the road, while the remaining third drive on the left. Over time, some countries have changed their driving orientation for various reasons.

The most recent country in continental Europe to make the switch was Sweden, which changed from left- to right-hand driving on 3 September 1967. The transition, which aimed to align Sweden with its neighbouring countries, involved an investment of over 100 million euros in public education and new road signage. Samoa made a shift in the opposite direction – changing from driving on the right to the left – in order to align with its closest neighbours, Australia and New Zealand. An additional motivation was the lower cost of importing vehicles from these left-hand-driving countries. Other countries that have switched from left- to right-hand driving include Iceland in 1968, Burma (now Myanmar) in 1970, and Ghana in 1974. Today, all countries follow a consistent national driving orientation, with the exception of China. Following the handovers of Hong Kong in 1997 and Macau in 1999 – both of which drive on the left – China now encompasses territories with differing driving rules.

According to Poddar and Singh (2021), countries that drive on the left share this cultural heritage primarily because of British colonial influence. These regions either were directly governed by the United Kingdom or are in such close proximity to former British colonies that aligning with their driving orientation proved more practical. Similarly, it can be argued that the influence of the United States and Napoleonic France led many countries to adopt right-hand traffic. An exception is Japan, which drives on the left because of a samurai tradition that prioritised movement from the left. In general, left-hand traffic is considered safer – not for cultural reasons, but because it corresponds better with the cognitive responses of the predominantly right-handed population (accounting for 80% globally).

Economic aspects

Wealthier countries tend to have fewer road fatalities per capita, despite having larger car fleets and higher traffic volumes than lower-income countries. However, wealth is also linked to a country's cultural context. Hofstede's widely used individualism/collectivism dimension, for instance, can statistically account for over half of the variation in road safety outcomes between countries (Van der Berghe et al., 2020). For example, comparative studies using Hofstede's cultural dimensions have found that countries with higher levels of power distance, collectivism, and short-term orientation generally exhibit higher traffic fatality rates. The conclusion can be drawn that economic growth leads to cultural change, yet this relationship is not merely a reflection of differences in wealth between countries. Van der Berghe and co-authors (2020) demonstrated that countries with higher levels of collectivism tend to experience more road fatalities than those with similar levels of prosperity but higher individualism. People in individualistic societies, even if initially sceptical about new rules that may restrict their freedom, often adopt a more holistic view. They consider the well-being of society as a whole, rather than focusing solely on their in-group, which is more characteristic of collectivistic societies. Additionally, they trust others to exercise sound judgement and feel less of a need to regulate society to prevent chaos, compared to their collectivistic peers. In more collectivistic societies, where the focus is on in-groups, people are often less likely to comply with new rules unless strict control mechanisms, such as fines or police enforcement, are in place. Collectivistic societies are typically more conformist – they tend to follow societal rules, but these are usually traditional rules (often tied to religion), rather than traffic laws, which are relatively new and not yet fully integrated into the cultural fabric. Minkov and co-authors (2017) support this notion, showing that the more religious a society is, the more likely it is to be collectivistic.

It is clear that national differences in traffic safety exist, with high-income countries (HICs) outperforming LMICs (Uzondu et al., 2020). The Global Status Report on Road Safety (WHO, 2018) indicates that the highest road traffic fatality rates are found in the African and Southeast Asian regions, despite Africa being the least motorised (2%) region in the world. Road traffic mortality rates per capita can be up to 30 times higher in developing countries than in high-income states. Developing countries thus bear a disproportionately high share of global road traffic accidents and fatalities; for example, 96% of all fatal child traffic accidents occur in developing countries. Unsurprisingly, the highest fatality rates are reported in Nigeria and South Africa (WHO, 2009, 2011, 2015). The best results are seen in HICs such as Norway, the Netherlands, and Sweden. These disparities stem from a combination of factors, including poorer transport infrastructure, lower levels of healthcare and emergency services, and inadequate legislation or enforcement of traffic regulations, as well as differing driving cultures and attitudes to risk (Peden et al., 2004).

In LMICs, the enforcement of traffic regulations is generally weak. Combined with underdeveloped barrier systems, this leaves greater room for individual and informal interpretations of what constitutes safe behaviour within the traffic system (Nordfjærn et al., 2014). In contrast, HICs benefit from more consistent police enforcement and targeted attitude campaigns, which together provide a clearer framework for behaviour and have had a positive impact on traffic safety outcomes (Elvik et al., 2009).

Wealth is also reflected in micro-level behaviours and their interpretations. In HICs, for example, the horn is primarily used to express annoyance at other drivers or to alert pedestrians, cyclists, and other motorists in potentially dangerous situations. In contrast, in LMICs, it is commonly used when overtaking other cars, serving as an alternative to visual signals such as turn indicators or headlights, as a way of asserting one's presence and social rights in traffic, and even to show appreciation for courteous behaviour by other drivers (Nordfjærn et al., 2014).

The influence of wealth is also evident in the context of economic growth or decline. For instance, during economic crises, lower accident rates are observed in those regions with the highest unemployment levels (Lloyd et al., 2015). Economic recession is associated with reduced traffic volumes, particularly in freight transport and commuting (Zámečník et al., 2017), and a shift towards more localised activities (Sivak & Schoettle, 2010). Higher unemployment has been linked to a lower predicted crash fatality rate, presumably because it reduces driving exposure and increases the perceived risk of a crash because of the higher subjective costs (Reinfurt et al., 1991). Conversely, technological advancements in passive safety and improvements in healthcare systems – both associated with national wealth – contribute to reductions in traffic fatalities (Cheung & McCartt, 2010)

Overall, it can be concluded that reducing road traffic injuries has a positive effect on a country's economy (World Bank, 2017). It is therefore essential for LMICs to recognise the importance of culture. Culturally oriented research clearly demonstrates that national culture is closely linked to the level of road safety. In other words, cultural values such as attitudes towards authority or individualism influence the degree to which people adhere to rules and foster risk prevention. This, in turn, affects accident rates, which have a direct impact on the economy.

Infrastructural aspects

Apart from the economy and education, driving behaviour and safety are also influenced by road infrastructure (Dixit et al., 2012; Hao et al., 2016). Infrastructure has a critical impact on road user safety (Jamroz, 2011), as altering and improving it can play a key role in reducing road traffic risks (Pérez, 2006). Appropriate road infrastructure can promote safer behaviours (Theeuwes & Godthelp, 1995; Martens, 2007), while improvements can reduce crashes by minimising opportunities for user errors or "forgiving" such errors (Almqvist & Hydén, 1994). Conversely, insufficient, poorly maintained, or misleading road markings and signs, degraded road surfaces, and obstructive street furniture can increase the likelihood of accidents (Stanton & Salmon, 2009). Compared to HICs, roads in LMICs are often constructed or reconstructed without adequate consideration for safety (Uzondu et al., 2018). Although infrastructure improvements have become a priority in road safety programmes and strategies, financial constraints in LMICs make their implementation difficult. Moreover, if drivers in LMICs have come to rely on and respond to cultural cues rather than infrastructural ones, investment in infrastructure may be misdirected and might not achieve the intended safety outcomes (Uzondu et al., 2020). It also appears that poor traffic safety culture can be resistant to change, even when infrastructure is improved. Improvements to infrastructure alone may not bring about significant changes in drivers' behaviour (Uzondu et al., 2020; Novaco, 2001; WHO, 2009), as demonstrated in the case of Nigeria (Uzondu et al., 2018). Musselwhite and co-authors (2010) found that habitual processes – formed through repeated interaction with infrastructure – play a more decisive role in shaping behaviour. Driving tasks become automated and can be performed without conscious attention (Boer & Hoedemaeker, 1998). Therefore, improving safety in countries such as Nigeria requires more than just upgrading road infrastructure (Uzondu et al., 2018).

Other studies focus directly on international comparisons, with an emphasis on traffic culture and infrastructure. For example, Atchley and co-authors (2014) looked into China, Japan, and the United States. They found that because of the rapid increase in automobile use and widespread construction, China faces several safety challenges related to its traffic infrastructure. The relatively short history of traffic culture in the country has contributed to a shortage of designers with expertise in traffic engineering. This has led to various issues, such as narrower spacing of urban expressway entrances compared to Japan or the United States, resulting in collisions. In addition, excessively wide urban roadways and traffic signal cycles that are too long have prompted unsafe pedestrian behaviours, including ignoring traffic signals, crossing roads on which there is active traffic, or jaywalking. Wang and Shi (2016) further note that inadequate infrastructure in China results in overcrowded roads, limited traffic control, and a high proportion of young drivers with insufficient training and experience. These conditions contribute to competitive driving behaviours such as speeding, crowding, and lane-hopping, often used by drivers in an attempt to save a few minutes during urban rush hours.

Unlike China, the United States has had a much longer period in which to make substantial progress in the field of road safety (Atchley et al., 2014). Compared to China, for instance, the United States of America has established effective policies and public attitudes regarding drink-driving. In comparison to Japan, however, the United States of America appears to be more reluctant to regulate vehicle safety. Moreover, the United States could benefit from Japan's experience in addressing the challenges of an ageing population and managing the impacts of natural disasters (Mwebesa et al., 2018).

A further cultural aspect of infrastructure is road signage. For example, in China, some local names are written using highly complex Chinese characters (Huang et al., 2006). This poses difficulties not only for Chinese people from other regions but even more so for foreigners. One of the key goals in traffic system design is to ensure comprehensibility. Traffic signs and in-vehicle controls should rely on internationally recognisable symbols. Text should be avoided wherever possible, as drivers may not understand the language that is used.

Social aspects

In addition to economic, infrastructural, and educational differences, traffic psychology can also identify divergences across many other dimensions. One such dimension of major significance is cultural variation in behaviour and risk perception: culture shapes drivers' attitudes towards safety, their understanding of rules, and their willingness to take risks. In some regions (often in developing countries), fatalistic attitudes are widespread. People tend to believe that accidents are a matter of fate or the will of a higher power, and therefore cannot be fully prevented (Hoffmann & Muttarak, 2017). Such beliefs reduce motivation to follow safety measures – people who believe that accidents are predestined are less likely to wear helmets or seat belts and are more prone to risk-taking (Oluwadiya et al., 2009). Research in West Africa has shown that drivers often attribute the causes of accidents to external factors (such as poor road conditions or lack of infrastructure) rather than to their own actions (Atubi, 2012).

The degree of religiosity has a significant impact on road safety (Melinder, 2007). Religious countries tend to exhibit a greater fatalistic attitude towards road traffic (Nordfjærn et al., 2014). A strong belief in fate is thus associated with a lower perceived need to use protective measures and even with more active risk-taking behaviour (Oluwadiya et al., 2009). In particular, optimism bias and fatalism may contribute to a lower willingness to use safety devices such as seat belts, and can result in a general passivity towards safety (Peltzer & Renner, 2003). However, when one compares religious countries with different GDP levels, such as Iran and Turkey (which has double the GDP of Iran), it becomes evident that in higher-income countries, people are more likely to attribute road traffic crashes to individual risk-taking behaviour rather than to fate, bad luck, or the neglect of religious practices. In addition, religious culture also has a statistically significant effect on car ownership rates on local levels (Newmark & Rearick, 2021).

It should be noted, however, that in certain cultural contexts, religiosity may in fact promote safer behaviour. For example, in Turkey, strongly religious drivers reported fewer risky violations (Korkmaz, 2020). Similarly, in Kenya and Bangladesh, a deep faith in divine protection was associated with more cautious attitudes, while in some Asian and European countries, the relationship was reversed (Mcilroy et al., 2019).

To some extent, risk perception is related to different traffic regulations when driving in a foreign country. Government campaigns can provide support to foreign drivers in adapting to a new traffic culture. These campaigns should be tailored to the specific cultural groups they are intended for, as people from different cultures vary in their responsiveness to different persuasive techniques. For example, Turkish drivers may behave more safely behind the wheel if attention is focused on the emotional aspects of their behaviour (Üzümcüoğlu et al., 2019).

Another important dimension is cultural variability within countries, with a typical example being the differences between urban and rural traffic cultures. Statistically, the risk of a fatal crash is significantly higher in rural areas than in cities (NHTSA, 2020). The causes include not only objective factors – such as poorer road quality, longer emergency response times, and higher travel speeds on open roads – but also differing attitudes and driving habits. Surveys show that rural drivers are more likely to take risks such as driving without seat belts on or under the influence, as they tend to perceive the risks of such behaviours as lower than urban drivers do (Rakauskas et al., 2009). Not wearing a seat belt may seem "less dangerous" to rural drivers, particularly on familiar local roads with light traffic. This perception is also reflected in their more sceptical attitudes towards safety measures. It was found that rural drivers in the United States of America place less importance on government-sponsored traffic safety initiatives compared to their urban counterparts (Rakauskas et al., 2009).

It is therefore key to adapt strategies to local conditions. In every country or community, preliminary research should be conducted to understand its specific characteristics before implementing measures (Mehar et al., 2022). Unique cultural features of the region should be taken into account when designing road safety programmes, whether they involve educational campaigns, traffic signage, or enforcement methods. Only context-sensitive approaches enable traffic psychology interventions to be effective across diverse environments. In this regard, the principle of participatory design is gaining momentum. Target groups, such as cycling communities, rural drivers, and young people, are involved in developing campaigns or interventions, thereby enhancing both the relevance and acceptability of such measures (Hamann et al., 2021). An approach that combines expert knowledge with an understanding of local culture and lived realities appears to be vital for improving road safety on a global scale.

Last but not least, there is an ethical dimension. The traffic policies of many developed countries today are based on the 4E approach, which combines *E*ducation of road users, *E*nforcement of traffic rules, *E*ngineering solutions, and *E*valuation of effectiveness (WHO, 2018; Wijnen et al., 2019). Additionally, the Graduated Driver Licensing (GDL) system, which gradually extends driving privileges as novice drivers gain experience, has been shown to reduce crashes significantly (Insurance Institute for Highway Safety, 2024). Recent technological advancements have caused these two fundamental frameworks to involve robust data collection which raises ethical concerns regarding privacy, surveillance, and fairness.

Modern vehicles and transport systems generate enormous amounts of data – from records of driving behaviour to footage from in-cabin cameras. Event data recorders (EDRs) have been mandatory in all new vehicles in the EU since 2022 (European Commission, 2019). However, in order to protect privacy, European legislation restricts the types of data that may be stored; for example, it must not include location information or the exact time of an accident. These limitations reduce the usefulness of such data for research purposes (European Commission, 2019).

Similarly, telematics systems used by insurance companies to assess risk raise questions concerning transparency, consent, and discrimination. Studies show that usage-based insurance programmes may disadvantage certain groups of drivers based on socio-economic factors like birthplace or gender – even when driving behaviour is similar (Fabris et al., 2021). There are also concerns about algorithmic bias: the training of models on unbalanced datasets can lead to skewed assessments of certain population groups (Barocas et al., 2023).

In some scenarios, it is theoretically possible to implement a system of "total enforcement" of traffic regulations, that is, continuous surveillance which automatically detects and penalises every violation. While this could significantly improve legal enforceability, it would also result in a loss of autonomy and privacy (Litman, 2020). The European Union is currently pursuing a more balanced approach – for instance, the intelligent speed assistance (ISA) system, which alerts drivers when they exceed the speed limit, does not actively intervene in driving and leaves final control with the human driver (European Commission, 2019). However, most data is not owned by the EU but by private companies, which can increase their profits substantially through the monetisation of transport and mobility data.

Key requirements for the use of traffic data are informed consent and transparent communication with drivers about what is being monitored, how, and for what purpose. Data collection must be proportionate, targeted, and anonymised wherever possible (Zarsky, 2017). An ethically responsible application of psychology in transport thus involves not only a deeper understanding of human behaviour but also the defence of individual dignity and rights in the face of every technological innovation.

Ethical issues associated with the collection of traffic-related data and the monitoring of drivers are not culturally neutral. Levels of trust in the government, perceptions of privacy, and acceptance of surveillance vary considerably across different parts of the world. In some Asian countries, including China and South Korea, digital surveillance is more commonly tolerated by the public as a means of ensuring collective safety whereas in Western liberal democracies (e.g. Germany), there appears to be a greater sensitivity to intrusions into privacy and stronger concerns about excessive state intervention (Kostka et al., 2023; Kostka & Habich-Sobiegalla, 2024). Cultural values such as individualism versus collectivism, high or low power distance, and trust in public institutions play a crucial role in shaping how societies perceive the legitimacy of surveillance technologies and the use of personal data (Hofstede, 2001; Pavlou & Chai, 2002). This is also reflected in road traffic: while some cultures may accept continuous telematic monitoring of driving as a rational safety measure, others may view it as an unacceptable limitation on driver autonomy.

Moreover, in developing countries where robust legal frameworks for data protection are lacking, the introduction of surveillance technologies may lead to abuse, including discrimination against drivers on the basis of social class or ethnicity, without access to legal recourse (Taylor & Broeders, 2015). In this context, ethical frameworks for traffic-related data should be sensitive to cultural and institutional settings, rather than uncritically adopting models from other parts of the world.

The above insights underline the fact that cultural influences are complex and cannot be reduced to a single pattern – each society may subscribe to specific values that shape traffic behaviour in distinct ways.

References

Almqvist, S., & Hydén, C. (1994). Methods for assessing traffic safety in developing countries. *Building issues (LCHS)*, *6*(1). www.lth.se/fileadmin/hdm/BI_Volume_06_1_1994_Methods_for_Assessing_Traffic_Safety_in_Development_Countries.pdf

Atchley, P., Shi, J., & Yamamoto, T. (2014). Cultural foundations of safety culture: A comparison of traffic safety culture in China, Japan and the United States. *Transportation Research Part F: Traffic Psychology and Behaviour*, *26*(Part B), 317–325. https://doi.org/10.1016/j.trf.2014.01.004

Atubi, A. O. (2012). Epidemiology of injuries from road traffic accidents in Lagos State, Nigeria. *AFREV STECH: An International Journal of Science and Technology*, *1*(2), 56–75. www.ajol.info/index.php/stech/article/view/106350

Barocas, S., Hardt, M., & Narayanan, A. (2023). *Fairness and machine learning*. https://fairmlbook.org

Boer, E. R., & Hoedemaeker, M. (1998, December 14–16). *Modeling driver behaviour with different degrees of automation: A hierarchical decision framework of interacting mental models*. 17th European Annual Conference on Human Decision Making and Manual Control, Valenciennes, France.

Cheung, I., & McCartt, A. T. (2010). Declines in fatal crashes of older drivers: Changes in crash risk and survivability. *Accident Analysis & Prevention*, *43*(3), 666–674. https://doi.org/10.1016/j.aap.2010.10.010

Dixit, V., Gayah, V., & Radwan, E. (2012). Comparison of driver behaviour by time of day and wet pavement conditions. *Journal of Transportation Engineering*, *138*(8), 1023–1029. https://doi.org/10.1061/(ASCE)TE.1943-5436.0000400

Edwards, J., Freeman, J., Soole, D., & Watson, B. (2014). A framework for conceptualising traffic safety culture. *Transportation Research Part F: Traffic Psychology and Behaviour: Part B*, 26, 293–302. https://doi.org/10.1016/j.trf.2014.03.002

Elvik, R., Høye, A., Vaa, T., & Sørensen, M. (2009). *The handbook of road safety measures*. Emerald Group Publishing Limited.

European Commission. (2019). Regulation (EU) 2019/2144 of the European Parliament and of the Council of 27 November 2019 on type-approval requirements for motor vehicles. *Official Journal of the European Union*. https://eur-lex.europa.eu/legal-content/EN/TXT/?uri=CELEX:32019R2144

Fabris, A., Mishler, A., Gottardi, S., Carletti, M., Daicampi, M., Susto, G. A., & Silvello, G. (2021). Algorithmic audit of Italian car insurance: Evidence of unfairness in access and pricing. In *Proceedings of the 2021 AAAI/ACM conference on AI, ethics, and society, 19–21 May 2021* (pp. 458–468). https://doi.org/10.1145/3461702.3462569

Factor, R., Mahalel, D., & Yair, G. (2007). The social accident: A theoretical model and a research agenda for studying the influence of social and cultural characteristics on motor vehicle accidents. *Accident Analysis & Prevention*, *39*(5), 914–921. https://doi.org/10.1016/j.aap.2006.12.015

Flora, C. B., Flora, J. L., Spears, J. D., Swanson, L. E., Lapping, M. B., & Weinberg, M. L. (1992). *Rural communities: Legacy and change*. Westview Press.

Giddens, A., Duneier, M., Appelbaum, R. P., & Carr, D. (2011). *Introduction to sociology* (8th ed.). W. W. Norton & Company.

Guldenmund, F. (2000). The nature of safety culture: A review of theory and research. *Safety Science*, *34*(1–3), 215–257. https://doi.org/10.1016/S0925-7535(00)00014-X

Hamann, C. J., Daly, E., Schwab-Reese, L., Askelson, N., & Peek-Asa, C. (2021). Community engagement in the development and implementation of a rural road safety campaign: Steps and lessons learned. *Journal of Transport & Health*, *23*, 101282. https://doi.org/10.1016/j.jth.2021.101282

Hao, W., Kamga, C., & Wan, D. (2016). The effect of time of day on drivers' injury severity at highway rail grade crossings in the United States. *Journal of Traffic and Transportation Engineering (English Edition)*, *3*(1), 37–50. https://doi.org/10.1016/j.jtte.2015.10.006

Hoffmann, R., & Muttarak, R. (2017). Learn from the past, prepare for the future: Impacts of education and experience on disaster preparedness in the Philippines and Thailand. *World Development*, *96*, 32–51. https://doi.org/10.1016/j.worlddev.2017.02.016

Hofstede, G. (2001). *Culture's consequences: Comparing values, behaviors, institutions and organizations across nations* (2nd ed.). Sage Publications.

Huang, Y.-H., Zhang, W., Roetting, M., & Melton, D. (2006). Experiences from dual-country drivers: Driving safely in China and the US. *Safety Science*, *44*, 785–795. https://doi.org/10.1016/j.ssci.2006.05.002

Hughes, B. P., Anund, A., & Falkmer, T. (2015). System theory and safety models in Swedish, UK, Dutch and Australian road safety strategies. *Accident Analysis & Prevention*, *74*, 271–278. https://doi.org/10.1016/j.aap.2014.07.017

Hussain, B., Miwa, T., Sato, H., & Morikawa, T., (2023). Subjective evaluations of self and others' driving behaviors: A comparative study involving data from drivers in Japan, China, and Vietnam. *Journal of Safety Research*, *84*, 316–329. https://doi.org/10.1016/j.jsr.2022.11.007

Insurance Institute for Highway Safety. (2024). *Teenagers*. www.iihs.org/research-areas/teenagers

Jamroz, K. (2011). *Method of risk management in highway engineering*. Gdansk University of Technology.

Korkmaz, S. (2020). Dindarlık, Empati ve Sürücü Davranışları Arasındaki İlişkiler [The relationships between religiosity, empathy, and driver behaviors]. *Cumhuriyet İlahiyat Dergisi – Cumhuriyet Theology Journal*, *24*(2), 631–643. https://doi.org/10.18505/cuid.694393

Kostka, G., & Habich-Sobiegalla, S. (2024). In times of crisis: Public perceptions toward COVID-19 contact tracing apps in China, Germany, and the United States. *New Media & Society*, *26*(4), 2256–2294. https://doi.org/10.1177/14614448221083285

Kostka, G., Steinacker, L., & Meckel, M. (2023). Under big brother's watchful eye: Cross-country attitudes toward facial recognition technology. *Government Information Quarterly*, *40*(1), 101761. https://doi.org/10.1016/j.giq.2022.101761

Lajunen, T., Gaygısız, E., & Özkan, T. (2017). *Analysis of cultural patterns in (road) traffic systems: A comprehensive research report*. Deliverable 2.3 for the project Traffic Safety Cultures and the Safe Systems Approach Towards a Cultural Change Research and Innovation Agenda for Road Safety [TraSaCu].

Leviäkangs, P. (1998). Accident risk of foreign drivers – the case of Russian drivers in SouthEastern Finland. *Accident Analysis & Prevention*, *30*(2), 245–254. https://doi.org/10.1016/s0001-4575(97)00077-8

Litman, T. (2020). *Autonomous vehicle implementation predictions: Implications for transport planning*. Victoria Transport Policy Institute. www.vtpi.org/avip.pdf

Lloyd, L., Wallbank, C., & Broughton, J. (2015). A collection of evidence for the impact of the economic recession on road fatalities in Great Britain. *Accident Analysis & Prevention*, *80*, 274–285. https://doi.org/10.1016/j.aap.2015.03.026

Martens, M. H. (2007). *The failure to act upon important information: Where do things go wrong?* [Doctoral dissertation, Free University Amsterdam].

Mayhew, D., Vanlaar, W., Lonero, L., Robertson, R., Marcoux, K., Wood, K., Clinton, K., & Simpson, H. (2017). Evaluation of beginner driver education in Oregon. *Safety*, *3*(1), 9. https://doi.org/10.3390/safety3010009

Mcilroy, R. C., Kokwaro, G. O., Wu, J., Jikyong, U., Hoài, N. V., Hoque, Md. S., Preston, J. M., Plant, K. L., & Stanton, N. A. (2019). How do fatalistic beliefs affect the attitudes and pedestrian behaviours of road users in different countries? A cross–cultural study. *Accident Analysis & Prevention*, *139*, 105491. https://doi.org/10.1016/j.aap.2020.105491

Mehar, A., Saini, N., & Goyal, P. (2022). A comprehensive review of traffic safety behavior: Influence of culture, gender, and socioeconomic context. *International Journal of Injury Control and Safety Promotion*, *29*(4), 451–466.

Melinder, K. (2007). Socio-cultural characteristics of high versos low risk societies regarding road traffic safety. *Safety Science*, *45*(3), 397–414. https://doi.org/10.1016/j.ssci.2006.07.004

Minkov, M., Dutt, P., Schachner, M., Morales, O., Sanchez, C., Jandosova, J., Khassenbekov, Y., & Mudd, B. (2017). A revision of Hofstede's individualism-collectivism dimension: A new national index from a 56-country study. *Cross Cultural & Strategic Management*, *24*(3), 386–404. https://doi.org/10.1108/CCSM-11-2016-0197

Morrow, S., & Coplen, M. (2017). *Safety culture: A significant influence on safety in transportation. Final report.* U.S. Department of Transportation. https://rosap.ntl.bts.gov/view/dot/32538

Musselwhite, C., Avineri, E., Fulcher, E., Goodwin, P., & Susilo, Y. (2010). *Understanding public attitudes to road-user safety – literature review: Final report* (Road Safety Research Report No. 112). Department for Transport.

Mwebesa, M. E., Yoh, K., Inoi, H., & Doi, K. (2018). A new approach to cross-sector cooperation in road safety through a comparison of policies in selected countries. *IATSS Research*, *42*(4), 197–206. https://doi.org/10.1016/j.iatssr.2018.08.001

National Highway Traffic Safety Administration (NHTSA). (2020). *Traffic safety facts 2018 data: Rural/urban comparison of traffic fatalities* (Report No. DOT HS 812 957). US Department of Transportation. https://crashstats.nhtsa.dot.gov/Api/Public/Publication/812957

Newmark, G. L., & Rearick, E. L. (2021). Religious culture and rural car ownership. *Journal of Transport Geography*, *93*, 103035. https://doi.org/10.1016/j.jtrangeo.2021.103035

Nordfjærn, T., Şimşekoğlu, Ö., Zavareh, M. F., Hezaveh, A. M., Mamdoohi, A. R., & Rundmo, T. (2014). Road traffic culture and personality traits related to traffic safety in Turkish and Iranian samples. *Safety Science*, *66*, 36–46. https://doi.org/10.1016/j.ssci.2014.02.004

Novaco, R. W. (2001). Psychology of transportation. In *International Encyclopaedia of the social & behavioural sciences* (pp. 15878–15882). Science direct.

Oluwadiya, K. S., Kolawole, I. K., Adegbehingbe, O. O., Olasinde, A. A., Agodirin, O., & Uwaezuoke, S. C. (2009). Motorcycle crash characteristics in Nigeria: Implication for control. *Accident Analysis & Prevention*, *41*(2), 294–298. https://doi.org/10.1016/j.aap.2008.12.002

Otto, J., Finley, K., & Ward, N. J. (2016). *An assessment of traffic safety culture related to engagement in efforts to improve traffic safety* (Final Report No. FHWA/MT-16–012/8882-309-03). Western Transportation Institute, Montana State University.

Özkan, T. (2006). *The regional differences between countries in traffic safety: A cross-cultural study and Turkish Case* [Academic dissertation, University of Helsinki].
Özkan, T., & Lajunen, T. (2011). Person and environment: Traffic culture. In B. Porter (Ed.), *Handbook of traffic psychology* (pp. 179–192). Elsevier Ltd. https://doi.org/10.1016/B978-0-12-381984-0.10014-1
Özkan, T., & Lajunen, T. (2015). *A general traffic (safety) culture system (G-TraSaCu-S)*. Deliverable 2.1 for the project Traffic Safety Cultures and the Safe Systems Approach Towards a Cultural Change Research and Innovation Agenda for Road Safety [TraSaCu].
Parker, D. (2007). *Traffic culture paradigm*. Presentation at the 5th Australasian College of Road Safety (ACSQHC).
Pavlou, P. A., & Chai, L. (2002). What drives electronic commerce across cultures? A cross-cultural empirical investigation of the theory of planned behavior. *Journal of Electronic Commerce Research*, *3*(4), 240–253.
Peden, M., Scurfield, R., Sleet, D., Mohan, D., Hyder, A. A., Jarawan, E., & Mathers, C. (Eds.). (2004). *World report on road traffic injury prevention*. World Health Organization. https://iris.who.int/bitstream/handle/10665/42871/9241562609.pdf
Peltzer, K., & Renner, W. (2003). Superstition, risk-taking and risk perception of accidents among South African taxi drivers. *Accident Analysis & Prevention*, *35*(4), 619–623. https://doi.org/10.1016/S0001-4575(02)00035-0
Pérez, I. (2006). Safety impact of engineering treatments on undivided rural roads. *Accident Analysis & Prevention*, *38*(1), 192–200. https://doi.org/10.1016/j.aap.2005.09.008
Poddar, P., & Singh, V. (2021). When left is 'right'! The impact of driving-side practice on road fatalities in Africa. *Transport Policy*, *114*, 225–232. https://doi.org/10.1016/j.tranpol.2021.10.004
Rakauskas, M. E., Ward, N. J., & Gerberich, S. G. (2009). Identification of differences between rural and urban safety cultures. *Accident Analysis & Prevention*, *41*(5), 931–937. https://doi.org/10.1016/j.aap.2009.05.008
Reinfurt, D. W., Stewart, J. R., & Weaver, N. L. (1991). The economy as a factor in motor vehicle fatalities, suicides, and homicides. *Accident Analysis & Prevention*, *23*(5), 453–462. https://doi.org/10.1016/0001-4575(91)90065-D
Sharda, S., Wang, Y., & Ward, N. J. (2017). *Mapping driver behaviors and identifying the geo-spatial correlates of the utility truck crashes: A spatial statistical approach*. TRB 96th Annual Meeting Compendium of Papers (No. 17–0621).
Shell, D. F., Newman, I. M., Córdova-Cazar, A. L., & Heese, J. M. (2015). Driver education and teen crashes and traffic violations in the first two years of driving in a graduated licensing system. *Accident Analysis & Prevention*, *82*, 45–52. https://doi.org/10.1016/j.aap.2015.05.011
Sivak, M., & Schoettle, B. (2010). Toward understanding the recent large reductions in U.S. road fatalities. *Traffic Injury Prevention*, *11*(6), 561–566. https://doi.org/10.1080/15389588.2010.520140
Stanojević, P., Jovanović, D., & Lajunen, T. (2013). Influence of traffic enforcement on the attitudes and behavior of drivers. *Accident Analysis & Prevention*, *52*, 29–38. https://doi.org/10.1016/j.aap.2012.12.019
Stanton, N. A., & Salmon, P. M. (2009). Human error taxonomies applied to driving: A generic driver error taxonomy and its implications for intelligent transport systems. *Safety Science*, *47*, 227–237. https://doi.org/10.1016/j.ssci.2008.03.006
Taylor, L., & Broeders, D. (2015). In the name of development: Power, profit and the datafication of the global South. *Geoforum*, *64*, 229–237. https://doi.org/10.1016/j.geoforum.2015.07.002
Theeuwes, J., & Godthelp, H. (1995). Self-explaining roads. *Safety Science*, *19*(2–3), 217–225. https://doi.org/10.1016/0925-7535(94)00022-U

Timmermans, C. P. M., Alhajyaseen, W. K. M., Ross, V., & Nakamura, H. (2020). Introducing a multi-variate classification method: Risky driving acceptance among different heterogeneous driver sub-cultures. *Journal of Safety Research*, *73*, 81–91. https://doi.org/10.1016/j.jsr.2020.02.009

Topolšek, D., Babić, D., & Fiolić, M. (2019) The effect of road safety education on the relationship between driver's errors, violations and accidents: Slovenian case study. *European Transport Research Review*, *11*(18). https://doi.org/10.1186/s12544-019-0351-y

Uzondu, C. C., Jamson, S., & Lai, F. (2018). Exploratory study involving observation of traffic behaviour and conflicts in Nigeria using the traffic conflict technique. *Safety Science*, *110*(Part A), 273–284. https://doi.org/10.1016/j.ssci.2018.08.029

Uzondu, C. C., Jamson, S., & Hibberd, D. (2020). Can infrastructure improvements mitigate unsafe traffic safety culture: A driving simulator study exploring cross cultural differences. *Transportation Research Part F: Traffic Psychology and Behaviour*, *73*, 205–221. https://doi.org/10.1016/j.trf.2020.06.022

Üzümcüoğlu, Y., Özkan, T., Wu, C., & Zhang, H. (2019). How drivers perceive traffic? How they behave in traffic of Turkey and China? *Transportation Research Part F: Traffic Psychology and Behaviour*, *64*, 463–471. https://doi.org/10.1016/j.trf.2019.06.006

Van den Berghe, W., Schachner, M., Sgarra, V., & Christie, N. (2020). The association between national culture, road safety performance and support for policy measures. *IATSS Research*, *44*(3), 197–211. https://doi.org/10.1016/j.iatssr.2020.09.002

Wang, H., & Shi, J. (2016). *Culture, crowding and social influence all tied to aggressive driving behavior*. Oregon State University. https://news.oregonstate.edu/news/culture-crowding-and-social-influence-all-tied-aggressive-driving-behavior

Ward, N., Otto, J., Swinford, S., & Borkowski, J. (2015). *Measuring Minnesota's traffic safety culture. Final report*. Minnesota Department of Transportation. www.lrrb.org/pdf/201513.pdf

Ward, N., & Özkan, T. (2014). In consideration of traffic safety culture. *Transportation Research Part F*, *26*, 291–292. https://doi.org/10.1016/j.trf.2014.09.004

Ward, N. J., Linkenbach, J., Keller, S. N., & Otto, J. (2010). *White paper on traffic safety culture No. 2 for: "Toward zero deaths: A national strategy on highway safety"*. www.towardzerodeaths.org/wp-content/uploads/2020/01/Draft_TZD_White_Paper_2_Safety_Culture.pdf

Whitley, E., Batty, G. D., Gale, C. R., Deary, I. J., Tynelius, P., & Rasmussen, F. (2010). Intelligence in early adulthood and subsequent risk of unintentional injury over two decades: Cohort study of 1 109 475 Swedish men. *Journal of Epidemiology & Community Health*, *64*(5), 419–425. https://doi.org/10.1136/jech.2009.100669

Wiegmann, D. A., von Thaden, T. L., & Gibbons, A. M. (2007). A review of safety culture theory and its potential application to traffic safety. In *Improving the traffic safety culture in the United States: The journey forward* (pp. 113–129). AAA Foundation for Traffic Safety.

Wijnen, W., Weijermars, W., & Wesemann, P. (2019). *An international comparison of the national road safety strategies and targets*. SWOV Institute for Road Safety Research. www.swov.nl/publicatie/international-comparison-national-road-safety-strategies-and-targets

Willemsen, J., Dula, C. S., Declercq, F., & Verhaeghe, P. (2008). The Dula dangerous driving index: An investigation of reliability and validity across cultures. *Accident Analysis & Prevention*, *40*(2), 798–806. https://doi.org/10.1016/j.aap.2007.09.019

World Bank. (2017). *The high toll of traffic injuries: Unacceptable and preventable*. World Bank. https://openknowledge.worldbank.org/handle/10986/29129

World Health Organization. (2009). *Global status report on road traffic. Time for action*. WHO Press. https://iris.who.int/bitstream/handle/10665/44122/9789241563840_eng.pdf

World Health Organization. (2011). *Global plan for the decade of action for road safety 2011–2020*. www.who.int/publications/m/item/global-plan-for-the-decade-of-action-for-road-safety-2011-2020
World Health Organisation. (2015). *Road safety in the African region*. WHO Regional Office for Africa. www.afro.who.int/sites/default/files/2017-06/Road_Safety_AFRO_for_web_0.pdf
World Health Organization. (2018). *Global status report on road safety 2018*. www.who.int/publications/i/item/9789241565684
Zaidel, D. M. (1992). A modeling perspective on the culture of driving. *Accident Analysis & Prevention*, *24*(6), 585–597. https://doi.org/10.1016/0001-4575(92)90011-7
Zámečník, P., Gabrhel, V., Kurečková, V., & Řezáč, P. (2017). 7 Years of experience with demerit point system in the Czech Republic: Is it effective? In N. Stanton, S. Landry, G. Di Bucchianico, & A. Vallicelli (Eds.), *Advances in human aspects of transportation. Advances in intelligent systems and computing* (Vol. 484). Springer. https://doi.org/10.1007/978-3-319-41682-3_23
Zarsky, T. Z. (2017). Incompatible: The GDPR in the age of big data. *Seton Hall Law Review*, *47*(4), 995–1020. https://scholarship.shu.edu/shlr/vol47/iss4/2

12 Road users and vulnerable road users

Lucie Hartmannová

Introduction

Applying the perspective of social and traffic psychology, this chapter discusses different groups of road users, with special attention being focused on vulnerable road users (VRUs). Road users can assume various roles in traffic situations. Their behaviour may be influenced by a range of factors, such as physical and mental characteristics, age, experience, the dynamics and structure of interactions with other road users, and the traffic system. Road users' behaviour is also shaped by the physical properties of the means of transport (e.g. speed, size, and weight) and the traffic infrastructure. This brings us to a key aspect of traffic, which is safety. As far as traffic safety is concerned, VRUs are at the highest risk. The vulnerability of road users is related to the concept of vulnerable transport planning, as regular transport planning often favours the needs of motorised traffic. The chapter concludes with an overview of projects dedicated to protective measures and the planning of infrastructure that is more accommodating to VRUs.

Who are (vulnerable) road users?

Road users include scooter riders, the drivers of passenger cars, lorries, and buses, cyclists, pedestrians, motorcyclists, and even animals such as guide dogs. This chapter focuses on what are referred to as VRUs. The literature provides several ways of defining VRUs. According to the OCDE/OECD (1998), the main criterion featured in all the definitions is the *lack of external protection*. The definitions often refer to limited abilities to handle traffic situations because of, for example, children's inexperience or the declining task capability and physical frailty of the elderly. The World Health Organization (WHO, 2019a) defines VRUs as pedestrians, cyclists, and riders of two-wheeled motor vehicles. The European Commission defines VRUs as non-motorised road users, including not only pedestrians, cyclists, and motorcyclists but also "persons with disabilities or reduced mobility and orientation" (European Parliament, 2018, p. 3).

DOI: 10.4324/9781003608394-12

Numerous studies (e.g. Mayou & Bryant, 2003; Peden et al., 2004; Razzak & Luby, 1998; Sewalkar & Seitz, 2019) indicate that this group of road users is at a higher risk of becoming victims of traffic accidents or experiencing permanent physical or psychological consequences as a result of an accident. VRUs lack external protection in the event of a collision (Wegman et al., 2008) and account for more than half of all road fatalities globally (Constant & Lagarde, 2010), a statistic also pointed out by the WHO (2019b).

Road user groups

A more detailed literature review yields many other definitions of VRUs. We therefore present an integrative diagram that illustrates the aforementioned considerations while accounting for individuals with specific needs. It is also necessary to consider the level of skills and age-related cognitive processes (Touron et al., 2004). Figure 12.1 provides a classification of road users hierarchised according to their vulnerability (Holländer et al., 2021).

This taxonomy is organised (along the vertical axis) into *four layers*, from the most general (Layer 1 – Road Users, RUs) to specific groups of road users (Layer 4 – Especially Vulnerable Road Users, EVRUs). Layer 2 represents a broader distinction between Vulnerable Road Users (VRUs) and Protected Road Users (PRUs), with the latter not being the centre of attention in this taxonomy. Layer 3 classifies VRUs according to whether they are in a motorised vehicle, using a non-motorised mode of transport, or walking. The horizontal axis (bottom continuum) arranges the subgroups from left (the slowest) to right (the fastest) according to their estimated maximum speed

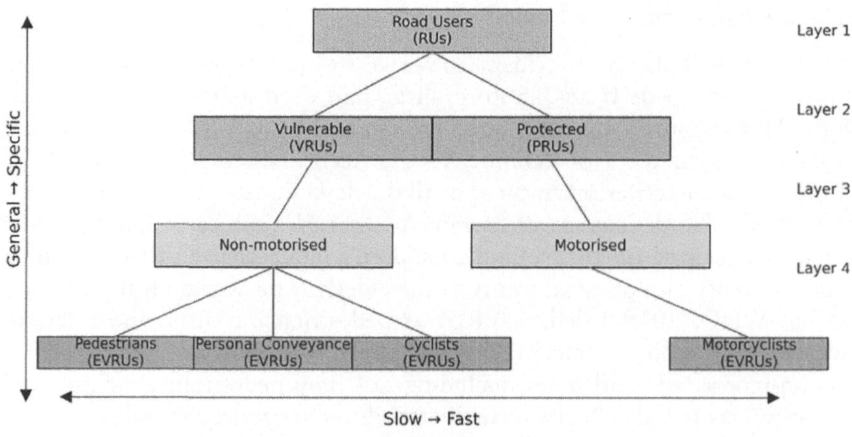

Figure 12.1 Taxonomy of VRUs (Holländer et al., 2021).

(Holländer et al., 2021). The three layers of RUs according to Holländer et al. (2021) are described as follows:

- *Layer 1* – RUs include every being (both people and animals) and all vehicles participating in road traffic.
- *Layer 2* – RUs are divided into vulnerable (VRUs) and protected (PRUs).
  - *VRUs* characteristics: they typically have lower weight, slower speeds, and are not protected against collisions.
  - *PRUs* characteristics: they travel in vehicles (cars, buses, trams, etc.) that provide them with protection (airbags, cabins, and restraint systems).
- *Layer 3* – VRUs are divided into non-motorised (powered by muscle strength) and motorised (powered by an engine).
  - *Non-motorised VRUs* include pedestrians, cyclists, and users of electric bicycles.
  - *Motorised VRUs* include users of mopeds and other light motor vehicles.

Especially vulnerable road users

The taxonomy of VRUs presented earlier, specifically Layer 4, can be expanded to include two additional layers, as shown in Figure 12.2. These cover specific attributes that are part of each EVRUs category and allow for a more precise description of the VRUs subgroups. Holländer et al. (2021) note that although the taxonomy does not provide for PRUs subgroups such

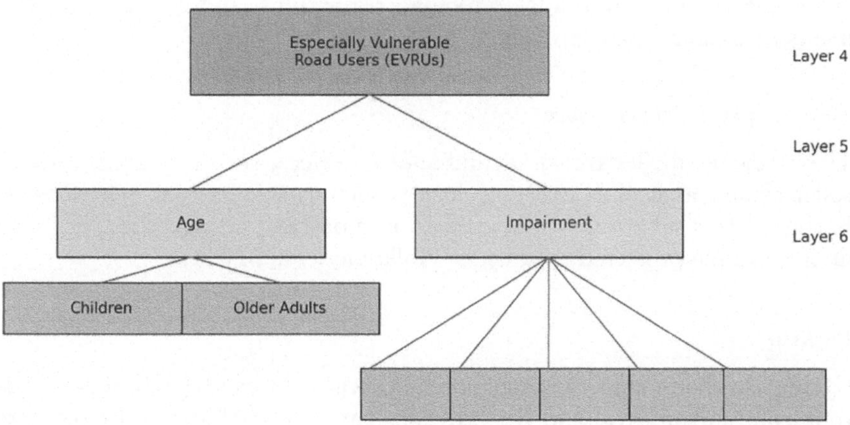

Figure 12.2 Three-layer representation of specific attributes for EVRUs (Holländer et al., 2021).

as drivers of passenger cars, lorries, and buses, both dimensions of EVRUs can also be applied to these road users.

Layer 4

This layer defines specific target groups of the most VRUs and is closely related to the attributes assigned to each group. A more detailed description of selected groups in this layer, namely pedestrians, cyclists, and motorcyclists, is provided as follows:

Pedestrians

Pedestrians rank among the most VRUs as they are not protected by a vehicle body or safety features (Tefft, 2013). Their safe movement requires visual scanning abilities, selective attention, and risk assessment (Barton & Schwebel, 2007; Thomson et al., 1996). Failures in these processes, particularly inattention, contribute to a high number of accidents (BESIP, 2021). Moreover, pedestrians often consciously ignore traffic rules, which increases the risk of collisions. On the other hand, drivers may cause accidents because of their incorrect assessment of a situation or limited visibility (BESIP, 2021).

Although roads in the EU are among the safest in the world, 99% of pedestrian deaths result from collisions with motor vehicles (ETSC, 2020). Between 2010 and 2018, at least 51,300 pedestrians were killed on European roads, with the number of fatalities decreasing by only 19% over the period (Adminaité-Fodor & Jost, 2020). The severity of injuries is influenced by the pedestrian's age, impact speed, and vehicle type (CDV, 2021). The most common consequences of collisions are head and limb injuries (Hirt & Vavera, 2022; Li et al., 2019). These facts highlight the necessity for pedestrians to exercise greater caution, adhere to road regulations, and use reflective elements to enhance their visibility.

Means of personal conveyance

They include smaller transport and sports devices, such as skateboards or roller skates, as well as assisting devices such as strollers and wheelchairs. Entities in this subgroup can be either non-motorised or motorised, but they do not include protected road users (Holländer et al., 2021).

Cyclists

Cycling is becoming increasingly popular because of its health benefits, environmental advantages, and the growing market for e-bikes and e-scooters (Budi et al., 2021; De Hartog et al., 2010; Peruzzi et al., 2020). Bicycle collisions often resemble pedestrian accidents, with the primary cause typically being distraction on the part of the cyclist or the driver of a motor vehicle

(ETSC, 2020). The most serious accidents are rear-end collisions, where the risk of fatal or severe injuries is approximately 47% (CDV, 2021).

Between 2010 and 2018, at least 19,450 cyclists were killed on EU roads, with almost no decrease in the annual mortality rate (Adminaité-Fodor & Jost, 2020). While the number of motorised road user deaths decreased by an average of 3.1% annually from 2010, the number of cyclist fatalities dropped by a mere 0.4% per year (ETSC, 2020). The slow decline reflects the growing popularity of cycling in the EU. A major concern is the non-use of bicycle helmets. Head injuries, in particular, are often fatal or lead to permanent disabilities (CDV, 2020). Therefore, it is vital that cyclists pay attention to safety, wear helmets, use reflective elements, and ensure proper lighting.

Motorcyclists

Motorcyclists are also especially VRUs. Their accidents are often associated with misjudging traffic situations, failing to adjust speed, or engaging in risky overtaking (CDV, 2021). Another significant factor is the personality trait of "sensation-seeking", that is, the need to look for intense experiences even at the cost of risk (Romero et al., 2019). Additionally, inadequate skills or training may lead motorcyclists to underestimate danger and increase the likelihood of an accident (Di Stasi et al., 2011).

In the European Union, motorcyclists account for an average of 18% of all road fatalities, with their numbers decreasing by 20% between 2011 and 2018 (Adminaité-Fodor & Jost, 2020; ETSC, 2020). Nevertheless, their safety remains a serious issue, as other drivers play a role in a significant proportion of these accidents by failing to pay enough attention to motorcyclists on the road. Consideration and mutual respect on the part of all road users are therefore key to reducing the number of fatal crashes.

Layers 5 and 6

The attributes of especially VRUs can be distinguished on the basis of limitations determined by age and individual impairments. When formulating these taxonomy layers, Holländer et al. (2021) were inspired by the categorisation proposed by Hanson et al. (2015), where a distinction is also drawn between cognitive, visual, auditory, and mobility impairments, as well as age-related limitations. These authors point out the linguistic aspect of different designations. For instance, they recommend avoiding the term "old person" in favour of the more appropriate "senior citizen", with "older adult" being the preferred term (Hanson et al., 2015).

Age

This category implies all the age-related limitations of the entire range from children to the elderly. An example is the ability to read, which children may not possess, or the potential difficulty senior citizens may have in understanding new technologies (Holländer et al., 2021). Regarding age-related risk behaviour in the pedestrian category, it has been shown that as road users get older, the proportion of accidents involving failure to comply with road regulations increases. On the other hand, the most common risk factor among younger age groups appears to be a pedestrian running out from behind an obstacle that blocks their view (CDV, 2021). Children and older adults are typically much more vulnerable and fragile and they may face greater difficulties in recovering from serious injuries.

Children

As regards the traffic environment, Holländer et al. (2021) categorise children as individuals aged from 0 to 15/17 years. Studies focusing on children as pedestrians often work with the age range from 5 to 14/15 years, in which road accidents are the second most common cause of death (Bart et al., 2008; Bou-Karroum et al., 2022). Young school-age children (5–9 years) are not yet sufficiently developed emotionally or cognitively, which manifests itself in their limited ability to focus attention, tendency to be easily distracted, and reduced inhibitory control, particularly among boys (Alonso et al., 2015). Cognitive abilities such as the integration of visual information improve around the age of 7 or 8 years, but even then, they have not reached adult levels (Bart et al., 2008). Children's short stature limits their field of vision, which, when combined with their lack of experience, may lead to misjudgements of vehicle speed and distance (Barton & Schwebel, 2007; Whitebread & Neilson, 2000).

Children often experience difficulties in applying road traffic rules in real-life situations, with the main factors contributing to accidents being inattention, impulsively stepping into the road, and inability to assess safe crossing points (CDV, 2021). International research indicates that motor vehicle-related accidents are the leading cause of death among children and adolescents in the United States (Cunningham et al., 2018). Children's safe behaviour in traffic environments can be significantly facilitated by parental supervision, which helps to reduce children's tendency to overestimate their abilities (Barton & Schwebel, 2007). On the other hand, excessive supervision or frequent car transport may delay the development of autonomy and result in children becoming less experienced pedestrians (Pfeffer et al., 2010). Therefore, improving child safety requires not only education and supervision but also fostering their independence in a safe environment.

Older adults

Older adults (individuals over the age of 65) represent a specific group of road users whose cognitive and motor abilities may vary (Brüchert et al., 2021; Daigneault et al., 2002; Hawkes et al., 2012). While ageing is an individual process, and old age in itself is not a medical condition, recovery from road traffic accidents can be more challenging for older adults (Fitzpatrick & O'Neill, 2017; Greene & Smith, 2019; Vágnerová, 2017). In the EU, an average of 5,180 pedestrian deaths and 2,160 cyclist deaths are recorded annually, with 47% of pedestrian fatalities and 44% of cyclist fatalities involving individuals aged 65 and over (Adminaité-Fodor & Jost, 2020).

The main causes of accidents involving pedestrians over the age of 65 include inattention, misjudgement of situations, and, less frequently, failure to observe traffic rules (CDV, 2021). The decline in psychomotor speed associated with increasing age can lead to poorer estimation of the time needed to cross the road (CDV, 2021; Mayhew et al., 2003). On the other hand, caution increases and risky behaviour becomes less frequent (CDV, 2021). The inevitable changes in cognitive and psychomotor abilities can be offset to a certain degree by experience, safe mobility attitudes, and a realistic assessment of one's own abilities (Daigneault et al., 2002; Rimmö & Hakamies-Blomqvist, 2002; Štikar et al., 2007).

Impairment

This category covers specific manifestations of increased vulnerability which are defined on the basis of various special needs that are not necessarily related to age (Holländer et al., 2021). It includes cognitive, visual, auditory, motor, and other impairments, which may be combined and vary in intensity. Before travelling, individuals with impairments need to verify whether their journey features easy access, as they find it more challenging to adapt to barriers. A seamless connection between different segments of the journey – referred to as the "chain of accessible routes" – is key (Park & Chowdhury, 2018; Zhang et al., 2011). This concept plays a crucial role in traffic environment planning (Zhang et al., 2011).

Cognitive impairments have effects on mental processes, including knowing, learning, and understanding (Holländer et al., 2021). The most common sign is a decline in the speed at which the nervous system processes sensory information. With increasing age, this may elevate the risk of falls and cause slower walking and greater difficulty with orientation or making quick decisions (Dommès et al., 2015; Salthouse, 1996, as cited in Tournier et al., 2016; Welmerink et al., 2010). Cognitive function disorders may significantly affect safety in the traffic environment, particularly when complex traffic situations need to be processed (Dommes et al., 2013; Kliegel et al., 2007; Salthouse & Siedlecki, 2007).

Motor, auditory, and visual impairments limit mobility and the ability to process sensory information (Holländer et al., 2021). Movement limitations may increase the risk of a person crossing the road away from a designated crossing (Tournier et al., 2016). Hearing impairments make it more difficult to localise vehicles spatially and may lead to risky street-crossing decisions (Rodrigues et al., 2012). Visual impairments increase the risk of falls, especially when there are environment barriers (Bromley et al., 2007; Rosenberg et al., 2013). The "other" subcategory includes less frequent impairments with a smaller impact on traffic safety, which may, however, require special modifications of the traffic environment (Holländer et al., 2021).

VRUs and vulnerable transport planning

This section offers a critical perspective on the concept of VRUs. Both research and practice show that vulnerability also lies within transport planning, which tends to focus more on the needs of motorised traffic. While it holds that pedestrians, bicyclists, motorcyclists, children, and older adults are limited by certain deficits, these are also a consequence of *transport planning* (referred to as "vulnerable transport planning") that makes the traffic environment riskier for other road users by prioritising motor vehicles (Khayesi, 2020). Such planning results in the omission or neglect of the needs of certain groups of road users whom we regard as vulnerable. This hypothesis is supported by several studies (e.g. Legacy, 2016; Mohan, 2008; Stone & Mees, 2010; Whitelegg, 2016). Traditional planning generating transport systems that fail to create equitable opportunities – in terms of accessibility, environmental, and safety issues – for all road users was already pointed out in 2001 by Vasconcellos (2001/2014).

Khayesi (2020) recommends that current research and practice should address the root causes of vulnerability rather than focusing primarily on the individuals affected by it. The latest studies and projects strive to work with both concepts – vulnerable planning and VRUs. Marina et al. (2022), for example, sought to identify urban features that determine the safety of VRUs. Their findings suggest that urban planning aimed at increasing walking and cycling shares is key to improving road safety for all users. Similarly, Brüchert et al. (2021) examined the preferences for various urban design features among older adults (65+) living in less densely populated areas. The planning of a transport system should take into account its complexity, which makes it difficult to meet the needs of all road users. Prioritising one group leads to sidelining the needs of others. Therefore, finding a balance is crucial.

Projects to promote the safety of VRUs

The previous section highlighted the need to focus not only on VRUs themselves but also on changes in transport system planning that consider the interests of more than just motorised vehicles. This part will outline selected projects that seek to address this issue.

Projects focused on safe transport for VRUs

Within the EU, legislation, financial support, and projects concerning road safety are overseen by the European Commission. Possessing extensive experience in relation to road safety decision-making (evidenced by the safest roads in the world), the European Commission and EU member states establish collaboration with non-European countries in order to support global progress in this field (EC, 2022). In addition, the European Transport Safety Council (2022) is committed to reducing road deaths and injuries. Major EU projects aimed at enhancing the safety of VRUs are described as follows.

The VRUITS project was implemented from 2013 to 2016. The name of the project is composed of two abbreviations: the *VRU* introduced earlier for "Vulnerable Road Users" and *ITS* standing for "Intelligent Transport Systems". The purpose of ITS was mainly to support the safe and efficient mobility of motorised vehicles. *VRUITS* sought to integrate the element of VRUs into ITS applications in such a way as to respect the needs of VRUs. After assessing ten innovative road safety systems, the *VRUITS* project presented what could work best to improve not only safety but also the comfort of travel for VRUs (CORDIS, 2017). Carried out between 2015 and 2018, the *InDeV* ("In-Depth understanding of accident causation for Vulnerable road users") project focused on a thorough analysis of the causes of traffic accidents involving VRUs. The goal was to understand these causes and develop a set of tools capable of linking the causal factors of traffic accidents involving VRUs with the risk of their occurrence. These tools are intended for the development of preventive measures and aim to facilitate the improvement of the safety of VRUs in Europe (CORDIS, 2019b).

PROSPECT ("PROactive Safety for PEdestrians and CyclisTs") was a project carried out from 2015 to 2018. Its main goal was to reduce the statistics of pedestrians and cyclists killed or seriously injured in road accidents by improving the effectiveness of active safety systems for VRUs in comparison to those available on the market at the time (CORDIS, 2019a). The *SENIORS* project ("Safety-ENhancing Innovations for Older Road userS"), implemented during the same period, addressed the safety and protection of older road users by also transferring the safety standards of today's younger generation. Additionally, the project focused on the road safety of individuals with obesity (CORDIS, 2022).

Although the EU has the safest roads in the world, many challenges remain, as road safety has not yet reached its full potential. It can therefore be expected that further projects and innovations will be developed, extended, and advanced. However, their integration into a broader framework will also be required.

References

Adminaité-Fodor, D., & Jost, G. (2020). *How safe is walking and cycling in Europe? PIN flash report* (No. 38). European Transport Safety Council. https://etsc.eu/wp-content/uploads/PIN-Flash-38_FINAL.pdf

Alonso, F., Pastor, J. C., Montoro, L., & Esteban, C. (2015). Driving under the influence of alcohol: Frequency, reasons, perceived risk and punishment. *Substance Abuse Treatment, Prevention, and Policy*, *10*(1), 11. https://doi.org/10.1186/s13011-015-0007-4

Bart, O., Katz, N., Weiss, P. L., & Josman, N. (2008). Street crossing by typically developed children in real and virtual environments. *OTJR Occupation Participation Health*, *28*(2), 42–46. https://doi.org/10.3928/15394492-20080301-01

Barton, B. K., & Schwebel, D. C. (2007). The roles of age, gender, inhibitory control, and parental supervision in children's pedestrian safety. *Journal of Pediatric Psychology*, *32*(5), 517–526. https://doi.org/10.1093/jpepsy/jsm014

BESIP. (2021). *Národní strategie bezpečnosti silničního provozu 2011–2020 – Informace o plnění v roce 2020* [National road safety strategy 2011–2020 – implementation report 2020]. https://besip.cz/getattachment/Pro-odborniky/Narodni-strategie-BESIP/Plneni-strategie/Informace-o-plneni-NSBSP-za-rok-2020.pdf?lang=cs-CZ

Bou-Karroum, L., El-Jardali, F., Jabbour, M., Harb, A., Fadlallah, R., Hemadi, N., & Al-Hajj, S. (2022). Preventing unintentional injuries in school-aged children: A systematic review. *Pediatrics*, *149*(Suppl. 6). https://doi.org/10.1542/peds.2021-053852J

Bromley, R. D., Matthews, D. L., & Thomas, C. J. (2007). City centre accessibility for wheelchair users: The consumer perspective and the planning implications. *Cities*, *24*(3), 229–241. https://doi.org/10.1016/j.cities.2007.01.009

Brüchert, T., Baumgart, S., & Bolte, G. (2021). Social determinants of older adults' urban design preference: A cross-sectional study. *Cities & Health*, 6(2), 1–15. https://doi.org/10.1080/23748834.2020.1870845

Budi, D. R., Widyaningsih, R., Nur, L., Agustan, B., Dwi, D. R. S., Qohhar, W., & Asnaldi, A. (2021). Cycling during COVID-19 pandemic: Sports or lifestyle? *International Journal of Human Movement and Sports Sciences*, *9*(4), 765–771. https://doi.org/10.13189/saj.2021.090422

Centrum dopravního výzkumu (CDV). (2020). *Bezpečnější cesty, bezpečnější města. Jak zlepšit bezpečnost silničního provozu ve městech v EU* [Safer roads, safer cities. How to improve urban road safety in the EU]. www.cdv.cz/tisk/jak-zlepsit-bezpecnost-silnicniho-provozu-ve-mestech-v-eu/

Centrum dopravního výzkumu (CDV). (2021). *Dílčí cíle NSBSP z pohledu Hloubkové analýzy dopravních nehod* [Components of the national road safety strategy from the perspective of the in-depth analysis of traffic accidents]. www.ibesip.cz/getattachment/Pro-odborniky/Narodni-strategie-BESIP/Plneni-strategie/19-Dilci-cile-NSBSP-z-pohledu-Hloubkove-analyzy-dopravnich-nehod.docx?lang=cs-CZ

Constant, A., & Lagarde, E. (2010). Protecting vulnerable road users from injury. *PLOS Medicine*, 7(3), e1000228. https://doi.org/10.1371/journal.pmed.1000228

CORDIS. (2017). *Moving in the right direction for the protection of vulnerable road users*. https://cordis.europa.eu/article/id/198035-moving-in-the-right-direction-for-the-protection-of-vulnerable-road-users

CORDIS. (2019a). *A future with zero road deaths – is it possible?* https://cordis.europa.eu/article/id/266022-a-future-with-zero-road-deaths-is-it-possible

CORDIS. (2019b). *How can we keep vulnerable road users safe on European roads?* https://cordis.europa.eu/article/id/293959-how-can-we-keep-vulnerable-road-users-safe-on-european-roads

CORDIS. (2022). *Safety-enhancing innovations for older road users*. https://cordis.europa.eu/project/id/636136

Cunningham, R. M., Walton, M. A., & Carter, P. M. (2018). The major causes of death in children and adolescents in the United States. *New England Journal of Medicine*, *379*(25), 2468–2475. https://doi.org/10.1056/NEJMsr1804754

Daigneault, G., Joly, P., & Frigon, J. Y. (2002). Executive functions in the evaluation of accident risk of older drivers. *Journal of Clinical and Experimental Neuropsychology*, 24(2), 221–238. https://doi.org/10.1076/jcen.24.2.221.993

De Hartog, J. J., Boogaard, H., Nijland, H., & Hoek, G. (2010). Do the health benefits of cycling outweigh the risks? *Environmental Health Perspectives*, *118*(8), 1109–1116. https://doi.org/10.1289/ehp.0901747

Di Stasi, L. L., Contreras, D., Cándido, A., Cañas, J. J., & Catena, A. (2011). Behavioral and eye-movement measures to track improvements in driving skills of vulnerable road users: First-time motorcycle riders. *Transportation Research Part F: Traffic Psychology and Behaviour*, *14*(1), 26–35. https://doi.org/10.1016/j.trf.2010.09.003

Dommes, A., Cavallo, V., & Oxley, J. (2013). Functional declines as predictors of risky street-crossing decisions in older pedestrians. *Accident Analysis & Prevention*, *59*, 135–143. https://doi.org/10.1016/j.aap.2013.05.017

Dommès, A., Le Lay, T., Vienne, F., Dang, N. T., Beaudoin, A. P., & Do, M. C. (2015). Towards an explanation of age-related difficulties in crossing a two-way street. *Accident Analysis & Prevention*, *85*, 229–238. https://doi.org/10.1016/j.aap.2015.09.022

European Commission (EC). (2022). *EU road safety policy*. https://road-safety.transport.ec.europa.eu/eu-road-safety-policy_en

European Parliament. (2018). *Consolidated version: Directive 2010/40/EU of the European Parliament and of the Council of 7 July 2010 on the framework for the deployment of Intelligent Transport Systems in the field of road transport and for interfaces with other modes of transport (text with EEA relevance)*. https://eur-lex.europa.eu/legal-content/CS/TXT/?uri=CELEX%3A02010L0040-20180109

European Transport Safety Council (ETSC). (2020). *Urgent action needed to tackle deaths of pedestrians and cyclists*. https://etsc.eu/urgent-action-needed-to-tackle-deaths-of-pedestrians-and-cyclists/

European Transport Safety Council (ETSC). (2022). *About us*. https://etsc.eu/about-us/

Fitzpatrick, D., & O'Neill, D. (2017). The older motorcyclist. *European Geriatric Medicine*, *8*(1), 10–15. https://doi.org/10.1016/j.eurger.2016.10.004

Greene, W. R., & Smith, R. (2019). Driving in the geriatric population. *Clinics in Geriatric Medicine*, *35*(1), 127–131. https://doi.org/10.1016/j.cger.2018.08.011

Hanson, V. L., Cavender, A., & Trewin, S. (2015). Writing about accessibility. *Interactions 22*(6), 62–65. https://doi.org/10.1145/2828432

Hawkes, T. D., Siu, K. C., Silsupadol, P., & Woollacott, M. H. (2012). Why does older adults' balance become less stable when walking and performing a secondary task? Examination of attentional switching abilities. *Gait & Posture*, *35*(1), 159–163. https://doi.org/10.1016/j.gaitpost.2011.09.001

Hirt, M., & Vavera, F. (2022). *Soudní lékařství pro právníky* [Forensic medicine for lawyers]. Grada Publishing a.s.

Holländer, K., Colley, M., Rukzio, E., & Butz, A. (2021). A taxonomy of vulnerable road users for HCI based on a systematic literature review. In *Proceedings of the 2021 CHI conference on human factors in computing systems* (pp. 1–13). https://doi.org/10.1145/3411764.3445480

Khayesi, M. (2020). Vulnerable road users or vulnerable transport planning? *Frontiers in Sustainable Cities*, 2, 25. https://doi.org/10.3389/frsc.2020.00025

Kliegel, M., Martin, M., McDaniel, M. A., & Phillips, L. H. (2007). Adult age differences in errand planning: The role of task familiarity and cognitive resources. *Experimental Aging Research*, *33*(2), 145–161. https://doi.org/10.1080/03610730601177395

Legacy, C. (2016). Transforming transport planning in the postpolitical era. *Urban Studies*, *53*(14), 3108–3124. https://doi.org/10.1177/0042098015602649

Li, G., Wang, F., Otte, D., & Simms, C. (2019). Characteristics of pedestrian head injuries observed from real world collision data. *Accident Analysis & Prevention*, *129*, 362–366. https://doi.org/10.1016/j.aap.2019.05.007

Marina, K., Laetitia, G., Michele, T., & Szell, M. (2022). Identifying urban features for vulnerable road user safety in Europe. *EPJ Data Science*, *11*(1), 27. https://doi.org/10.1140/epjds/s13688-022-00339-5

Mayhew, D. R., Simpson, H. M., & Pak, A. (2003). Changes in collision rates among novice drivers during the first months of driving. *Accident Analysis & Prevention*, *35*(5), 683–691. https://doi.org/10.1016/s0001-4575(02)00047-7

Mayou, R., & Bryant, B. (2003). Consequences of road traffic accidents for different types of road user. *Injury*, *34*(3), 197–202. https://doi.org/10.1016/S0020-1383(02)00285-1

Mohan, D. (2008). Traffic safety and city structure: Lessons for the future. *Salud Pública de México*, *50*(Suppl. 1), 93–100. www.scielo.org.mx/scielo.php?pid=S0036-36342008000700014&script=sci_arttext

Organisation for Economic Co-Operation and Development (OCDE/OECD). (1998). *Safety of vulnerable road users*. OCDE/OECD.

Park, J., & Chowdhury, S. (2018). Investigating the barriers in a typical journey by public transport users with disabilities. *Journal of Transport & Health*, *10*, 361–368. https://doi.org/10.1016/j.jth.2018.05.008

Peden, M., Scurfield, R., Sleet, D., Mohan, D., Hyder, A. A., Jarawan, E., & Mathers, C. (Eds.). (2004). *World report on road traffic injury prevention*. World Health Organization. https://iris.who.int/bitstream/handle/10665/42871/9241562609.pdf

Peruzzi, M., Sanasi, E., Pingitore, A., Marullo, A. G., Carnevale, R., Sciarretta, S., Sciarra, L., Frati, G., & Cavarretta, E. (2020). An overview of cycling as active transportation and as benefit for health. *Minerva Cardioangiologica*, *68*(2), 81–97. https://doi.org/10.23736/S0026-4725.20.05182-8

Pfeffer, K., Fagbemi, H. P., & Stennet, S. (2010). Adult pedestrian behavior when accompanying children on the route to school. *Traffic Injury Prevention*, *11*(2), 188–193. https://doi.org/10.1080/15389580903548576

Razzak, J. A., & Luby, S. P. (1998). Estimating deaths and injuries due to road traffic accidents in Karachi, Pakistan, through the capture-recapture method. *International Journal of Epidemiology*, *27*(5), 866–870. https://doi.org/10.1093/ije/27.5.866

Rimmö, P. A., & Hakamies-Blomqvist, L. (2002). Older drivers' aberrant driving behaviour, impaired activity, and health as reasons for self-imposed driving limitations. *Transportation Research Part F: Traffic Psychology and Behaviour*, *5*(1), 47–62. https://doi.org/10.1016/S1369-8478(02)00005-0

Rodrigues, J., Pinto, M., Dommes, A., Cavallo, V., & Vienne, F. (2012). Simulated traffic and auditory information: The impact on street crossing in young and old adults. *Actes INRETS*, 95–104.

Romero, D. L., de Barros, D. M., Belizario, G. O., & Serafim, A. D. P. (2019). Personality traits and risky behavior among motorcyclists: An exploratory study. *PLOS One*, *14*(12), e0225949. https://doi.org/10.1371/journal.pone.0225949

Rosenberg, D. E., Huang, D. L., Simonovich, S. D., & Belza, B. (2013). Outdoor built environment barriers and facilitators to activity among midlife and older adults with mobility disabilities. *The Gerontologist*, *53*(2), 268–279. https://doi.org/10.1093/geront/gns119

Salthouse, T. A., & Siedlecki, K. L. (2007). Efficiency of route selection as a function of adult age. *Brain and Cognition*, *63*(3), 279–286. https://doi.org/10.1016/j.bandc.2006.09.006

Sewalkar, P., & Seitz, J. (2019). Vehicle-to-pedestrian communication for vulnerable road users: Survey, design considerations, and challenges. *Sensors*, *19*(2), 358. https://doi.org/10.3390/s19020358

Štikar, J., Hoskovec, J., & Šmolíková, J. (2007). *Bezpečná mobilita ve stáří* [Safe mobility in old age]. Karolinum.

Stone, J., & Mees, P. (2010). Planning public transport networks in the post-petroleum era. *Australian Planner*, *47*(4), 263–271. https://doi.org/10.1080/07293682.2010.526550

Tefft, B. C. (2013). Impact speed and a pedestrian's risk of severe injury or death. *Accident Analysis & Prevention*, *50*, 871–878. https://doi.org/10.1016/j.aap.2012.07.022

Thomson, J., Tolmie, A., Foot, H., & McLaren, B. (1996). *Child development and the aims of road safety education: A review and analysis* (Road Safety Research Report No. 1). The Stationery Office.
Tournier, I., Dommes, A., & Cavallo, V. (2016). Review of safety and mobility issues among older pedestrians. *Accident Analysis & Prevention*, *91*, 24–35. https://doi.org/10.1016/j.aap.2016.02.031
Touron, D. R., Hoyer, W. J., & Cerella, J. (2004). Cognitive skill learning: Age-related differences in strategy shifts and speed of component operations. *Psychology and Aging*, *19*(4), 565–580. https://doi.org/10.1037/0882-7974.19.4.565
Vágnerová, M. (2017). *Vývojová psychologie dětství a dospívání* [Developmental psychology of childhood and adolescence] (2nd ed.). Karolinum.
Vasconcellos, E. A. (2014). *Urban transport environment and equity: The case for developing countries*. Routledge.
Wegman, F., Aarts, L., & Bax, C. (2008). Advancing sustainable safety: National road safety outlook for the Netherlands for 2005–2020. *Safety Science*, *46*(2), 323–343. https://doi.org/10.1016/j.ssci.2007.06.013
Welmerink, D. B., Longstreth, W. T., Jr., Lyles, M. F., & Fitzpatrick, A. L. (2010). Cognition and the risk of hospitalization for serious falls in the elderly: Results from the cardiovascular health study. *Journals of Gerontology Series A: Biomedical Sciences and Medical Sciences*, *65*(11), 1242–1249. https://doi.org/10.1093/gerona/glq115
Whitebread, D., & Neilson, K. (2000). The contribution of visual search strategies to the development of pedestrian skills by 4-11 year-old children. *British Journal of Educational Psychology*, *70*(4), 539–557. https://doi.org/10.1348/000709900158290
Whitelegg, J. (2016). *Mobility: A new urban design and transport planning philosophy for a sustainable future*. Createspace Independent Publishing Platform.
World Health Organization (WHO). (2019a). *Global status report on road safety 2018*. www.who.int/publications/i/item/9789241565684
World Health Organization (WHO). (2019b). *Protecting vulnerable road users*. www.who.int/china/activities/protecting-vulnerable-road-users
Zhang, Y., Bauernfeind, B., & Fokdal, J. (2011). Barrier-free transport facilities in Shanghai: Current practice and future challenges. *Bridging Urbanities: Reflections on Urban Design in Shanghai and Berlin*, *135*.

13 Public health issues

Matúš Šucha and Ralf Risser

Introduction

Currently, the following points are more or less part of the general knowledge base in industrialised countries: walking strengthens the heart, lowers the risk of diabetes and cancer, helps with weight loss, prevents dementia, improves physical fitness, boosts vitamin D levels (since one is outside), supports youthfulness by enhancing circulation and increasing oxygen supply, has an anti-depressive effect, and, importantly, causes no harm to the environment. Medical professionals may not be impressed by the layperson's phrasing in this list, but we all understand what these expressions mean. Walking – regularly and over somewhat longer distances (some recommend 10,000 steps a day, others suggest that 3,000 or even just half an hour is sufficient) – is healthy and makes us feel good, or at the very least, better. Moreover, in addition to its health benefits, walking enhances social competence: people who walk more tend to perform better in communication and other social skills. And, last but not least, walking helps save money.

With a pinch of salt, similar arguments apply to cycling, the other major active mode of transport. Micromobility is not discussed here, as research into the health impacts of these modes is ongoing and their role in relation to public health is not yet fully understood.

The tasks of psychology in this context are threefold: (1) to provide a comprehensive summary of the aforementioned benefits; (2) to understand which of these benefits are actually known to the population, both among those who already use active travel modes and those who do not (while bearing in mind that driving a car or riding a motorcycle probably contributes little, if anything, to public health). In this regard, gaps in awareness of these health benefits must be identified, and on this basis, targeted awareness-raising measures should be developed and implemented; and (3) the third task – though by no means the least important – is to make these advantages tangible for those segments of the population who remain unmoved by the facts, even when they are aware of them. We should remember that there are many physicians who smoke, which clearly illustrates that knowledge alone is not enough to produce sensible behaviour. Psychological strategies, as discussed

DOI: 10.4324/9781003608394-13

in other chapters of this book, need to be applied to ensure that people not only accept public health arguments on an intellectual level but are also motivated to act accordingly.

Public health implications of transport: noise

This section outlines the main categories of public health impacts, apart from traffic safety issues, that are significantly influenced by the choice of travel mode.

For instance, transport is the primary source of noise pollution in Europe, with road traffic being the main cause of human exposure to noise, except for those living near airports or railway lines. Ambient noise levels have increased steadily as a result of the growing number of road journeys and kilometres travelled by motor vehicles, higher driving speeds, and the increased frequency of air travel and use of larger aircraft. There is strong evidence demonstrating the harmful effects of noise on communication, academic performance, sleep, mood, and cardiovascular health, as well as its contribution to hearing impairment.

Vehicle pollution exposure

Motor vehicles emit a range of pollutants that can cause health problems and ecological damage, including contributing to *climate change* (Litman, 2009). Although emission control technologies have reduced the amount of pollution per vehicle-kilometre, motor vehicle emissions remain a significant health risk, partly because the benefits of reduced emission rates are offset by increasing volumes of vehicle travel (Health Effects Institute, 2010). The number of premature deaths attributed to air pollution from motor vehicles is probably comparable to those caused by road traffic crashes.

Traffic-calming and *speed reduction* strategies, such as lower speed limits and more effective speed enforcement, not only improve safety by reducing the frequency and severity of collisions but in some cases also help reduce overall vehicle travel (Bellefleur & Gagnon, 2012). Their impact on pollution levels varies with the context; emission rates per kilometre typically decrease when speeds are reduced from high to moderate levels, but may rise again at very low speeds (below 20 km/h) and with increased stop-and-go driving (e.g. where speed bumps are used). However, reducing traffic speeds generally tends to improve conditions for walking and cycling, which would lead to lower per capita emissions, increased physical activity, and better access to essential services (Mackett & Brown, 2011).

Physical activity and fitness

A further major category of health impacts pertains to the influence of transport planning decisions on physical activity and fitness. Public health authorities are increasingly concerned about *declining physical fitness*, rising rates of *overweight* and *obesity*, and the consequent *increase in illnesses* such as diabetes. Current recommendations suggest that adults should engage in at least 150 minutes per week (approximately 22 minutes per day) of moderate-intensity physical activity, while children should achieve around three times that amount (US Department of Health and Human Services, 2008). A meta-analysis of 22 cohort studies found that, compared to no reported physical activity, 2.5 hours of moderate activity per week is associated with a 19% reduction in mortality, while 7 hours per week is associated with a 24% reduction (Woodcock et al., 2011). There are many ways to be physically active, but most, such as going to the gym or participating in organised sports, require additional time, skills, and financial resources. These barriers discourage consistent, lifelong engagement in physical activity, particularly among vulnerable groups such as people with low incomes or those who are currently sedentary and overweight. Research has shown that car use is positively associated with sedentary lifestyles and increased body weight, while walking and cycling are linked to lower rates of obesity and related conditions such as high blood pressure and diabetes. As most journeys by public transport involve walking to and from stops, the use of *public transport* also *tends to increase physical activity* (Lachapelle et al., 2011). Many experts conclude that one of the most practical ways of improving the population's fitness and health is to promote walking and cycling for both utilitarian and recreational purposes. However, conventional transport planning often fails to account for the significant effects that policy and planning decisions can have on public health and fitness.

Improvements to walking and cycling infrastructure, such as better pavements, pedestrian crossings, and dedicated pathways, can reduce the risk of crashes for these modes. As previously discussed, increasing the use of active transport tends to lower overall crash rates thanks to the "safety in numbers" effect. In a typical case, doubling the distance walked and cycled in an area leads to a 32% increase in pedestrian and cyclist injuries, which means a decrease per capita, but reduces the risks for other road users, resulting in an overall decline in traffic casualties, as well (Jacobsen, 2003). Shifts from driving to active modes can produce proportionately greater reductions in air pollution emissions, as these modes typically replace short urban vehicle trips, which tend to have high per-kilometre emission rates because of cold engine starts and traffic congestion. Consequently, a 1% shift from motorised to non-motorised modes typically results in a 2–4% reduction in emissions (Grabow et al., 2012). Finally, improvements for walking also tend to promote and enhance access to amenities and services – where there are people out walking shops and gastronomy do better business.

Improvements to public transport services, such as additional routes, longer operating hours, more frequent services, more comfortable vehicles and stations and stops, grade separation, and enhanced user information, can attract discretionary users (i.e. those who would otherwise drive), thereby reducing overall crash rates and pollution emissions. Public transport travel has a traffic fatality rate an order of magnitude lower than that of car travel, and high-quality services often promote further reductions in vehicle use by acting as a catalyst for transit-oriented development. Such development fosters neighbourhoods in which residents tend to own fewer cars, drive less, and rely more on walking, cycling, and public transport, resulting in additional health and safety benefits (Bailey et al., 2008). These findings help to explain why per capita traffic fatalities tend to fall significantly as per capita transit ridership increases within a community.

Public transport also has other health-related implications. By reducing congestion delays, features such as bus priority lanes and signal controls can lower bus emission rates. When combined with supportive land use policies, public transport improvements can lead to transit-oriented development, producing further reductions in car use and the associated benefits. As most public transport journeys involve walking or cycling segments, improvements in services tend to support greater physical fitness. They also enhance basic accessibility and can reduce commuter stress, particularly where the services are of high quality (Wener & Evans, 2007).

Mental health implications

Transport planning decisions can also affect mental health in various ways. Improvements to walking and cycling infrastructure and public transport services tend to support mental well-being by increasing physical activity and *fostering community cohesion*, that is, encouraging positive interactions among neighbours. Higher levels of neighbourhood walkability are associated with *reduced symptoms of depression* in older men and a *lower incidence of dementia* in both women and men (Larson et al., 2006). In a study of 299 older adults in the United States (mean age 78), Erickson et al. (2010) found significantly *higher grey matter volume* and *cognitive ability* among those who previously walked more than 72 blocks per week. High-quality public transport can also *reduce commuting stress* in comparison with car travel.

The number of motor vehicle accidents and the physical injuries and deaths resulting from them are closely monitored in most countries. However, the long-term psychological effects frequently experienced by accident survivors, even if they suffered minor or no physical injury, are less consistently tracked. Studies have shown that 14% of survivors develop diagnosable post-traumatic stress disorder (PTSD) (Goldberg & Gara, 1990), 25% display psychiatric disorders one year after an accident, and one-third exhibit clinically significant symptoms at an 18-month follow-up (Green et al., 1993). *PTSD* is a debilitating condition involving symptoms such as * re-experiencing the trauma through nightmares, flashbacks, or uncontrollable intrusive recollections; * avoidance behaviours, including steering clear of situations that trigger recollections of the event, blocking feelings, and becoming detached and estranged from others; and * heightened arousal, resulting in sleep disturbances, poor concentration and memory, hypervigilance, and an exaggerated startle response (De Horne, 1995).

As previously noted, traffic noise has been shown to provoke nervousness, depression, sleeplessness, and excessive irritability. However, other aspects of the transport system can also contribute to psychological distress. Regular exposure to traffic congestion has been linked to poorer health outcomes, impaired psychological adjustment, diminished work performance, and reduced overall life satisfaction (Novaco et al., 1990). Congestion-related constraints are conducive to raised blood pressure and a reduced ability to cope with frustration. This not only affects well-being but can also lead to aggressive behaviour and an increased risk of involvement in traffic accidents (Mayer & Treat, 1977).

Excessive automobile use has impacted on people's social lives. While the car has enabled people to move away from cities and settle in suburban areas, many of these neighbourhoods have been developed around the needs of the automobile rather than the psychological and social needs of residents. Close-knit communities have been replaced by neighbourhoods that do not foster social interaction, resulting in increased social isolation (Cassidy, 1997). These new areas rarely include local schools, small shops, or other venues that might facilitate community engagement. Instead, shopping and services have relocated to large, impersonal out-of-town centres. Additionally, the growing use of cars has affected social contact through what is known as the *community severance effect* – the dividing impact that roads and traffic can have on local communities. A seminal study of the influence of traffic on three residential streets in San Francisco demonstrated how traffic volume and speed affect the way people use streets for purposes other than transport (Appleyard & Lintell, 1972). The three streets were similar in all respects except for traffic volume: one street had 2,000 vehicles per day (Light Street), another 8,000 (Moderate Street), and the third 16,000 (Heavy Street). Residents were asked about their perceptions of their neighbourhoods, and researchers observed behaviours such as pedestrian delay times, the number of closed windows, drawn blinds, parked cars, flower boxes, and the amount of litter. Those living on Light Street had three times as many friends and twice as many acquaintances among their neighbours compared to the residents of Heavy Street. Light Street was perceived as friendly, and families with children felt relatively safe from traffic hazards. In contrast, Heavy Street showed little to no street-level activity and was described as "used solely as a corridor between the sanctuary of individual homes and the outside world" (Appleyard & Lintell, 1972). The decline in environmental quality on Heavy Street led to a process of environmental selection and adaptation: the street's residential composition changed significantly over time in response to the hostile traffic conditions. The residents tended to withdraw from the public realm and keep to themselves, and there was little sense of community (Appleyard & Lintell, 1972; Okeke et al., 2020).

How to convince people that all this is relevant

All the issues discussed so far affect people's and communities' lives rigorously. However, there does not seem any strong tendency to strive for a change. One assumption is of course that people have become used to the situation and changes are anticipated as creating disturbance – even if the change would be to the better from a bird's-eye view. The COVID-19 pandemic has demonstrated that effective public health management requires not only epidemiological measures but also a thorough understanding of how effective communication with specific population groups should look like in order to work. Psychology plays a key role in this process by enabling the segmentation of the population on the basis of attitudes, values, and behavioural patterns, thereby allowing communication strategies to be tailored to different target groups (Betsch et al., 2020; Risser & Šucha, 2020). A marketing approach informed by behavioural science helps to understand why certain population groups reject recommended measures (such as mask-wearing or vaccination) and to design targeted campaigns that address these barriers in both emotional and rational terms, substantially increasing the likelihood of desirable behavioural change.

Finally, trust in public institutions and sources of information has been shown to be a crucial factor in following recommendations during pandemic crises. Psychology can help ensure that communication is transparent, consistent, and adapted to diverse value orientations within the population, thereby increasing its effectiveness (Van Bavel et al., 2020). Approaches such as *social marketing* (Lee & Kotler, 2011) draw on insights into motivations and behavioural barriers to propose positive interventions that support healthy decision-making. Successful strategies must therefore combine scientific evidence with special regard for emotional factors, cultural specifics, and value systems in order to influence behaviour effectively, and not only during times of crisis.

References

Appleyard, D., & Lintell, M. (1972). The environmental quality of city streets: The resident's viewpoint. *Journal of the American Institute of Planners*, *38*, 84–101.

Bailey, L., Mokhtarian, P. L., & Little, A. (2008). *The broader connection between public transportation, energy conservation and greenhouse gas reduction*. ICF International. https://onlinepubs.trb.org/onlinepubs/tcrp/docs/TCRPJ-11Task3-FR.pdf

Bellefleur, O., & Gagnon, F. (2012). *Urban traffic calming and health: A literature review*. National Collaborating Centre for Healthy Public Policy. www.ncchpp.ca/docs/ReviewLiteratureTrafficCalming_En.pdf

Betsch, C., Wieler, L. H., & Habersaat, K. (2020). Monitoring behavioural insights related to COVID-19. *The Lancet*, *395*(10232), 1255–1256. https://doi.org/10.1016/S0140-6736(20)30729-7

Cassidy, T. (1997). *Environmental psychology – behaviour and experience in context.* Psychology Press.
De Horne, L. D. J. (1995). Detection and assessment of PTSD by health and legal professions. *Psychiatry, Psychology and Law*, 2(1), 65–73.
Erickson, K. I., Raji, C. A., Lopez, O. L., Becker, J. T., & Rosano, C. (2010). Physical activity predicts gray matter volume in late adulthood: The cardiovascular health study. *Neurology*, *75*, 1415–1422. https://doi.org/10.1212/WNL.0b013e3181f88359
Goldberg, L., & Gara, M. A. (1990). A typology of psychiatric reactions to motor vehicle accidents. *Psychopathology*, *23*(1), 15–20. https://doi.org/10.1159/000284632
Grabow, M. L., Spak, S. N., Holloway, T., Stone, B., Mednick, A., & Patz, J. A. (2012). Air quality and exercise-related health benefits from reduced car travel in the midwestern United States. *Environmental Health Perspectives*, *120*(1), 68–76. https://doi.org/10.1289/ehp.1103440
Green, M. M., McFarlane, A. C., Hunter, C. E., & Grigg, M. (1993). Undiagnosed post-traumatic stress disorder following motor vehicle accidents. *Medical Journal of Australia*, *159*(8), 529–534. https://doi.org/10.5694/j.1326-5377.1993.tb138006.x
Health Effects Institute. (2010). *Traffic-related air pollution: A critical review of the literature on emissions, exposure, and health effects* (HEI Special Report No. 17). Health Effects Institute. www.healtheffects.org/system/files/SR17TrafficReview.pdf
Jacobsen, P. L. (2003). Safety in numbers: More walkers and bicyclists, safer walking and bicycling. *Injury Prevention*, *9*(3), 205–209. https://doi.org/10.1136/ip.9.3.205
Lachapelle, U., Saelens, B. E., Sallis, J. F., & Conway, T. L. (2011). Commuting by public transit and physical activity: Where you live, where you work, and how you get there. *Journal of Physical Activity & Health*, *8*(Suppl. 1), S72–S82. https://doi.org/10.1123/jpah.8.s1.s72
Larson, E. B., Wang, L., Bowen, J. D., McCormick, W. C., Teri, L., Crane, P., & Kukull, W. (2006). Exercise is associated with reduced risk for incident dementia among persons 65 years of age and older. *Annals of Internal Medicine*, *144*(2), 73–81. https://doi.org/10.7326/0003-4819-144-2-200601170-00004
Lee, N. R., & Kotler, P. (2011). *Social marketing: Influencing behaviors for good.* SAGE publications.
Litman, T. (2009). *Transportation cost and benefit analysis: Techniques, estimates and implications* (2nd ed.). Victoria Transport Policy Institute. www.vtpi.org/tca/tca01.pdf
Mackett, R. L., & Brown, B. (2011). *Transport, physical activity and health: Present knowledge and the way ahead.* Centre for Transport Studies, University College London. www.ipenproject.org/documents/publications_docs/UCL%20T%2C%20PA%20%26%20H%20report%20Dec%2011.pdf
Mayer, R. E., & Treat, J. R. (1977). Psychological, social and cognitive characteristics of high-risk drivers: A pilot study. *Accident Analysis and Prevention*, *9*, 1–8. https://doi.org/10.1016/0001-4575(77)90002-1
Novaco, R. W., Stokols, D., & Milanesi, L. (1990). Objective and subjective dimensions of travel impedance as determinants of commuting stress. *American Journal of Community Psychology*, *18*(2), 231–257. https://doi.org/10.1007/BF00931303
Okeke, F. O., Okosun, A. E., Udeh, C. A., & Okekeogbu, C. J. (2020). Cities for people: The dependency & impact of automobile in the life of city dwellers. *European Journal of Sustainable Development*, *9*(3), 157. https://doi.org/10.14207/ejsd.2020.v9n3p157
Risser, R., & Šucha, M. (2020). Start walking! How to boost sustainable mode choice – psychological measures to support a shift from individual car use to more sustainable traffic modes. *Sustainability*, *12*(2), 554. https://doi.org/10.3390/su12020554

US Department of Health and Human Services. (2008). *Physical activity guidelines for Americans*. USDHHS. http://health.gov/paguidelines/pdf/paguide.pdf

Van Bavel, J. J., Baicker, K., Boggio, P. S., Capraro, V., Cichocka, A., Cikara, M., Crockett, M. J., Crum, A. J., Douglas, K. M., Druckman, J. N., Drury, J., Dube, O., Ellemers, N., Finkel, E. J., Fowler, J. H., Gelfand, M., Han, S., Haslam, S. A., Jetten, J., . . . & Willer, R. (2020). Using social and behavioural science to support COVID-19 pandemic response. *Nature Human Behaviour*, *4*(5), 460–471. https://doi.org/10.1038/s41562-020-0884-z

Wener, R. E., & Evans, G. W. (2007). A morning stroll: Levels of physical activity in car and mass transit commuting. *Environment and* Behavior, *39*(1), 62–74. https://doi.org/10.1177/0013916506295571

Woodcock, J., Franco, O. H., Orsini, N., & Roberts, I. (2011). Non-vigorous physical activity and all-cause mortality: Systematic review and meta-analysis of cohort studies. *International Journal of Epidemiology*, *40*(1), 121–138. https://doi.org/10.1093/ije/dyq104

14 Traffic mode choice

Elisabeta Drimlová

Introduction

A wide range of transport options is available to us, thanks to technological advancements. European society is accustomed to overusing cars, which is not sustainable. We need to determine what influences our choice of transport mode and then apply psychological strategies through which individual car use can be reduced. This chapter describes transport mode choice from the perspective of four selected theories. We will introduce each theory and describe how they can be used to explain transport mode choice, as well as discussing their limitations. In addition, we will address the role of habit in choosing a specific transport mode and examine what life without a car might look like. Finally, recommendations for reducing individual car use will be outlined.

Is the car the only option?

In today's world, there are many means of transport to choose from for travelling. The possibility of getting from one place to another under given circumstances is referred to as *mobility*. It is closely linked to quality of life – the higher the mobility, the better the quality of life. While mobility is achieved through transport, we aim to minimise the volume of transport, as it has a negative impact on the environment. The less impact transportation has, the more sustainable it is considered to be. *Sustainable modes of transport* are divided into active and passive. Active transport modes, such as walking, have more benefits for both the environment and individuals, as they involve the use of our own physical power. Public transport is also regarded as sustainable, given the number of passengers it can carry.

The least sustainable mode of transport is individual car use. Many people could hardly imagine their lives without cars, and it comes as no surprise that the term car-dependent society has become widely used (Newman & Kenworthy, 1989). This is demonstrated by roads, parking spaces, the number of cars, and car-centric urban planning. Driving a car can entail pleasure and feelings of comfort, effortlessness, and control. The car is a symbol of

DOI: 10.4324/9781003608394-14

freedom and speed (Selzer & Lanzendorf, 2022). The adverse effects of transport include its impact on the environment (the transport sector is responsible for 23% of global greenhouse gas emissions contributing to climate change), land consumption and public space taken up by transport infrastructure, road fatalities and injuries, public health concerns (e.g. noise, pollution, less physical activity, and increased levels of stress and frustration), and economic costs (Conley, 2016; Kirschner, n.d.; McCarthy et al., 2018; World Health Organization, 2018). Despite efforts to mitigate the negative effects of car use (e.g. by improving engine technology), it is evident that human behaviour related to car use needs to be influenced through psychological approaches (Abrahamse et al., 2009; World Health Organization, 2022).

Theory of planned behaviour (TPB)

The *theory of planned behaviour* (TPB) is a general social theory that builds on the theory of reasoned action (Ajzen, 1991). It proposes that human behaviour is determined by *intention*, which is shaped by *attitudes*, *subjective norms*, and *perceived behavioural control*. Attitudes refer to the positive or negative evaluation of choosing a particular transport mode (e.g. driving a private car is perceived negatively when an individual believes that one should behave in an environmentally friendly manner). Subjective norms are shaped by social pressure exerted towards a specific behaviour (e.g. to what extent car use is approved by one's social environment). They reflect people's motivation to align their behaviour with the expectations of reference groups (Lanzini & Khan, 2017). Perceived behavioural control refers to the degree to which an individual finds it easy or difficult to engage in a certain behaviour; for example, someone who finds driving challenging will perceive a low level of control over such behaviour. TPB assumes that we behave pro-environmentally because we have rationally evaluated all the relevant advantages and disadvantages.

Bamberg et al. (2003) used TPB to investigate travel-related behaviour among college students. TPB was found to be more effective in explaining car use than the norm activation model. In their study of university students' bus use, Heath and Gifford (2006) found that attitudes, social norms, and perceived behavioural control could account for up to 70% of the variance in behaviour. In addition, a descriptive norm (i.e. more people engage in a particular behaviour) was found to enhance the predictive power of TPB in relation to bus use. Similarly, the interaction between intention and perceived behavioural control (i.e. when greater control is perceived, the link between intention and behaviour strengthens) significantly improved the power of TPB to predict bus use (see also Javid et al., 2021).

The most common theories used in research were the *theory of planned behaviour* (TPB) and the *norm activation model* (NAM). Perceived behavioural control, intention, and habit consistently generated higher effect sizes on average. Attitudes, as conceptualised by TPB, can also be effective in predicting travel mode choice depending on the type of journey. For example, attitudes may be stronger predictors of a travel mode chosen for commuting (Hoffmann et al., 2017). Lanzini and Khan (2017) concluded that the best predictors of transport mode choice were intention, habit, and past travel behaviour.

Despite the good applicability of TPB, research should be expanded to include moral norms and the human–car relationship, as the dominance of automobiles in our culture and our psychological attachment to cars have a significant influence on our reluctance to give up individual car use (Anable, 2005). No aspect of TPB covers moral reasoning, and indeed, personal norms have been shown to enhance the explanation of both intention and past behaviour beyond the scope of TPB (Harland et al., 1999).

Abrahamse et al. (2009) integrated TPB and personal norms into their research. Car use was explained by perceived behavioural control and attitudes, while personal norms were better predictors of the intention to reduce it. Perceived behavioural control was found to moderate the relationship between personal norms and behavioural intention: when our personal norms are strong and our perceived behavioural control is low, our intentions to engage in certain behaviour grow stronger. In other words, when we see few opportunities to reduce individual car use, we feel an even greater moral obligation to do so. Other limitations of TPB include its reduction of the situational influences on perceived behavioural control and its weaker prediction of repetitive behaviour (Klöckner & Blöbaum, 2010).

Norm activation model (NAM)

The *norm activation model* (NAM) explains altruistic behaviour. It understands it as behaviour in which a person voluntarily forgoes personal benefits for the sake of others. Pro-environmental behaviour can be considered altruistic too: for example, we may reduce our comfort by giving up commuting to work by car so that other people could benefit from this behaviour.

NAM assumes that *moral reasoning* is important for behaviour. The feelings of moral obligation are reflected in *personal norms*. Two factors that play key roles in the norm activation process are (1) *awareness of the consequences of our behaviour* and (2) a *sense of personal responsibility*. If our behaviour is in accord with our personal norm, we may feel pride. If our behaviour does not follow our personal norm, we feel guilt. When we recognise our personal responsibility, our behaviour can become altruistic (Schwartz, 1977; Schwartz & Howard, 1981). If we believe, for example, that driving short distances is harmful, we will feel a greater moral obligation to reduce car use. A sense of personal responsibility will help us act in ways that contribute to reducing individual car travel (Abrahamse et al., 2009). NAM perceives personal norms as a direct predictor of human behaviour (Klöckner & Blöbaum, 2010).

The predictive power of NAM (and TPB) is supported by an Indonesian research project in which personal norms explained 62.7% variance of the intention to use cars (Setiawan et al., 2014). NAM has also proved useful in the study of tourists' travel behaviour: a sense of personal responsibility for environmental protection, awareness of the consequences of one's actions, and personal ecological norms explained 44% of holiday travelling behaviours (Vaske et al., 2015).

A Pakistani study applied NAM to identify attitudes towards public transport during the COVID-19 pandemic. Awareness of consequences and responsibility was found to be in positive correlation with personal norms, while attitudes towards public transport correlated negatively with personal norms. Attitudes were also shown to be strong predictors of public transport preferences (Javid et al., 2021). However, NAM (similarly to TPB) limits the situational influence on behaviour to perceived behavioural control (Klöckner & Blöbaum, 2010).

Value-belief-norm theory

The *value-belief-norm* (VBN) theory can be seen as a younger sibling of the norm activation model (NAM). Drawing on NAM, it integrates Schwartz's value theory and the *new environmental paradigm*. Its authors propose a chain of variables – *values* and *environmental concern* – that influence behaviour by activating personal norms (Stern et al., 1999).

The VBN theory is capable of predicting the acceptance of transport policies, such as road pricing, intended to reduce excessive car use (De Groot & Steg, 2008). Similarly, the VBN theory is supported by Eriksson et al. (2006, 2008), who suggest that the acceptability of various travel demand management measures is related to environmental sustainability orientation, problem awareness, and personal norms. The theory appears to hold across different cultural contexts, for example, in Argentina (Jakovcevic & Steg, 2013) and Norway (Lind et al., 2015). Values and personal norms were found to predict the acceptance of transport policies and the intention to reduce car use. Car-using respondents felt a greater sense of responsibility for averting the negative environmental impacts of cars in comparison to frequent users of public transport (Lind et al., 2015). However, the VBN theory was found to be a weak predictor of the acceptance of disincentives to driving and active transport mode choices among Chinese parents. Situational variables played an important role: car ownership was found to have the strongest association with lower acceptance of driving disincentives, while longer walking times to school were associated with a lower probability of active travel mode preferences (Nordfjærn & Zavareh, 2017).

Comprehensive action determination model

The *comprehensive action determination model* (CADM) seeks to explain pro-environmental behaviour in broader terms. While recognising the strong predictive power of both TPB and NAM, Klöckner and Blöbaum (2010) argue that these models fail to incorporate other aspects of pro-environmental behaviour. For example, TPB emphasises intention, NAM highlights personal norms, and VBN focuses on values, but these theories seem to disregard habits and situational variables, which also influence our behaviour.

Klöckner and Blöbaum (2010) propose that behaviour is determined by intention, situation, and habits. Social and personal norms are present in both intention and habit. Intention itself, which, as in TPB, precedes our behaviour, is shaped by norms. Additionally, it is influenced by attitudes and perceived behavioural control. These sources of behaviour interact with one another. Personal norms refer to our value system and adapt to situational variables. They are generated by the awareness of need, awareness of the consequences of behaviour, and perceived behavioural control. Habit is understood as stable, repetitive behaviour.

Klöckner and Friedrichsmeier (2011) studied CADM in terms of its ability to explain driving as a preferred travel mode for getting to school, work, leisure activities, and shops in a university student sample. CADM accounted for 62% of the variance, while the remaining situational variables included in the statistical model as predictors (e.g. car availability, weather, or daylight) explained another 48% of the variance. CADM is also validated by means of studies exploring ecological behaviour. Joanes et al. (2020) looked into the reduction of clothing consumption. Personal norms were shown to be a significant predictor of waste prevention behaviours (Balundė et al., 2020).

The role of habit in travel mode choice

Our decision-making is not influenced only by the attributes of the aforementioned theories; habit and automated behaviour also play important roles. While habit appears in some theories, it never features as a central motive for travel mode choice. Habit is a repeated behaviour under similar conditions. It is therefore logical that, in commuting, where spatiotemporal and social conditions do not vary much, we tend to stop looking for new solutions to familiar situations. The ability of attitude and habit to explain travel mode choice was tested by Verplanken et al. (1994). When the habit of using a car was strong, the explanatory power of attitude in predicting travel mode choice was weak. Moreover, the habit of using a particular mode of transport tends to be generalised and independent of the purpose of the journey (Verplanken et al., 1994).

The functionality of habit lies in its "simplicity". Habit limits the supply of information needed to reconsider our choice of travel mode by reducing cognitive efforts in the conditions that trigger habitual behaviour (Engel, 1993; Verplanken et al., 1994). Verplanken and Roy (2016) refer to habit as *tunnel vision* resulting in superficial decision-making. While experimental intervention may have an effect on the decision-making process, its degree is not sufficient to weaken the relationship between habit and car use (Garvill et al., 2003).

How can habit explain travel mode choice?

It is useful to study the influence of habit, attitudes, and norms on travel mode choice. It helps to determine when the choice of travel mode is conscious (and, accordingly, where there is greater potential for change) rather than involving automatised behaviour.

One of the best predictors of travel mode choice is past behaviour (Lanzini & Khan, 2017). The habit of driving was the most common and relevant predictor of all instances of car use, regardless of the purpose of the journey. Habit was the strongest predictor of travel mode choice for leisure activities and shopping and was the only factor that could explain car use for childcare purposes. The habit of using a private car is influenced by the place of residence: people in urban areas drive less because they have access to a dense public transport network (Ramos et al., 2020).

Disruption of habit as an opportunity for change

An opportunity for change is created by disrupting habits, which increases our sensitivity to previously overlooked information (Šucha et al., 2019; Verplanken & Roy, 2016). Habits are naturally disrupted during certain life events, such as residential relocation. Such key events can have a direct impact (e.g. obtaining a driving licence) or an indirect impact (e.g. the birth of a child) on the availability of transport modes (Lanzendorf, 2010). People with weaker habits in choosing a particular travel mode were found to be more likely to have experienced at least one life change during the study period (Klöckner & Matthies, 2004). This is referred to as the habit discontinuity hypothesis (Ramos et al., 2020; Verplanken & Roy, 2016). Hoffmann et al. (2017) supported this hypothesis by concluding that identifying opportunities for disrupting habits may be just as beneficial as using persuasive interventions to promote reduced car use.

Habits can be disrupted by *incentives*. Common examples of those are discounts for students or pensioners. Discounts can also be granted as a reward for certain activities, such as blood donation. Another opportunity for disrupting habits may be a research *experiment*.

Car-less and car-free urban areas

Private car use has been recognised as an issue for over 20 years. Our mobility is influenced by the place in which we live. This is why large European cities have developed what are referred to as car-less or car-reduced neighbourhoods, where efforts are made to minimise car use, and car-free neighbourhoods, where the goal is to do without cars completely. These areas are designed so that all essential daily life amenities are accessible within 15 minutes by sustainable transport modes. The community element within car-less and car-free neighbourhoods plays a crucial role as a source of cooperation and mutual support (Baehler & Rérat, 2020a). The longer a person lives in a car-less neighbourhood, the more likely they are to give up private car use, as living without a car becomes "normal" there (Selzer & Lanzendorf, 2022).

Nearly half of the residents in car-less/car-free neighbourhoods in Switzerland and Germany are *families with children*. A German study exploring mobility practices in a car-reduced housing development found that a third of the residents hold a university degree, and 84% had a driving licence and could afford to own a car. Their *decision to live without a car was deliberate and voluntary*. While environmental reasons played an important role in their choice, these alone would not have convinced the majority of the respondents to give up cars. The availability of amenities and services within the neighbourhood was key (Baehler & Rérat, 2020b).

The phenomenon described earlier, known as *residential self-selection (RSS)*, has been studied since 2008. The RSS concept proposes that people choose their place of residence according to transport conditions, available services, and personal needs (Mokhtarian & Cao, 2008). It was found that certain segments of the population – mainly younger people and those without children – did indeed select their residential location on the basis of their travel mode preferences. Areas with sustainable, efficient, and diverse mobility options were typically found in city centres or densely populated neighbourhoods with a mixed-use built environment (Humphreys & Ahern, 2019; Wolday et al., 2019). Families with children were less likely to live in city centres because of their greater need for green areas but less money to purchase a family home (Wolday et al., 2019). Stability and change in travel behaviour should be considered in the context of RSS and an individual's life stage (Scheiner, 2014).

People with a generally positive stance towards travel tend to be more satisfied with their journeys and are more likely to live in suburban neighbourhoods (spending more time travelling). In contrast, "travel haters" have been shown to be more inclined to reside in the city centre and are estimated to exhibit a higher degree of RSS (Vos & Witlox, 2016). In a study conducted in Alexandria, Egypt, "availability of transportation" was reported as a major factor influencing residential location choices. However, this did not apply to the respondents who owned a car (Masoumi et al., 2021).

To conclude this section, we will outline the findings of experiments involving limited car use. Families participating in a six-month Finnish experiment were asked to give up their car ownership (with 7 out of 11 families left with one vehicle). As a result, the average distance travelled weekly by car dropped from 210 km to 100 km. The participants made greater use of alternative transport modes and travelled more efficiently. Responsibility for child-related errands shifted towards the adult with a car. Some participants occasionally felt socially excluded. While some viewed the loss of a car as a loss of independence, others learned to make use of the time spent on the bus or walking (e.g. reading or taking walking as exercise). Some experienced a sense of joy from acquiring new competences and felt liberated from taking care of the car, while others felt a loss of independence (having associated car ownership with ease, convenience, and privacy (Laakso, 2017).

The participants in a Czech experiment (ten families) tried life without a car for three months. The need to plan their journeys ahead was found inconvenient. The majority complained about inefficient public transport services. Those who had already considered reducing their car use before the experiment were able to manage without a car during the study, and their reduced car use continued until a year later. Their overall experience was more positive: they appreciated the relief from having to look for a parking place and the health benefits of active travel modes. However, these advantages did not generally outweigh the benefits of car use, such as flexibility, freedom, and time savings (Šucha et al., 2019).

Practical recommendations for reducing private car use

While people are aware of the negative environmental impact of private car use when it comes to commuting, they tend to overlook these adverse effects for journeys undertaken for other purposes. Therefore, campaigns pointing out the impact of car use for different purposes are needed (Ramos et al., 2020).

In order to adopt effective measures for reducing private car use, incentives, campaigns, and information need to be targeted properly (Anable, 2005; Ramos et al., 2020). The most effective approach appears to be aiming at *dissatisfied motorists*, because they feel a moral responsibility when using a car and would rather opt for alternative modes of transport but perceive barriers to doing so (Anable, 2005). It is more effective to *highlight the comfort and time saving* associated with the use of public transport rather than pointing out the costs of car ownership and the health benefits of alternative transport modes (Šucha et al., 2019).

A comprehensive approach to *promoting walking* in society is proposed by Risser and Šucha (2020). They recognise that to "sell" the idea of walking, it must be presented as a valuable activity (intelligent people's travel option rather than an inferior transport mode for the less affluent) that benefits both society and individuals. The emphasis should be placed on the advantages that one can see and *feel*.

Tailored campaigns can be effective, but it is essential that favourable preconditions for alternative modes and supportive infrastructure are created. Urban areas should focus on active modes of transport such as walking and cycling. The concepts of "walkability" and "cycleability" may be seen as reflecting to what degree cities are friendly towards pedestrians and cyclists. Such cities have well-developed networks of pedestrian and cycling zones, cycle paths, and bike lanes. Streets feature traffic-calming measures such as speed bumps and attention is also given to the aesthetic quality of streets. Key preconditions include wide pavements with smooth surfaces, benches for rest, and urban greenery.

Cities support efficient public transport systems. Heath and Gifford (2006) recommend introducing *student bus passes*, as they enhance positive beliefs about public transport use. Through the improvement of conditions for desirable behaviour, people will begin to act accordingly and adjust their beliefs. Strategies to reduce car use are more effective where a habit has not developed (Ramos et al., 2020).

If an opportunity for change arises, a poor experience with alternative transport modes may result in a *boomerang effect* (Šucha et al., 2019). Negative experience may reinforce our belief that alternative travel modes are "terrible". Positive travel attitudes generally have a strong positive impact on commuting satisfaction (Ye & Titheridge, 2017). The reduction of private car use has also been facilitated by *digitalisation*, for example, through smartphone.

References

Abrahamse, W., Steg, L., Gifford, R., & Vlek, C. (2009). Factors influencing car use for commuting and the intention to reduce it: A question of self-interest or morality? *Transportation Research Part F: Traffic Psychology and Behaviour*, *12*(4), 317–324. https://doi.org/10.1016/j.trf.2009.04.004

Ajzen, I. (1991). The theory of planned behavior. *Organizational Behavior and Human Decision Processes*, *50*(2), 179–211. https://doi.org/10.1016/0749-5978(91)90020-T

Anable, J. (2005). 'Complacent car addicts' or 'aspiring environmentalists'? Identifying travel behaviour segments using attitude theory. *Transport Policy*, *12*(1), 65–78. https://doi.org/10.1016/j.tranpol.2004.11.004

Baehler, D., & Rérat, P. (2020a). Beyond the car. Car-free housing as a laboratory to overcome the "system of automobility". *Applied Mobilities*, *7*(3), 1–18. https://doi.org/10.1080/23800127.2020.1860513

Baehler, D., & Rérat, P. (2020b). Between ecological convictions and practical considerations – profiles and motivations of residents in car-free housing developments in Germany and Switzerland. *Geographica Helvetica*, *75*(2), 169–181. https://doi.org/10.5194/gh-75-169-2020

Balundė, A., Jovarauskaitė, L., & Poškus, M. S. (2020). Exploring adolescents' waste prevention via value-identity-personal norm and comprehensive action determination models. *Journal of Environmental Psychology*, *72*, 101526. https://doi.org/10.1016/j.jenvp.2020.101526

Bamberg, S., Ajzen, I., & Schmidt, P. (2003). Choice of travel mode in the theory of planned behavior: The roles of past behavior, habit, and reasoned action. *Basic and Applied Social Psychology*, *25*(3), 175–187. https://doi.org/10.1207/S15324834BASP2503_01

Conley, J. (2016). *Car troubles: Critical studies of automobility and auto-mobility*. Routledge.

De Groot, J. I. M., & Steg, L. (2008). Value orientations to explain beliefs related to environmental significant behavior: How to measure egoistic, altruistic, and biospheric value orientations. *Environment and Behavior*, *40*(3), 330–354. https://doi.org/10.1177/0013916506297831

Engel, C. (1993). Real exchange rates and relative prices: An empirical investigation. *Journal of Monetary Economics*, *32*(1), 35–50. https://doi.org/10.1016/0304-3932(93)90034-D

Eriksson, L., Garvill, J., & Nordlund, A. M. (2006). Acceptability of travel demand management measures: The importance of problem awareness, personal norm, freedom, and fairness. *Journal of Environmental Psychology*, *26*(1), 15–26. https://doi.org/10.1016/j.jenvp.2006.05.003

Eriksson, L., Garvill, J., & Nordlund, A. M. (2008). Acceptability of single and combined transport policy measures: The importance of environmental and policy specific beliefs. *Transportation Research Part A: Policy and Practice*, *42*(8), 1117–1128. https://doi.org/10.1016/j.tra.2008.03.006

Garvill, J., Marell, A., & Nordlund, A. (2003). Effects of increased awareness on choice of travel mode. *Transportation*, *30*(1), 63–79. https://doi.org/10.1023/A:1021286608889

Harland, P., Staats, H., & Wilke, H. A. M. (1999). Explaining proenvironmental intention and behavior by personal norms and the theory of planned behavior. *Journal of Applied Social Psychology*, *29*, 2505–2528. https://doi.org/10.1111/j.1559-1816.1999.tb00123.x

Heath, Y., & Gifford, R. (2006). Extending the theory of planned behavior: Predicting the use of public transportation. *Journal of Applied Social Psychology*, *32*, 2154–2189. https://doi.org/10.1111/j.1559-1816.2002.tb02068.x

Hoffmann, C., Abraham, C., White, M. P., Ball, S., & Skippon, S. M. (2017). What cognitive mechanisms predict travel mode choice? A systematic review with meta-analysis. *Transport Reviews*, *37*(5), 631–652. https://doi.org/10.1080/01441647.2017.1285819

Humphreys, J., & Ahern, A. (2019). Is travel based residential self-selection a significant influence in modal choice and household location decisions? *Transport Policy*, *75*, 150–160. https://doi.org/10.1016/j.tranpol.2017.04.002

Jakovcevic, A., & Steg, L. (2013). Sustainable transportation in Argentina: Values, beliefs, norms and car use reduction. *Transportation Research Part F: Traffic Psychology and Behaviour*, *20*, 70–79. https://doi.org/10.1016/j.trf.2013.05.005

Javid, M. A., Abdullah, M., Ali, N., & Dias, C. (2021). Structural equation modeling of public transport use with COVID-19 precautions: An extension of the norm activation model. *Transportation Research Interdisciplinary Perspectives*, *12*, 100474. https://doi.org/10.1016/j.trip.2021.100474

Joanes, T., Gwozdz, W., & Klöckner, C. A. (2020). Reducing personal clothing consumption: A cross-cultural validation of the comprehensive action determination model. *Journal of Environmental Psychology*, *70*, 101396. https://doi.org/10.1016/j.jenvp.2020.101396

Kirschner, W. (n.d.). *The impact of car dependent policy on standard of living in the US*. https://scholarworks.arcadia.edu/cgi/viewcontent.cgi?article=1522&context=showcase

Klöckner, C. A., & Blöbaum, A. (2010). A comprehensive action determination model: Toward a broader understanding of ecological behaviour using the example of travel mode choice. *Journal of Environmental Psychology*, *30*(4), 574–586. https://doi.org/10.1016/j.jenvp.2010.03.001

Klöckner, C. A., & Friedrichsmeier, T. (2011). A multi-level approach to travel mode choice – how person characteristics and situation specific aspects determine car use in a student sample. *Transportation Research Part F: Traffic Psychology and Behaviour*, *14*(4), 261–277. https://doi.org/10.1016/j.trf.2011.01.006

Klöckner, C. A., & Matthies, E. (2004). How habits interfere with norm-directed behaviour: A normative decision-making model for travel mode choice. *Journal of Environmental Psychology*, *24*(3), 319–327. https://doi.org/10.1016/j.jenvp.2004.08.004

Laakso, S. (2017). Giving up cars – the impact of a mobility experiment on carbon emissions and everyday routines. *Journal of Cleaner Production*, *169*, 135–142. https://doi.org/10.1016/j.jclepro.2017.03.035

Lanzendorf, M. (2010). Key events and their effect on mobility biographies: The case of childbirth. *International Journal of Sustainable Transportation*, *4*(5), 272–292. https://doi.org/10.1080/15568310903145188

Lanzini, P., & Khan, S. A. (2017). Shedding light on the psychological and behavioral determinants of travel mode choice: A meta-analysis. *Transportation Research Part F: Traffic Psychology and Behaviour*, *48*, 13–27. https://doi.org/10.1016/j.trf.2017.04.020

Lind, H. B., Nordfjærn, T., Jørgensen, S. H., & Rundmo, T. (2015). The value-belief-norm theory, personal norms and sustainable travel mode choice in urban areas. *Journal of Environmental Psychology*, *44*, 119–125. https://doi.org/10.1016/j.jenvp.2015.06.001

Masoumi, H., Ibrahim, M. R., & Aslam, A. B. (2021). The relation between residential self-selection and urban mobility in middle eastern cities: The case of Alexandria, Egypt. *Urban Forum*, *32*, 261–287. https://doi.org/10.1007/s12132-020-09414-4

McCarthy, C. P., Vaduganathan, M., & Song, Z. (2018). Moving beyond the hospital. *Circulation*, *138*(20), 2169–2171. https://doi.org/10.1161/CIRCULATIONAHA.118.036391

Mokhtarian, P. L., & Cao, X. (2008). Examining the impacts of residential self-selection on travel behavior: A focus on methodologies. *Transportation Research Part B: Methodological*, *42*(3), 204–228. https://doi.org/10.1016/j.trb.2007.07.006

Newman, P., & Kenworthy, J. (1989). *Cities and automobile dependence: An international sourcebook*. Avebury Technical.

Nordfjærn, T., & Zavareh, M. F. (2017). Does the value-belief-norm theory predict acceptance of disincentives to driving and active mode choice preferences for children's school travels among Chinese parents? *Journal of Environmental Psychology*, *53*, 31–39. https://doi.org/10.1016/j.jenvp.2017.06.005

ProDarce. (n.d.). Doprava. www.prodarce.cz/kategorie/doprava/

Ramos, É., Bergstad, C., & Nässén, J. (2020). Understanding daily car use: Driving habits, motives, attitudes, and norms across trip purposes. *Transportation Research Part F: Traffic Psychology and Behaviour*, *68*, 306–315. https://doi.org/10.1016/j.trf.2019.11.013

Risser, R., & Šucha, M. (2020). Start walking! How to boost sustainable mode choice – psychological measures to support a shift from individual car use to more sustainable traffic modes. *Sustainability*, *12*(2), 554. https://doi.org/10.3390/su12020554

Scheiner, J. (2014). Residential self-selection in travel behavior: Towards an integration into mobility biographies. *The Journal of Transport and Land Use*, *7*, 15–29. https://doi.org/10.5198/jtlu.v7i3.439

Schwartz, S. H. (1977). Normative influences on altruism. *Advances in Experimental Social Psychology*, *10*, 221–279. https://doi.org/10.1016/s0065-2601(08)60358-5

Schwartz, S., & Howard, J. A. (1981). A normative decision-making model of altruism. In J. P. Rushton & R. M. Sorrentino (Eds.), *Altruism and helping behavior: Social, personality, and developmental perspectives* (pp. 189–211). Lawrence Erlbaum Associates.

Selzer, S., & Lanzendorf, M. (2022). Car independence in an automobile society? The everyday mobility practices of residents in a car-reduced housing development. *Travel Behaviour and Society*. *28*, 90–105. https://doi.org/10.1016/j.tbs.2022.02.008

Setiawan, R., Santosa, W., & Sjafruddin, A. (2014). Integration of theory of planned behavior and norm activation model on student behavior model using cars for traveling to campus. *Civil Engineering Dimension*, *16*(2). https://doi.org/10.9744/ced.16.2.117-122

Stern, P. C., Dietz, T., Abel, T., Guagnano, G. A., & Kalof, L. (1999). A value-belief-norm theory of support for social movements: The case of environmentalism. *Human Ecology Review*, 81–97. https://humanecologyreview.org/pastissues/her62/62sternetal.pdf

Šucha, M., Viktorová, L., & Risser, R. (2019). Can an experience with no car use change future mode choice behavior? *Sustainability*, *11*(17). https://doi.org/10.3390/su11174698

Vaske, J. J., Jacobs, M. H., & Espinosa, T. K. (2015). Carbon footprint mitigation on vacation: A norm activation model. *Journal of Outdoor Recreation and Tourism*, *11*, 80–86. https://doi.org/10.1016/j.jort.2015.05.002

Verplanken, B., Aarts, H., van Knippenberg, A., & van Knippenberg, C. (1994). Attitude versus general habit: Antecedents of travel mode choice. *Journal of Applied Social Psychology*, *24*(3), 285–300. https://doi.org/10.1111/j.1559-1816.1994.tb00583.x

Verplanken, B., & Roy, D. (2016). Empowering interventions to promote sustainable lifestyles: Testing the habit discontinuity hypothesis in a field experiment. *Journal of Environmental Psychology*, *45*, 127–134. https://doi.org/10.1016/j.jenvp.2015.11.008

Vos, J., & Witlox, F. (2016). Do people live in urban neighbourhoods because they do not like to travel? Analysing an alternative residential self-selection hypothesis. *Travel Behaviour and Society*, *4*, 29–39. https://doi.org/10.1016/j.tbs.2015.12.002

Wolday, F., Næss, P., & X. Cao, J. (2019). Travel-based residential self-selection: A qualitatively improved understanding from Norway. *Cities*, *87*, 87–102. https://doi.org/10.1016/j.cities.2018.12.029

World Health Organization. (2018). *Cop24 special report health and climate*. https://apps.who.int/iris/bitstream/handle/10665/276405/9789241514972-eng.pdf

World Health Organization. (2022). *Compendium of WHO and other UN guidance on health and environment: 2022 update* (WHO/HEP/ECH/EHD/22.01). www.who.int/publications/i/item/WHO-HEP-ECH-EHD-22.01

Ye, R., & Titheridge, H. (2017). Satisfaction with the commute: The role of travel mode choice, built environment and attitudes. *Transportation Research Part D: Transport and Environment*, *52*, 535–547. https://doi.org/10.1016/j.trd.2016.06.011

15 Sustainability and sustainable transport modes

Matúš Šucha and Ralf Risser

Introduction

Sustainable transport makes a positive contribution to the social, environmental, and economic sustainability of the communities it serves (Poboon & Kenworthy, 1995). The term *sustainable transport* refers to travel modes and planning systems that align with the broader notion of sustainability. According to the definition by the Council of Transport Ministers of the European Union (Litman, 2009), a transport system is considered sustainable if it (1) enables individuals, businesses, and civil society to meet their essential needs for access and development in ways that are safe and compatible with human health and the health of ecosystems and support an equitable consideration of the needs of different generations (*social aspect*); (2) is affordable, operates fairly and efficiently, provides a range of transport mode choices, and supports a competitive economy as well as balanced regional development (*economic aspect*); and (3) reduces emissions and waste to levels the planet can absorb, uses renewable resources at or below their rate of regeneration, and uses non-renewable resources at or below the rate at which sustainable substitutes can be developed, while minimising land take and noise generation (*environmental aspect*). A balance between all three aspects is crucial for sustainable transport, or mobility.

The downsides of transport

Transport systems have a significant impact on the environment. Transport accounts for 20–25% of global energy consumption and carbon dioxide emissions. Greenhouse gas emissions from transport are increasing faster than those of any other energy-consuming sector. Road traffic is also a major source of local air pollution and smog formation (World Energy Council, 2007). The social costs of transport include traffic accidents, air and noise pollution, the absence of physical activity, time spent commuting away from family, due both to congestions and long distances to working places enhanced by the possibility to cover such distances "easily" by car (Korsu & Le Néchet, 2017), dependence on volatile oil prices, and others (Maier et al.,

DOI: 10.4324/9781003608394-15

2023). Traffic congestion also incurs economic costs through lost time and delays in the delivery of goods and services.

Many of these negative impacts disproportionately burden the very social groups who do not own or use a car. Traditional transport planning has focused on improving mobility, primarily for motor vehicles, and may not take sufficient account of wider impacts. However, the true purpose of mobility is to ensure access – to work, education, goods and services, friends, and family – and we already have proven techniques for ensuring better access while reducing environmental and social impacts (Litman, 2003). Communities that have succeeded in increasing the sustainability of their transport networks have done so as part of broader programmes aimed at creating vibrant and livable cities.

Sustainable transport and transport modes

Sustainability rests on three pillars: the economy, the environment, and social life. Consequently, cities that claim to pursue sustainability should adopt policies and measures that foster appropriate development across all three areas. In this context, *mode choice* becomes a key issue to address. For instance, promoting active modes of transport can be considered a meaningful step towards sustainability. Take walking as an example. From an economic perspective, walking is inexpensive in terms of the infrastructure required, and if adequate walking infrastructure encourages more people to walk, it also proves to be cost-effective. Recent studies have shown that an increase in the number of pedestrians consistently stimulates the local economy. Walking also offers clear environmental benefits: it produces hardly any emissions and generates minimal noise. As up to 20% of all car journeys in Europe cover distances that could easily be walked (>1 km; Gorges & Mingardo, 2025; Rudner & Malone, 2011), there is considerable potential for replacing ecologically harmful car use with walking – provided that the right measures are implemented. Walking also contributes positively to social life. One might say that where there are pedestrians, there is life. People tend to feel safer in places with a visible presence of others, thanks to the resulting informal social control. Moreover, walking enables individuals to be with or among other people without necessarily having to socialise. Last but not least, walking promotes health (see Chapter 13), which can be considered an advantage from all three perspectives: economic, environmental, and social (Risser & Šucha, 2020).

In any case, *walking* is, in many respects, the optimal form of mobility in the urban environment and therefore deserves special attention. To meet daily needs, short-distance walks (for work, shopping, leisure, and other purposes) must be encouraged and promoted. As already mentioned, up to 20% of all car journeys in Europe are shorter than one kilometre, which equates to a 10- to 15-minute walk. Sixty-six percent of children live less than one kilometre from their school, and 80% live within two kilometres (a 20- to 30-minute walk). Yet many of these trips are made by car (Frey, 2015). If larger segments of the population are to be motivated and inspired to walk more frequently, it is essential to study the motivations, attitudes, and characteristics of different population groups: what they enjoy and what deters them from walking. On the basis of this knowledge, a sensitive yet forward-thinking, road user-oriented approach must be taken when creating (new) urban spaces – one that places value on good design, functionality, and easy accessibility.

Walking, along with cycling, displays considerable potential for contributing to more sustainable transport. *Public transport* is also generally regarded as an environmentally preferable alternative to car use. A recent large-scale study conducted in Vienna (Füssl et al., 2019) demonstrated that using public transport involves a significant amount of physical activity. In this study, public transport in Vienna was even referred to as a "fitness course" because of the physical effort associated with walking to and from stations or stops, boarding vehicles, occasionally standing during the journey, or rushing to catch connections. Such activities all constitute physical exercise, certainly more than is typically involved in using a private car. In addition, people often walk on escalators rather than simply standing still, or choose to take the stairs instead of using the lift.

When it comes to the costs of infrastructure for pedestrians or cyclists, or for the provision and maintenance of public transport, there is often criticism that "those things" are too costly for taxpayers. However, it has been shown that, in comparison to public transport, car traffic imposes significantly higher public costs per capita each year ("external costs") on top of the personal expenses associated with car ownership and use (Umweltbundesamt, 2010).

Today, when sustainability is discussed in relation to car transport, it is often argued that *electromobility* will solve many of the existing problems. However, it is questionable whether such improvements will be sufficient. One major issue with current car use is the demand for space – a problem that will not be substantially mitigated by the use of electric vehicles. Furthermore, it remains uncertain whether there will be more or fewer cars on the roads in the future. The perception of driving an environmentally friendly vehicle may in fact encourage more people to purchase and use cars, rather than choosing other modes of transport. Another issue concerns the source of the electricity that is used: whether it is generated from renewable and clean energy sources such as wind, solar power, or "clean" waste incineration, or from coal- or oil-powered plants. Additionally, most of the arguments in favour of electromobility tend to focus solely on the ecological aspects. The question remains whether electromobility will also address the economic and social problems associated with car use.

It is evident that greater emphasis must be placed on promoting modes of mobility that require minimal external energy and resources (i.e. those not derived from the individual) and occupy the least possible space per person.

For shorter distances, walking is an obvious candidate. When combined with the use of public transport, walking could bring about significant improvements in the sustainability of human mobility, particularly if a large number of car journeys were replaced in this way. In a slightly different but still comparable way, the same holds true for cycling. However, appropriate conditions must be created to make such forms of mobility acceptable in terms of quality.

Interventions for achieving change

Taking a psychological perspective on walking, cycling, or public transport use in discussions of interventions aimed at achieving change always leads us to focus on the individual road user – or, more precisely, the individual human being. Striving for a change of transport and mobility in their character ultimately means that, in one way or another, many individuals' behaviour needs to be modified. While legal and enforcement measures may be necessary to some extent, democratic societies will primarily consider informing and persuading citizens. Efforts to promote behavioural change often involve awareness-raising and social marketing strategies (Kotler et al., 2016). Individuals should understand and be reminded of – or, ideally, feel – the many advantages which the adoption of more sustainable mobility habits brings both, to society AND themselves personally. The benefits of active travel modes that an individual understands – and finds convincing enough – can take various forms. One may have internalised the economic, ecological, and social values associated with walking, cycling, or using public transport, and may be motivated by these values, for instance, out of idealism or a sense of civic responsibility. Alternatively, one may find that walking or cycling meets other personal needs – those that are not always consciously recognised without frequent experience of such modes. Numerous studies show that walking and cycling can provide feelings of freedom and independence, physical fitness and strength, relaxation, and more such positive feelings. However, such feelings rarely become genuine motivations for action simply by hearing about them from others. It is important to experience them personally in order to really understand.

Thus, sustainable policies, whether implemented by public authorities, employers, and schools or other educational institutions, should include incentives that encourage people to use active travel modes, enabling them to experience the benefits for themselves. A guiding principle, particularly in the case of walking, could be: "We should not have to walk – we should love to walk." This, first and foremost, implies that people should not be forced to walk distances, or under conditions that are unacceptable, just because there are no alternatives. At the same time, walking should not be seen as something one *has to* do. Those who respond negatively even to short walks should begin to *understand* – and *feel* – the personal benefits that walking can offer. Finally, there is no harm if they also come to realise that increased walking is positive for society as a whole. In fact, something so beneficial might well deserve to be loved all the more.

Exposure to facts and arguments pointing out the advantages of active travel modes might help alleviate some of the burden experienced by those who have to walk, cycle, or use public transport under challenging conditions. In a sense, they may come to see some positive aspects associated with the "limitations" on their mobility that they otherwise experience. However, in order to avoid any cynical connotations of such an approach, it is imperative to develop serious and sustained efforts to consider improvements in the conditions for walking, making it accessible and attractive for as many people as possible.

Conditions for walking and cycling

When the aim is to promote sustainable modes of transport in the daily routines of citizens and to make active travel an integral part of everyday mobility patterns, it becomes necessary to make decisions that have complex consequences and long-term impacts on traffic planning, traffic management, and infrastructure design. The outcomes of these decisions, that is, the resulting conditions for walking or cycling, must meet the needs of those who already walk (or cycle), as well as those who would be willing to do so under certain circumstances.

In the EU project WALCYNG (Hydén et al., 1998), the following *quality dimensions* – relevant to the subjective well-being of road users and to their choice of transport mode – were identified. These were later elaborated on by Bein et al. (2004); see also Amann et al. (2006):

- *Safety*: low risk of accidents, low risk of criminal offences (theft and assault) – both objectively and subjectively felt.
- *Flow and progress*: uninterrupted movement; frequent stops are undesirable, and continuous progress is valued more than speed.
- *Comfort*: physical and mental well-being (appropriate temperature, shelter from weather, amenities such as kiosks, toilets, and benches).
- *Aesthetics*: the visual, auditory, and olfactory quality of the environment – do we like what we see, hear, and smell (or do not smell)?
- *Equity and respect*: being treated with respect during the journey (by other pedestrians, cyclists, motorists, the police, etc.).
- *Social climate*: the nature of interactions in the social environment (communication, courtesy, mutual consideration, etc.).
- *Costs*: perceived financial and time costs – travel time tends to feel shorter if the above factors are experienced positively.

As can be seen, these quality dimensions encompass both objective and subjective aspects. Anyone seeking to promote active travel modes must take these into careful consideration. Particularly, the subjective elements require thorough attention. This means that suitable methods must be employed to assess people's needs and how well these are being met. What constitutes a pleasant social climate? Under what conditions do parents feel confident about the safety of the areas where their children play or walk? And so on. It is essential to understand the users' perspectives; without this, planning measures are unlikely to succeed. Professionals working in the fields of traffic, mobility, and urban planning must recognise that people – or at least most of them – will only accept measures they experience as, or believe to be, *contributing to an improvement in their own quality of life*. Therefore, it is crucial that planners and practitioners maintain ongoing dialogue with the public in order to understand their views and, ultimately, gain their cooperation. It is important to highlight the more or less immediate benefits of walking or cycling while providing citizens with opportunities to enjoy these advantages for themselves.

The *psychological effects of climate change* and environmental transformation are increasingly influencing individuals' decisions regarding modes of transport. The phenomenon known as *climate anxiety* is particularly evident among younger generations, manifesting itself as a heightened sense of responsibility for one's environmental footprint, including the choice of transport (Clayton & Karazsia, 2020). Studies show that people who experience climate anxiety are more likely to prefer active travel modes, such as walking, cycling, or public transport, while being less likely to use cars for short journeys (Pihkala, 2020). Climate anxiety thus acts as a psychological motivator that supports environmentally sustainable decisions and increases people's willingness to adopt alternative forms of mobility. The shortcoming of an anxiety-driven way is however that it lacks positive reinforcement which is so important for well-being. Trying, under all conditions, to market the rewarding aspects of a change in mobility behaviour is therefore to be prioritised before blaring out the threats that we face and preaching to renounce in things. Renouncing is not connected to any positive motivation in the first place; and as discussed earlier, a change of mobility habits does in many respects not represent any renouncement at all.

Finally, environmental changes, such as more frequent extreme weather events or declining air quality, affect the practical feasibility of choosing certain transport modes and thus contributing to shifts in behaviour. For example, extreme temperatures or heavy rainfall may temporarily discourage walking or cycling. In the long term, adaptation to such changes may lead to an adaptation of infrastructure conditions that facilitates walking, cycling, and the use of micromobility vehicles under such changing conditions. Moreover, a development of more robust, collective, and environmentally friendly forms of transport that are less vulnerable to climate extremes may be expected – low-energy micro four-wheelers with a roof, driverless buses and trams, etc. (Bamberg, 2013). Integrated urban mobility planning must therefore not only promote the use of environmentally responsible transport modes – the task of psychology – but also respond to changing environmental conditions in order to sustain citizens' motivation and reduce the barriers associated with their adaptation to new climatic realities – the task of whomever this may concern.

References

Amann, A., Reiterer, B., Risser, R., & Haindl, G. (2006). *SIZE – life quality of senior citizens in relation to mobility conditions*. University of Vienna, Institute of Sociology, & FACTUM Chaloupka & Risser OG.

Bamberg, S. (2013). Changing environmentally harmful behaviors: A stage model of self-regulated behavioral change. *Journal of Environmental Psychology*, *34*, 151–159. https://doi.org/10.1016/j.jenvp.2013.01.002

Bein, N., Petrjánošová, M., Plichtová, J., Risser, R., & Ståhl, A. (2004). *Final report of the EU-project HOTEL: How to analyse life quality*. EU Fifth Framework Programme Key Action Improving the Socio Economic Knowledge Base. HPSE-2002-60057. Vienna: FACTUM Chaloupka & Risser OG.

Clayton, S., & Karazsia, B. T. (2020). Development and validation of a measure of climate change anxiety. *Journal of Environmental Psychology*, *69*, 101434. https://doi.org/10.1016/j.jenvp.2020.101434

Frey, H. (2015). *Wien zu Fuß 2015: Daten und Fakten zum Fußverkehr*. Mobilitätsagentur Wien GmbH.

Füssl, E., Jaunig, J., & Titze, S. (2019). ROUTINE: The development of a physical activity promoting journey planner web app. *Social Sciences*, *8*(3), 82. https://doi.org/10.3390/socsci8030082

Gorges, H., & Mingardo, G. (2025). The potential of active modes to reduce short car trips. A data-driven approach. *Transport Policy*, *168*, 1–14. https://doi.org/10.1016/j.tranpol.2025.03.015

Hydén, C., Nilsson, A., & Risser, R. (1998). *WALCYNG – how to enhance walking and cycling instead of shorter car trips and to make these modes safer: Final report. Transport RTD programme of the 4th framework programme*. Lund Institute of Technology, Department of Technology & Society.

Korsu, E., & Le Néchet, F. (2017). Would fewer people drive to work in a city without excess commuting? Explorations in the Paris metropolitan area. *Transportation Research Part A: Policy and Practice*, *95*, 259–274. https://doi.org/10.1016/j.tra.2016.10.030

Kotler, P., Armstrong, G., Harris, L. C., & Piercy, N. (2016). *Principles of marketing* (7th ed.). Pearson Education Limited.

Litman, T. (2003). Measuring transportation: Traffic, mobility and accessibility. *ITE Journal*, *73*(10), 28.

Litman, T. (2009). *Transportation cost and benefit analysis*. Victoria Transport Policy Institute. www.vtpi.org/tca/tca01.pdf

Maier, R., Posch, A., Proß, Ch., Plakolb, S., & Steininger, K. W. (2023). Cutting social costs by decarbonizing passenger transport. *Transportation Research Part D: Transport and Environment*, *122*, 103878. https://doi.org/10.1016/j.trd.2023.103878

Pihkala, P. (2020). Anxiety and the ecological crisis: An analysis of eco-anxiety and climate anxiety. *Sustainability*, *12*(19), 7836. https://doi.org/10.3390/su12197836

Poboon, C., & Kenworthy, J. R. (1995). *Towards a sustainable traffic solution*. Urban Habitat Conference, Delft.

Risser, R., & Šucha, M. (2020). *Psychological perspectives on walking: Interventions for achieving change*. Routledge.

Rudner, J., & Malone, K. (2011). Childhood in the suburbs and the Australian dream: How has it impacted children's independent mobility? *Global Studies of Childhood*, *1*(3), 150–164. https://doi.org/10.2304/gsch.2011.1.3.207

Umweltbundesamt. (2010). *PKW-Maut in Deutschland? Eine umwelt- und verkehrspolitische Bewertung*. www.umweltbundesamt.de/sites/default/files/medien/publikation/long/3929.pdf

World Energy Council. (2007). *Survey of energy resources*. http://hussonet.free.fr/resourc7.pdf

16 Systemic approach and Vision Zero

Petr Zámečník

Introduction

When considering a systemic approach and the Vision Zero strategy in transport, it is necessary to start with the nature of the (transport) system these concepts aim to reshape and cultivate. Over the past 100 years, the dominant element in transport has been the car, and consequently, automobilism has become the prevailing discourse associated with mobility. The technical and social interconnection with the industry – encompassing everything from spare parts, accessories, dealerships, and repair shops to road construction and maintenance, advertising, and suburban expansion – has created an exceptionally powerful complex with a significant influence on not only transport and industry in general but also society and ways of life. This complex has produced a dominant global form of pseudo-private mobility that subordinates other public mobility options, transforming how people access opportunities and navigate the constraints of work, family life, and leisure. *Automobilism* has thus created a dominant culture that sustains prevailing discourses on what constitutes a good life and what a proper mode of mobility requires (Sheller & Urry, 2000).

The history of automobilism and the myth of individual responsibility for road safety

Automobilism has become an extraordinarily powerful and enduring force in collective thinking and public discourse, primarily because it has met strong needs, both societal and individual. Significantly, safety was not among these needs for a long time. Gartman (2004) shows that, over the course of the 20th century, cars fulfilled three major needs. Initially, they satisfied a status need, as only the wealthiest could afford them because of their high cost. Cars became an essential accessory of the leisure class, with driving symbolising the ostentation and arrogance of the most affluent elites (Gartman, 2004). Subsequently, cars began to fulfil (the illusion of) mass individuality, as the advent of mass automobile production made cars accessible to much wider social strata. Finally, in the second half of the 20th century, cars helped to

DOI: 10.4324/9781003608394-16

satisfy the need for subcultural differentiation, serving as a means of expressing identity within an increasingly balanced and pluralistic consumer culture. However, their mass adoption also led to major problems concerning safety and the environment. In the 1960s, the United States, for example, responded to such issues by adopting key pieces of legislation such as the Motor Vehicle Air Pollution Control Act of 1965 and the National Traffic and Motor Vehicle Safety Act of 1966.

These developments significantly shaped the social perspective of cars and transport. Cars have always been at the centre of attention of both transport systems and society as a whole. Since cars were owned by the privileged classes from the outset, they were also viewed as a privileged mode of transport. As a result, road safety measures developed in line with this historical perspective of the car as a privileged object. In the first half of the 20th century, restrictions on drivers were minimal, and responsibility for road crashes was, in systemic terms, attributed to individual error, or to the "risky" behaviour of "second-class" road users (such as pedestrians and cyclists). Although mid-20th-century American newspapers competed in dramatic portrayals of pedestrian deaths, referring to them as "traffic terror in the cities", "carnage", or "massacre of innocent pedestrians under the wheels of trolleybuses", outrage over road collisions was largely directed towards large public transport vehicles.

In Canada, the road safety system of the early 20th century likewise reflected the dominance and privilege of private car use. Fines and warnings issued to drivers lacked the potential to reverse the rising tide of road violence, as the social and cultural framing of responsibility, shaped by the myth of personal responsibility, was directed at road users, particularly non-drivers, rather than the system. In 1955, for example, the "Elmer" campaign, which urged children to take responsibility for their own safety in traffic, was evaluated as a success: a 14% rise in children's injuries on Toronto's roads was still considered a positive outcome. While absurd from today's perspective, this conclusion was logical in the context of the car use culture of the time. Decades of subsequent research have shown that appeals to individual responsibility do not have the potential to protect people from reckless drivers or other traffic-related risks. On the contrary, the *myth of personal responsibility* has contributed to the worsening of road safety conditions (VZC, 2023).

In some European countries, a certain degree of reflection on the myth of personal responsibility occurred as early as the first half of the 20th century. In the United Kingdom, for example, criticism of a traffic safety system centred around individual responsibility had already begun to emerge in the 1940s. The first clear manifestation of the idea of systemic responsibility (later articulated as Vision Zero) can be found in an article written in 1946 by Eric Arthur Blair (better known by his pen name George Orwell). The piece was part of his *As I Please* series, published in the British newspaper *Tribune* between 1943 and 1947. Orwell's article offered a critical reflection on the British "Keep Death off the Roads" campaign, which, like Canada's

"Elmer the Safety Elephant", was driven by the same destructive logic of automobilism and contributed little of value to traffic safety. By that time, Orwell had already gained experience in evaluating campaigns countering Nazi propaganda, and he demonstrated a remarkable ability to capture the essence of the futility and absurdity of a road safety management system based on individual responsibility, as well as of its promotion. What emerged was a timeless text that could be published today without losing any of its relevance. So what did Orwell write?

> One interesting example of our unwillingness to face facts and our consequent readiness to make gestures which are known in advance to be useless, is the present campaign to Keep Death off the Roads.
>
> The newspapers have just announced that road deaths for September dropped by nearly 80 as compared with the previous September. This is very well so far as it goes, but the improvement will probably not be kept up – at any rate, it will not be progressive – and meanwhile everyone knows that you can't solve the problem while our traffic system remains what it is. Accidents happen because on narrow, inadequate roads, full of blind corners and surrounded by dwelling houses, vehicles and pedestrians are moving in all directions at all speeds from three miles an hour to 60 or 70. If you really want to keep death off the roads, you would have to replan the whole road system in such a way as to make collisions impossible. Think out what this means (it would involve, for example, pulling down and rebuilding the whole of London), and you can see that it is quite beyond the power of any nation at this moment. Short of that you can only take palliative measures, which ultimately boil down to making people more careful.
>
> But the only palliative measure that would make a real difference is a drastic reduction in speed. Cut down the speed limit to 12 miles an hour in all built-up areas, and you would cut out the vast majority of accidents. But this, everyone will assure you, is 'impossible'. Why is it impossible? Well, it would be unbearably irksome. It would mean that every road journey took twice or three times as long as it takes at present. Besides, you could never get people to observe such a speed limit. What driver is going to crawl along at 12 miles an hour when he knows that his engine would do fifty? It is not even easy to keep a modern car down to 12 miles an hour and remain in high gear – and so on and so forth, all adding up to the statement that slow travel is of its nature intolerable.
>
> In other words we value speed more highly than we value human life. Then why not say so, instead of every few years having one of these hypocritical campaigns (at present it is 'Keep Death off the Roads' – a few years back it was 'Learn the Kerb Step'), in the full knowledge that while our roads remain as they are, and present speeds are kept up, the slaughter must continue?

Orwell could thus be seen as an initiator of a paradigm shift, as by the late 1940s, efforts such as his were also endorsed by the United Nations. In 1948, the Universal Declaration of Human Rights was adopted by the UN, with Article 3 stating that everyone has the right to life, liberty, and personal security. This forms an ethical and legal basis for a paradigm shift, as the right to life and personal safety has by no means been adequately observed in the field of transport. While passive safety improved significantly in the second half of the 20th century, this largely benefited car users. Measures favouring other road users, such as reducing speed, were not pursued to the same degree. Sadly, Orwell's appeal thus went largely unanswered for another half-century, and the number of deaths and injuries soared. It was not until the 1990s that the groundwork for a new way of thinking re-emerged in Europe, and 50 years after Orwell's article, a paradigm shift towards a safe system materialised.

Vision Zero and Sustainable Safety as the realisation of the Safe System paradigm

So what exactly is the Safe System, and what does the paradigm shift involve? The Safe System can be understood as a comprehensive approach to managing road safety, based on addressing all components of the system, including road environments, speed regimes, vehicle safety, and post-crash interventions. The principal shift in thinking lies in moving away from blaming individuals and towards focusing on the system in its entirety and on what it makes (or does not make) possible for its users. The ultimate goal is to achieve zero fatalities and serious injuries on the roads. The *Safe System approach* was conceived and subsequently implemented in Sweden (in parallel with the Netherlands) in the 1990s and has since spread to other countries (Tingvall, 1998). The intellectual development from traditional to systems thinking is described by Shi et al. (2021) and Bliss and Breen (2009), for example. Placing safety and quality of life at the forefront of how we think about transport planning, design, and implementation (Kristianssen et al., 2018), this paradigm shift is a necessary part of the humanisation of both society and mobility. When Sweden and the Netherlands radically changed their national road safety policies by independently developing their own systemic approaches – known, respectively, as *Vision Zero* and Sustainable Safety (Tingvall & Haworth, 1999; Wegman & Elsenaar, 1997) – they achieved remarkable improvements in traffic safety. In the Netherlands, for example, the risk of dying in a road crash is now the same for pedestrians as for vehicle occupants (Shi et al., 2021).

Historically, the presumed causes of road traffic crashes have been divided into several phases. Bliss and Breen (2009), for example, summarised the evolution of thinking about road safety management in four chronological phases. The first phase, concerned with the human factor, covers the 1950s and 1960s. They point out the way safety was addressed in academic and professional discourse at the time. The literature focused on accident prevention in terms of the human factor – driver education and law enforcement – and profiles of "dangerous" drivers prone to crashes were looked for (Shaw & Sichel, 1971). Bliss and Breen thus reflected on the broader historical context: from the 1930s to the 1960s, the automobile industry produced campaigns and marketing materials that emphasised speed and performance (Mendoza et al., 2017). Marketing went so far as to portray higher speeds as safer (White, 2022). With the massive growth in car ownership and the spread of invasive marketing that encouraged risky behaviour, traditional approaches to traffic safety struggled to keep pace. Before long, the focus on the human factor proved to be of limited effectiveness (Froggatt & Smiley, 1964; Haight, 2004), as it underlined the *myth of individual responsibility* and contributed to the victimisation of crash victims (Goddard et al., 2019). One positive outcome of this line of thinking was the development of rehabilitation (therapeutic, remedial, or refresher programmes) for drivers, while one of the negative consequences was a worsening of mobility inequalities (Brown et al., 2016).

The next phase, labelled the "Three E's" approach[1] (*E*ngineering, *E*ducation, and *E*nforcement), spanned the 1970s and 1980s. According to Bliss and Breen (2009), attention during this time expanded beyond the human factor to include many other aspects. Since the 1970s, safety has been understood as a human–vehicle–environment triad (Haddon, 1968). This phase also saw the emergence of studies addressing vehicle energy in collisions (Björnstig, 1985) and passive safety features (Agent & Deen, 1975). While helping to enhance road safety, this approach still rested on the myth of personal responsibility, with the discourse of institutional responsibility waiting to be (re)discovered. Institutional responsibility did not begin to be discussed until the third phase, lasting from the late 1980s to the 1990s, when traffic safety indicators began to be established and public authorities assumed accountability for them (Bliss & Breen, 2009). An influential study on safety indicators (Elvik, 1993) demonstrated that countries with measurable, quantifiable targets were more successful in reducing crash rates.

The fourth and final phase – the Safe System approach – took hold in the late 1990s, when system-based road safety policies began to be implemented in countries such as Sweden and the Netherlands. Both countries embraced a discourse of institutional responsibility and a change in focus from erring or vulnerable road users to a flawed and vulnerable system (Khayesi, 2020). This shift resulted in unprecedented improvements in traffic safety (Kristianssen et al., 2018; Weijermars & Wegman, 2011).

The systemic approach

This conceptual shift in the 1990s proved beneficial for all road users, including drivers. Although tendencies to question or attack the systemic approach persist, decades of research have clearly demonstrated its positive impact (Belin et al., 2012; Johansson, 2009; Kristianssen et al., 2018; Tingvall & Haworth, 1999). Traditional systems often reflect a zero-sum logic, where one party's gain is assumed to come at another's expense; for example, that infrastructure improvements for pedestrians and cyclists must necessarily compromise conditions for car users. In fact, the opposite is true: high-quality infrastructure for micromobility can attract some drivers off the roads, thereby easing congestion and alleviating parking problems. Moreover, the happiest drivers are typically found in places with inclusive infrastructure (Chew, 2017).

The Vision Zero Canada (2023) report summarises the differences between the old and new paradigms. The traditional approach to a road safety system is characterised by the following mindset:

1. Responsibility for preventing crashes, injuries, and deaths rests with individuals.
2. Focus is placed on what causes "accidents".
3. Individual errors are allowed to kill and harm.
4. 90% of the problem is people and driver error.
5. The effects of single road safety interventions are studied one at a time.
6. Improvements can only be justified on the basis of a cost-benefit analysis.
7. It only works on problems with large numbers or high-crash locations.
8. It believes in the need for further "study" – waits for crashes and coroners' reports to identify problems.
9. It ignores exposure to motor vehicles as an injury risk factor and the carbon and pollution by-products of transport.

By contrast, the new, systemic approach is based on the following principles:

1. Responsibility for preventing crashes, injuries, and deaths rests with system designers.
2. The focus is on what causes safety.
3. It is unethical to allow individual failures to lead to death or serious injury.
4. 90% of the solutions involve speed, roads, and vehicles.
5. Road safety interventions work best when implemented together or in "bundles".
6. The default is to make the motor vehicle and the road system safe.
7. The system is safe everywhere.
8. It is recognised that there is enough evidence to act. Action is taken proactively using data, crash testing, simulations, physics, etc.
9. Multimodal transport is embraced to enhance safety and environmental sustainability.

This comparison shows how the systemic approach differs from the non-systemic one and how it should be applied. Many current interventions contradict the systemic approach and Vision Zero principles. These include campaigns that emphasise individual responsibility, such as the "Keep Death off the Roads" initiative criticised by Orwell (bullet 1), or various systematic analyses of road crashes that feed the illusion that reports informed by their results can make the old system safe (bullet 8). They just keep the "steam engine" running – maintaining a system of traffic safety which is outdated, ineffective, and unjust (and therefore unethical). The primary focus on the causes of road crashes is sometimes mistakenly presented as part of the systemic approach because of a fundamental misunderstanding of its core idea. In fact, this perspective belongs to the old paradigm, which reinforces the myth of individual responsibility and does not lead to genuine improvement in the traffic safety system. The key idea is that the cause of road crashes is systemic failure, not individual failure. Therefore, looking for the causes of individual error leads nowhere. At most, it confirms what has been known since Plato: that humans behave like humans – irrationally, with self-centred motivations, making mistakes, and disobeying rules. We already know why crashes occur; we do not need to wait for more of them in order to start improving safety.

Vision Zero

Philosophically, *Vision Zero* emerged as a materialisation of a shift towards life and health as primary societal values (see, e.g. Tingvall, 1998), representing an ethical and political stance on how the transport system should function. This standpoint is rooted in a philosophical principle that "man is the measure of all things" and refers to the Universal Declaration of Human Rights, which guarantees the right to personal security. The statement "man is the measure of all things" implies that the system and its components must accommodate human error and limitations rather than assuming that people will modify their behaviour. As the most vulnerable element of the system, the human being should define its parameters to such an effect that failure must not result in serious injury or death (Tingvall, 1998). Responsibility for fulfilling this ethical imperative then lies with all the actors who create and influence the transport system (Kristianssen et al., 2018). Since humans are fallible, it is necessary to design systems that account for errors. At the same time, the aim is to create a system with the lowest possible socio-economic costs (Kopits & Cropper, 2003).

In this philosophical spirit, Tingvall (1998) articulated four core postulates of Vision Zero, which he expanded in 2007 to include three additional principles to enhance drivers' understanding:

1. Ethics: human life and health are of the highest importance and take precedence over mobility and other objectives of the road transport system.
2. Responsibility: road transport system providers and regulators share equal responsibility with system users.

3. Safety: the road transport system should accommodate human fallibility and minimise both the likelihood of errors and their potential consequences.
4. Mechanisms for change: providers and regulators must do everything within their power to ensure the safety of all citizens and cooperate with road users, and all parties must be prepared to implement changes in pursuit of safety.
5. Road deaths and injuries are preventable and thus unacceptable at any level.
6. People will always make mistakes – the transport system should be designed in such a way as to ensure that such mistakes do not result in fatalities.
7. Safety must be the primary consideration in all transport-related decisions. Road safety must be addressed in comprehensive terms.

Vision Zero is commonly said to have originated in Sweden as the first attempt to implement the principles of the Safe System approach in 1995, and was subsequently endorsed by the Swedish government in 1997. However, its intellectual foundations were already laid in the late 1980s, when the first studies called for traffic crashes to be viewed as a social problem rather than an issue specific to risky drivers. This shift in perspective led to the creation of a multidisciplinary working group in Sweden (McAndrews, 2013). This team developed the central ideas of Vision Zero, with a particular emphasis on speed reduction (McAndrews, 2013; Mendoza et al., 2017), which were then presented to the Swedish government (Belin et al., 2012). Concrete implementations in Sweden included the introduction of median and side barriers, segregated pedestrian and cycle paths with vehicle-proof barriers, and lower speed limits where interactions between vehicles and other road users could occur (Johansson, 2009). Since its implementation, Vision Zero has been regularly evaluated (Belin et al., 2012; ITF, 2022; Johansson, 2009; Nihlén-Fahlquist, 2006; Rosencrantz et al., 2007; Tingvall & Haworth, 1999; Wegman, 2017; Weijermars & Wegman, 2011). It is evident that policies grounded in the Safe System approach are significantly more effective than those based on traditional models. One particularly noteworthy finding is the benefit–cost ratio of 4:1 (Weijermars & Wegman, 2011), a figure consistent with prevention/intervention ratios found in many other fields of human activity.

On the basis of data evaluations and effectiveness, Vision Zero has come to be recognised globally as the best approach to road safety system management. Its core principle is articulated in what is now considered Tingvall's classic text: while accidents cannot be entirely eliminated, the road transport system must be designed to accommodate human error (Tingvall, 1998; Tingvall & Haworth, 1999). Road users must comply with the rules, but the system must also reflect their capabilities in such a way that even non-compliance does not result in death or serious injury. Human life and health are therefore the key requirements for the design and operation of the road transport system, and all measures must be directed towards this end (Johansson, 2009). Vision Zero thus sets a long-term goal according to which no one should be killed or seriously injured in traffic crashes (Tingvall, 1998; Tingvall & Haworth, 1999), with the road transport system being devised and evolving in such a way as to meet this requirement (Kristianssen et al., 2018). The responsibility lies with those who design and plan the system (Johansson, 2009). Vision Zero thus represents a conceptual shift in thinking, where the main cause of road crashes is not human error, but an inadequate system of traffic safety management based on road users' unrealistic expectations (Langford, 2009; ITF, 2022).

Sustainable Safety

The concept of *Sustainable Safety* (SWOV, 2006, 2018), too, is underpinned by the philosophy of "man being the measure of all things". Similarly to Vision Zero, it prioritises a safe system over pressure on individual responsibility. Its primary focus is on mitigating or eliminating the potential risks embedded in traffic systems (SWOV, 2018). Sustainable Safety promotes the development of a system that recognises the human being as its most vulnerable element. Its goal is to prevent crashes, and where this is not possible, to at least reduce the likelihood of death or serious injury. Sustainable Safety takes into account both the physical fragility of the human body and the human tendency to make mistakes (whether deliberate or unintentional). The traffic system must therefore match the parameters of a fallible and fragile human. However, responsibility does not lie solely with the system; road users are expected to be properly informed about the rules of operation and subsequent enforcement. Together, this should lead to a safe system and safe behaviour. The result is a reduction in both system-induced and human errors, ultimately eliminating deaths and serious injuries. Sustainable Safety shifts the paradigm towards a model where traffic safety becomes almost independent of individual decisions, with the parameters of the system siding with the most vulnerable rather than the average user. To fulfil these assumptions and create such a system, Sustainable Safety is guided by five essential principles (SWOV, 2006, 2018):

1. Functionality
2. Homogeneity
3. Predictability
4. Forgivingness of both infrastructure and road users
5. Awareness (among road users)

The principle of functionality demands a hierarchically structured road network consisting of three basic categories of monofunctional roads: through-roads, distributor roads, and access roads. Within such a hierarchised road system, it is necessary to ensure homogeneity in key movement parameters: equality in speed, direction, and mass (weight) at medium and high speeds. But such a homogenised traffic flow must also be understandable. This is ensured by predictability, which makes it possible to anticipate the movement of traffic elements, the operation of traffic flow, and road users' behaviour by means of a traffic environment, particularly an adequate infrastructure, designed in line with users' needs and expectations. Even within such a well-structured system, errors will still occur, and safety buffers must be in place to mitigate their consequences. This is where the forgiving infrastructure and road users come into play, ensuring that mistakes do not inevitably result in crashes. To prevent such mistakes, Sustainable Safety emphasises the final principle: awareness. This involves educating road users towards self-reflection, enabling them to assess their own competence and capacity to manage traffic tasks (SWOV, 2006).

The difference made by system-based approaches can be clearly illustrated using the example of vulnerable road users (VRUs). The term is widely used in the discourse on road safety and transport in general. It refers to the road users (typically older adults, pedestrians, and cyclists) who are more vulnerable in the event of a collision. This is because they move through spaces dominated by cars, and thus face a higher risk of injury or death compared to automobilists. The concept of VRUs arises from a historical notion of the car as a privileged mode of transport and is grounded in the myth of personal responsibility. The vulnerability of these road users stems from a poorly designed system, not from the users themselves. The real issue – or real vulnerability – lies in an unfair, car-centric approach to transport planning, which strongly favours car traffic by prioritising its needs. Traditional transport systems have thus historically been non-inclusive, as they enabled and even encouraged the neglect of various communities, including pedestrians, cyclists, children, people with disabilities, and older people. In contrast, a systemic approach does not plan transport around vulnerable individuals (and accept their deaths as a given). Instead, it supports research and practice aimed at addressing the root causes of vulnerability. It focuses on developing inclusive transport that is safe for everyone, shifting attention to the vulnerable points in the system rather than to the people affected by those vulnerabilities.

The dissemination of Vision Zero

Sweden

Ten years after the introduction of Vision Zero, the Swedish Parliament adopted EU Directive 2008/96/EC (2008) on road infrastructure safety management. This piece of EU legislation formally regulates the safety of the backbone road infrastructure, which comprises two-thirds of Sweden's transport network (Belin & Tillgren, 2013). In 2010, Sweden passed a new Road Traffic Safety Act (*Vägsäkerhetslag*), which obliges national road authorities to systematically and continuously adopt the necessary measures to prevent road deaths and serious injuries, with such measures being granted priority in implementation. This was followed by the launch of numerous infrastructural innovations (Belin et al., 2012; Kristianssen et al., 2018). Alongside infrastructure, detailed regulations were also introduced to address the parameters of both passive and active vehicle safety, including safety features in automated vehicles (GOS, 2016). Finally, in 2017, Sweden adopted the Paris Agreement, adding another layer of measures to the systemic approach.

All these changes were grounded in well-formulated strategic documents developed during the first decade of implementation which clearly and understandably state that system designers are responsible for taking further steps to ensure that errors made by road users do not result in fatal or serious injuries (Larsson et al., 2010). These system designers are entities that shape the planning and operation of the road transport system – including road authorities, the automotive industry, transport service providers, and the organisations responsible for support systems such as the police, driving schools, emergency services, and healthcare professionals (Hysing, 2021). In addition to infrastructure-specific measures, Sweden also implements social ones. All state and governmental authorities are expected to set an example for society and use only those modes of transport that meet high standards of health (safety) and environmental sustainability. This involves, for example, the installation of alcohol interlocks for motor vehicles (including government vehicles, taxis, and school buses) (Belin & Tillgren, 2013; Belin et al., 2012; Magnusson et al., 2011). As a result, nearly 100% of public transport buses are equipped with these devices (Sveriges Bussföretag, 2018).

Other countries

The countries that have successfully implemented the Safe System approach include Australia and New Zealand. They adopted the system responsibility discourse in 2003 by linking Vision Zero with Sustainable Safety. Given its geographical and demographic specifics, Australia focused on identifying locations with high crash rates, changing infrastructure design, and reducing speeds. High technological and safety standards for vehicles, as well as rehabilitation and educational programmes aimed at changing driver behaviour, have also been identified as priorities (ACRS, 2016). According to ACRS, the systems approach in Australia and New Zealand focuses on four key areas:

1. Human performance limits (e.g. intersection design, fatigue, and fitness to drive)
2. Human tolerance limits to mechanical and dynamic forces (speed and vehicle design)
3. Safe road use (shared responsibility and speed)
4. Forgiving road infrastructure (a system tolerant in terms of physical laws).

Another European country to join in was Germany. One of the first Vision Zero strategic plans was developed in 2004 by the German automobile club, and in 2005, the federal state of North Rhine-Westphalia incorporated Vision Zero into its road safety programme (VCD, 2023). On the national level, activities linked to Vision Zero have been ongoing since 2007. The legal foundation is the constitutionally guaranteed right to life and physical integrity, which is the responsibility of public authorities. One of the first measures taken as part of the implementation of Vision Zero was accompanied driving, introduced in 2011 (Hell et al., 2022).

In Central Europe, the Czech Republic can serve as an example. Although the Czech Republic was not among the first European countries to endorse the vision (DfT, 2000), it joined during the global Decade of Action for Road Safety. Vision Zero formed the basis of the national strategy for 2011–2020 and the subsequent documents and their revisions (BESIP, 2021, 2023). The Czech Republic has long embraced the principles of Vision Zero, and in terms of strategic documents, this commitment appears strong. However, there are significant gaps in robust implementation and evaluation. There is no assessment of the fulfilment of the vision: only accident statistics are available, not data on the implementation of Vision Zero or analyses of any transition away from a non-systemic approach. The shift towards the Safe System approach thus remains largely declarative, with traditional methods still in place (Safarpour et al., 2020). That said, rhetorical commitment is not unique to Central Europe (see, for example, ETSC PIN Reports, 2022, 2023). For instance, a decade ago, Vision Zero was adopted by politicians and transport authorities in many parts of the United States of America, albeit without effective application of its principles or a genuine effort being made to fulfil them. In fact, a systemic approach requires strong political and financial support, which is something that ministries or other implementing institutions often lack.

In the case of post-Soviet countries in transition, Belarus offers a telling example. Growing motorisation in the country has put pressure on improving road safety. While the number of road accidents, including fatal ones, has been steadily declining over the past 20 years, Belarus still ranks among the European countries with the highest relative accident rates. Kapsky et al. (2020) attribute the start of the downward trend in crash rates to the adoption of the Road Traffic Safety Concept in 2006. However, since 2011, the decline has slowed, and no significant improvement is expected because Belarus relies mainly on engineering measures focused on road maintenance. Given the financial and political constraints, the implementation of Vision Zero in Belarus appears unlikely (Kapsky et al., 2020).

As regards developing countries, Ethiopia may serve as a good example. According to an analysis of the implementation of Vision Zero (Abebe, 2022), road fatalities and injuries continue to be a growing problem, despite governmental efforts to address road safety. Moreover, the WHO estimates that for countries such as Ethiopia the actual number of accidents may be six times higher than what national statistics report. Some progress has been made, such as the development of the first transport policy in the capital city in 2012, followed by a national strategy for non-motorised transport for 2019–2028, which aimed to provide a safe, efficient, and accessible network for pedestrians and cyclists and to improve access to mobility for all residents. However, these policy documents do not mandate system adjustments responding to road users' limitations as the primary means of preventing fatalities and serious injuries. Road safety problems are still viewed mainly as issues of individual road users, and real implementation of Vision Zero – as in Belarus – remains beyond the country's political and financial capacity (Abebe, 2022).

Recommendations for implementation and further development

In countries where there is an understanding of the Safe System philosophy and where political and financial support exists, efforts are under way to consider how the original concept can be advanced and improved. Over the past decade, a number of studies (Kristianssen et al., 2018; Mendoza et al., 2017; Safarpour et al., 2020; Shahum, 2015) have explored potential future directions. Their findings can be summarised into four key aspects (as outlined by Shahum, 2015).

The first and most important aspect is, under all circumstances, speed management (including speed differentials). The countries with the best road safety outcomes emphasise that managing speed is the number one factor in improving safety. Sweden, the Netherlands, and Germany have all changed their approaches to speed by systematically introducing lower maximum speeds and opting for speed homogeneity. The road environment must correspond to the composition of road users – in mixed traffic conditions, vehicle speeds need to be reduced to prevent collisions involving participants with differing mass and speed or mitigate consequences if such collisions do occur (SWOV, 2018).

Following speed reduction, road layouts must be physically redesigned to enhance safety for all users. Subsequently, "physical" or engineering interventions should be prioritised over education and enforcement, as street design and speed management have a greater impact than other areas. The fourth key aspect is engaging the private sector: collaboration with car manufacturers and insurance companies can yield better results, including private-sector investment in safety. An emerging fifth aspect involves the Safe System approach increasingly addressing cross-cutting social concerns, including environmental sustainability (Kristianssen et al., 2018), transport justice, and equity-based safety (Lubitow et al., 2020; Piatkowski & Marshall, 2020).

Note

1 The "Three E's" approach may have originated as early as the 1920s, but its initial application was in industry (Mendoza et al., 2017). It would take another 40 years before it was adopted in the field of transport.

References

Abebe, H. G. (2022). Road safety policy in Addis Ababa: A Vision Zero perspective. *Sustainability*, *14*(9), 5318. https://doi.org/10.3390/su14095318

Agent, K., & Deen, R. (1975). Relationship between roadway geometrics and accidents. *Transportation Research Record*, *541*, 1–11.

Australasian College of Road Safety (ACRS). (2016). *Safe system approach*. ACRS.

Belin, M. Å., & Tillgren, P. (2013). Vision Zero: How a policy innovation is dashed by interest conflicts, but may prevail in the end. *Scandinavian Journal of Public Administration*, *16*(3), 83–102. https://doi.org/10.58235/sjpa.v16i3.16258

Belin, M. Å., Tillgren, P., & Vedung, E. (2012). Vision Zero – a road safety policy innovation. *International Journal of Injury Control and Safety Promotion*, *19*(2), 171–179. https://doi.org/10.1080/17457300.2011.635213

BESIP. (2021). Národní strategie bezpečnosti silničního provozu 2021–2030 [National road safety strategy 2021–2030]. https://besip.gov.cz/Besip/media/Besip/data/web/Strategie-BESIP-2021-2030.pdf

BESIP. (2023). Akční plán Strategie BESIP 2021–***2030 pro období*** 2023–2024 [Action plan of the national road safety strategy 2021–2030 for 2023–2024]. www.mojevizenula.cz/app/webroot/files/editor/files/MV0%206%20dokumenty/AKCNI-PLAN-STRATEGIE-BESIP-pro-obdobi-2023-2024%20%281%29.pdf

Björnstig, U. (1985). *Snowmobile, motorcycle and moose-car accidents: Aspects on injury control* [Doctoral dissertation, Umeå universitet]. DiVa Portal. www.diva-portal.org/smash/get/diva2:1149799/FULLTEXT01.pdf

Bliss, T., & Breen, J. (2009). *Country guidelines for the conduct of road safety management capacity reviews and the specification of lead agency reforms, investment strategies and safe system projects*. The World Bank Global Road Safety Facility. https://thedocs.worldbank.org/en/doc/782371434470876828-0190022009/render/RoadSafetyManagementCapacityReviewGuidelinesOriginal329pg.pdf

Brown, C., Harvey, E., & Sinclair, J. (2016). *Understanding barriers to bicycle access & use in Black and Hispanic communities in New Jersey*. Alan M. Voorhees Transportation Center, Rutgers University. https://njbikeped.org/wp-content/uploads/2022/09/Minority-Report-Draft_FINAL.pdf

Chew, J. (2017). *Netherlands is the best country to drive in. Driver satisfaction survey*. Waze.

Department for Transport (DfT). (2000). *Tomorrow's roads: Safer for everyone*. https://masseguridadvial.com/FILES/Tomorrow's_Road_EN.pdf

Directive 2008/96/EC of the European Parliament and of the Council of 19 November 2008 on road infrastructure safety. (2008). https://eur-lex.europa.eu/legal-content/EN/TXT/?uri=CELEX:32008L0096

Elvik, R. (1993). Quantified road safety targets: A useful tool for policy making? *Accident Analysis & Prevention*, *25*(5), 569–583. https://doi.org/10.1016/0001-4575(93)90009-L

European Transport Safety Council (ETSC). (2022). *16th Annual road safety performance index (PIN) report*. ETSC.

European Transport Safety Council (ETSC). (2023). *17th Annual road safety performance index (PIN) report*. ETSC.

Froggatt, P., & Smiley, J. A. (1964). The concept of accident proneness: A review. *British Journal of Industrial Medicine*, *21*(1), 1–12. https://doi.org/10.1136/oem.21.1.1

Gartman, D. (2004). Three ages of the automobile: The cultural logics of the car. *Theory, Culture & Society*, *21*(4–5), 169–195. https://doi.org/10.1177/0263276404046066

Goddard, T., Ralph, K., Thigpen, C. G., & Iacobucci, E. (2019). Does news coverage of traffic crashes affect perceived blame and preferred solutions? *Transportation Research Interdisciplinary Perspectives*, *3*, 100073. https://doi.org/10.1016/j.trip.2019.100073

Government Offices of Sweden (GOS). (2016). *Renewed commitment to Vision Zero*. https://visaozero2030.pt/wp-content/uploads/Renewed_Commitment_Vision_Zero_Intensified_efforts_transport_safety_Sweden.pdf

Haddon, W., Jr. (1968). The changing approach to the epidemiology, prevention, and amelioration of trauma. *American Journal of Public Health*, *58*(8), 1431–1438. https://doi.org/10.2105/AJPH.58.8.1431

Haight, F. A. (2004). Accident proneness: The history of an idea. In T. Rothengatter & R. D. Huguenin (Eds.), *Traffic and transport psychology: Theory and application* (pp. 49–64). Elsevier.

Hell, W., Bodewig, K., Hammer, U., Kellner, C., Klinke, C., Mück, M., Schreiner, M., Walz, F., & Zielke, G. (2022). Vision Zero in Germany. In K. Edvardsson Björnberg, M. Å. Belin, S. O. Hansson, & C. Tingvall (Eds.), *The Vision Zero handbook* (pp. 1–21). Springer. https://doi.org/10.1007/978-3-030-23176-7_13-1

Hysing, E. (2021). Responsibilization: The case of road safety governance. *Regulation & Governance*, *15*(2), 356–369. https://doi.org/10.1111/rego.12288

International Transport Forum (ITF). (2022). *Zero road deaths and serious injuries: Leading a paradigm shift to a safe system*. OECD Publishing. https://doi.org/10.1787/9789282108055-en

Johansson, R. (2009). Vision Zero – implementing a policy for traffic safety. *Safety Science*, *47*(6), 826–831. https://doi.org/10.1016/j.ssci.2008.10.023

Kapsky, D., Bogdanovich, S., & Volynets, A. (2020). Implementation of the road traffic safety concept in Belarus. In *Vision Zero for sustainable road safety in Baltic sea region* (pp. 110–119). Springer. https://doi.org/10.1007/978-3-030-22375-5_13

Khayesi, M. (2020). Vulnerable road users or vulnerable transport planning? *Frontiers in Sustainable Cities*, *2*, 25. https://doi.org/10.3389/frsc.2020.00025

Kopits, E., & Cropper, M. (2003). *Traffic fatalities and economic growth* (Policy Research Working Paper No. 3035). World Bank. https://doi.org/10.1596/1813-9450-3035

Kristianssen, A. C., Andersson, R., Belin, M. Å., & Nilsen, P. (2018). Swedish Vision Zero policies for safety – a comparative policy content analysis. *Safety Science*, *103*, 260–269. https://doi.org/10.1016/j.ssci.2017.11.005

Langford, J. (2009). *Towards zero: Understanding a safe system approach to road safety (Fact sheet)*. Curtin – Monash Accident Research Centre. https://cdn-nrspp-s3-aus.s3.ap-southeast-2.amazonaws.com/wp-content/uploads/sites/4/2017/03/21124023/CMARC_Fact_Sheet1.pdf

Larsson, P., Dekker, S. W., & Tingvall, C. (2010). The need for a systems theory approach to road safety. *Safety Science*, *48*(9), 1167–1174. https://doi.org/10.1016/j.ssci.2009.10.006

Lubitow, A., Abelson, M. J., & Carpenter, E. (2020). Transforming mobility justice: Gendered harassment and violence on transit. *Journal of Transport Geography*, *82*, 102601. https://doi.org/10.1016/j.jtrangeo.2019.102601

Magnusson, P., Jakobsson, L., & Hultman, S. (2011). Alcohol interlock systems in Sweden. *American Journal of Preventive Medicine*, *40*(3), 378–379. https://doi.org/10.1016/j.amepre.2010.11.015

McAndrews, C. (2013). Road safety as a shared responsibility and a public problem in Swedish road safety policy. *Science, Technology, & Human Values*, *38*(6), 749–772. https://doi.org/10.1177/0162243913493675

Mendoza, A. E., Wybourn, C. A., Mendoza, M. A., Cruz, M. J., Juillard, C. J., & Dicker, R. A. (2017). The worldwide approach to Vision Zero: Implementing road safety strategies to eliminate traffic-related fatalities. *Current Trauma Reports*, *3*, 104–110. https://doi.org/10.1007/s40719-017-0085-z

Nihlén-Fahlquist, J. (2006). Responsibility ascriptions and Vision Zero. *Accident Analysis & Prevention*, *38*(6), 1113–1118. https://doi.org/10.1016/j.aap.2006.04.020

Orwell, G. (1946, November 8). As I please. *Tribune*. www.telelib.com/authors/O/OrwellGeorge/essay/tribune/AsIPlease19461108.html

Piatkowski, D. P., & Marshall, W. E. (2020). Not all crashes are created equal. *Journal of Transport & Health*, *16*, 100842. https://doi.org/10.1016/j.jth.2020.100842

Rosencrantz, H., Edvardsson, K., & Hansson, S. O. (2007). Vision Zero – is it irrational? *Transportation Research Part A: Policy and Practice*, *41*(6), 559–567. https://doi.org/10.1016/j.tra.2006.11.002

Safarpour, H., Khorasani-Zavareh, D., Soori, H., Bagheri-Lankarani, K., Ghomian, Z., & Mohammadi, R. (2020). Vision Zero: Evolution history and developing trend. *Trauma Monthly*, *25*(6), 275–286. https://doi.org/10.30491/TM.2020.244740.1166

Shahum, L. (2015). *Notes from abroad: Five key lessons from Europe's Vision Zero success*. Vision Zero Network. https://visionzeronetwork.org/notes-from-abroad-five-key-lessons-from-europes-vision-zero-success/

Shaw, L., & Sichel, H. S. (1971). *Accident proneness: Research in the occurrence, causation, and prevention of road accidents*. Pergamon.

Sheller, M., & Urry, J. (2000). The city and the car. *International Journal of Urban and Regional Research*, *24*(4), 737–757. https://doi.org/10.1111/1468-2427.00276

Shi, G., Methoxha, V., Atkinson-Palombo, C., & Garrick, N. W. (2021). Sustainable safety in the Netherlands. *Transportation Research Record*, *2675*(11), 792–803. https://doi.org/10.1177/03611981211019736

Sveriges Bussföretag. (2018). *Statistik om bussbranschen 2018*. www.transportforetagen.se/globalassets/rapporter/buss/statistik-om-bussbranschen-2018-03.pdf

SWOV. (2006). *Advancing sustainable safety: National road safety outlook for 2005–2020*. https://swov.nl/system/files/publication-downloads/advancing_sustainable_safety.pdf

SWOV. (2018). *Sustainable safety* (3rd ed.). https://swov.nl/system/files/publication-downloads/dv3_en_kort_rapport.pdf

Tingvall, C. (1998). *The Swedish 'Vision Zero' and how parliamentary approval was obtained*. Australasian Road Safety Research, Policing and Education Conference, Wellington, New Zealand.

Tingvall, C., & Haworth, N. (1999). *Vision Zero – an ethical approach to safety and mobility*. ITE International Conference Road Safety & Traffic Enforcement, Melbourne. www.monash.edu/muarc/archive/our-publications/papers/visionzero

Verkehrsclub Deutschland (VCD). (2023). *Zeit für null Verkehrstote*. www.vcd.org/themen/verkehrssicherheit/vision-zero/

Vision Zero Canada (VZC). (2023). *Vision Zero principles*. https://visionzero.ca

Wegman, F. (2017). The future of road safety: A worldwide perspective. *IATSS Research*, *40*(2), 66–71. https://doi.org/10.1016/j.iatssr.2016.05.003

Wegman, F., & Elsenaar, P. (1997). *Sustainable solutions to improve road safety in the Netherlands*. SWOV. https://swov.nl/system/files/publication-downloads/d-97-08.pdf

Weijermars, W., & Wegman, F. (2011). Ten years of Sustainable Safety in the Netherlands: An Assessment. *Transportation Research Record*, *2213*(1), 1–8. https://doi.org/10.3141/2213-01

White, R. (2022). *Auto safety history*. America on the Move: National Museum of American History.

17 Transport infrastructure

Jiří Ambros, Kateřina Bucsuházy, and Ralf Risser

Introduction

According to the World Road Association dictionary, transport infrastructure (hereinafter "infrastructure") comprises all the fixed facilities and works necessary for the circulation of vehicles and the operation of transport systems (PIARC, 2023). In addition to roads and motorways, the term therefore includes various road network features and equipment such as delineator posts, safety barriers (guard rails), traffic signs, traffic signal devices, and noise barriers.

Each of these elements exerts a specific influence on road users (including drivers, cyclists, and pedestrians). As outlined in the Diamond Model (Risser, 2004), infrastructure therefore constitutes one of the five key components that shape the transport system.

In terms of perception and behaviour (and consequently safety), the surroundings of the roadway (which may be collectively referred to as the "traffic environment"), including vegetation and advertising structures, can also be considered influential elements.

In general, the influence of infrastructure is primarily reflected in mobility (traffic volume and capacity) and safety (the level of risk and crash rates). This chapter focuses particularly on safety.

The following sections describe:

1. how transport infrastructure affects road users
2. how these factors have influenced the development of approaches to infrastructure design

How transport infrastructure affects road users

This section first describes the methods which may be used to assess the influence of infrastructure on road user behaviour (and consequently on road safety) and to identify risk factors. Then the main risk factors are outlined.

DOI: 10.4324/9781003608394-17

Methods

Approaches can be categorised according to various criteria, for example:

- the type of research
  - qualitative
  - quantitative
- the degree of control over extraneous variables
  - observational studies
  - experimental studies
- the temporal perspective
  - *proactive* (prospective): using indirect safety indicators
  - *reactive* (retrospective): using direct safety indicators derived from crash occurrence.

The validity and interpretability of each method are important when comparing these approaches; see Figure 17.1.

In line with Figure 17.1, the following description is divided into three groups of approaches:

- crash-based analyses
- observations
- experiments.

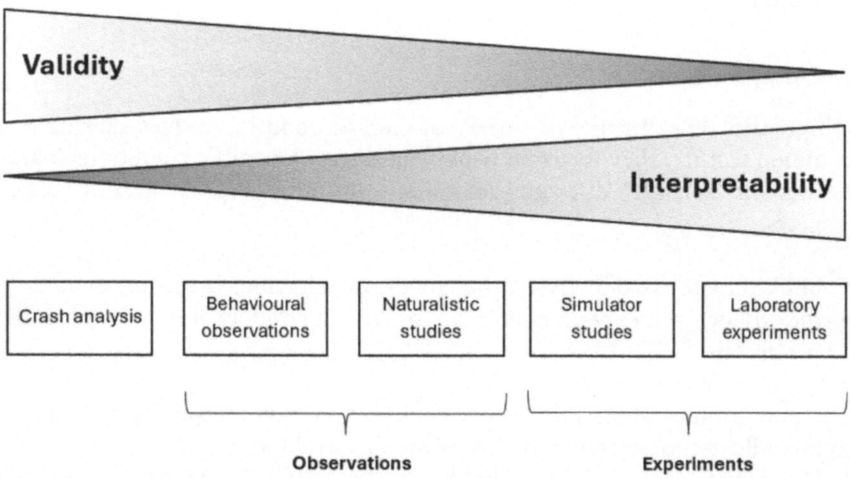

Figure 17.1 Schematic representation of the relationship between the validity and interpretability of individual methods (adapted from van der Horst, 2006).

Crash-based analyses

Traffic crashes are a traditional indicator of road safety. Depending on the scope of the crash sample under examination, several approaches can be distinguished.

Crash analysis deals with individual crashes. A typical example is the so-called *in-depth road crash investigation*, which focuses on a detailed examination of the mechanisms leading to the crash, its course, and its consequences. Selected crashes are typically analysed directly at the crash scene. All the relevant variables related to the crash are recorded for a comprehensive analysis, with road infrastructure design, human factors, and vehicle characteristics being taken into account. Various concepts of in-depth crash investigation analysis have been implemented in different countries all over the world, often depending on the type of road user or the injury severity. Examples include the Czech Republic (*Czech In-Depth Accident Study* (Centrum dopravního výzkumu, v. v. i., 2022)), Germany (GIDAS, n.d.), the United Kingdom (TRL, 2012), Australia (Centre for Automotive Safety Research, 2021), and the United States of America (National Highway Traffic Safety Administration, n.d.).

By contrast, crash statistics analysis deals with a larger number of crashes – that is, crashes (their frequency and severity) within selected spatial or temporal units. However, this type of analysis may be limited in cases of low statistical crash frequency – for instance, in short-term and/or localised evaluations, or in cases of crashes not recorded by the traffic police.

Moreover, since crash data is collected post-event (a reactive approach), any associated analyses provide only limited information on pre-crash behaviour.

Observations

Observation encompasses a variety of methods and techniques sharing one common feature: they focus on behaviour in regular traffic conditions (*naturalistic studies*). This distinguishes them from other approaches in two key respects:

- unlike crash-based analyses, they focus on behaviour preceding crashes
- they differ from experiments, as these are not conducted in real-life traffic conditions

The limitation of observational methods, however, lies in their susceptibility to the influence of extraneous (undesirable) variables.

Van Haperen et al. (2019) divide observational studies into naturalistic driving and behavioural observation.

They further highlight the difference between naturalistic and behavioural studies with respect to road users being informed about their participation in a research project:

- The aim of a naturalistic driving study is to collect data on natural behaviour in real traffic. The participants are informed that they are being monitored, and the monitoring occurs continuously, increasing the likelihood of behavioural adaptation.
- In behavioural observation studies, on the contrary, the participants are not informed (either beforehand or afterwards), and the observation occurs at a specific location (e.g. an intersection). As a result, it can be expected that, not being influenced by the knowledge that he is being monitored, the participants will behave naturally, including, for instance, engaging in risky or aggressive driving.

A special category of behavioural studies is the *observation of traffic conflicts* (i.e. "near-misses"), which is used to assess the safety of a particular location (most often an intersection). Numerous traffic conflict observation methods exist, with the Swedish approach being the best-known (Laureshyn & Várhelyi, 2018).

Experiments

The defining feature of experiments is that they are not potentially biased by the influence of undesirable variables. This distinguishes them from observations of driver behaviour in real-life traffic, which are always affected by factors such as weather, other vehicles, or the road environment. In experimental settings (e.g. using a driving simulator or during interviews), these external factors can be eliminated.

Experiments thus make it possible to simulate specific conditions in order to elicit desired responses – for example, creating a scenario for a driving simulator or directing questions towards specific situations during interviews. Contrarily, observational studies in real traffic are reliant on actual occurrences.

There are many approaches that can be classified as experimental. Van Haperen et al. (2019), for example, distinguish between:

- simulation-based approaches (e.g. driving simulators, microsimulations, and virtual reality) and
- communication-based approaches (e.g. questionnaires, interviews, and focus groups).

The three most common categories – driving simulators, microsimulations, and questionnaires/interviews – are described as follows.

Driving simulators

Driving simulators are now a commonly used tool for experimental research into driving behaviour. This is due to the simplicity and low cost of data collection, as well as its repeatability: a study that would take a disproportionately long time to carry out in real traffic conditions can be completed in a relatively short period. They can also be used to create scenarios that would be too risky in real traffic. However, simulators also have their limitations. A particularly significant one concerns validity; participants may behave differently than they would in real traffic. Another limitation is the restricted capacity to study learning and adaptation processes – for instance, it is possible to simulate repeated exposure to a traffic sign, but participant behaviour is unlikely to reflect long-term adaptation experienced in real traffic conditions (Carsten & Hamish Jamson, 2011). Another important factor to consider is the phenomenon referred to as simulator sickness (Fisher et al., 2011).

Microsimulations

The aim of computer-based microsimulations is to replicate road user behaviour using computer models calibrated against real-world behavioural and environmental data. These models can then be used to examine responses to infrastructural changes (e.g. intersection redesigns) or to assess the safety of entire modelled networks (e.g. following the construction of a road bypassing an urban area).

Microsimulation models are typically used to determine capacity, delays, and travel times. In recent years, applications have also emerged that focus on safety assessment, using principles and indicators similar to those utilised in traffic conflict studies. More information can be found in the review articles by Young et al. (2014) and Sohel Mahmud et al. (2019), for example.

While microsimulations have mostly focused on motor traffic, recent developments have included applications for pedestrians, cyclists, and motorcyclists; for more details, see, for example, Das and Maurya (2017) and Babojelić and Novačko (2020).

Questionnaires and interviews

Questionnaires and interviews are the most common methods used to obtain self-reported data on road user behaviour. Questionnaires are particularly popular, as they make possible the relatively fast collection of large datasets; a frequently used tool is the Driving Behaviour Questionnaire (DBQ). However, it is important to note that the validity of self-reported data has been questioned by a number of studies (Useche et al., 2022).

Risk factors

There are many types of behaviour that increase the risk and likelihood of road traffic crashes. Significantly, many of them are influenced by road infrastructure, as some of its features may serve as risk factors. This is particularly true in relation to driving speed, which is itself one of the major risk factors: higher speeds are associated with both an increased likelihood of a crash and increased crash severity (OECD, 2018).

It is therefore necessary to understand how the characteristics of the road environment influence the behaviour of road users. Infrastructural characteristics should not confuse drivers or lead them to inappropriate behaviour. The understanding of fundamental cognitive regulatory processes – motivation, capabilities, and information processing – needs to be considered. One core principle thus applies: the road environment should be both *calming* and *self-explaining* (Risser et al., 2020). The meaning of these concepts, along with the characteristics that contribute to them, is outlined below. Selected risk factors and infrastructure-related topics are introduced and their impact on behaviour and safety discussed. As mentioned in the introduction, both the road and its immediate surroundings are considered in this context.

The information presented is based on two groups of sources:

- crash-based analyses, including selected reviews and meta-analyses (Elvik et al., 2009; Papadimitriou et al., 2019; Roshandel et al., 2015; Theofilatos & Yannis, 2014; Wang et al., 2013) and data from in-depth crash investigations,
- naturalistic and simulator studies (on the basis of, e.g. Singh & Kathuria, 2021 and Bobermin et al., 2021), and
- other sources referenced as follows.

Horizontal curves

- The radius size is inversely proportional to crash rates: a higher number of crashes occurs on curves with a smaller radius (i.e. sharper curves). On smaller-radius curves, drivers are also more likely to adopt a trajectory with a larger radius ("cutting" the corner).
- Speed consistency is also important: along a route composed of straight sections and horizontal curves, there should be no significant speed changes. It is well established that speed differentials between straight segments and curves correlate with crash frequency. Critical curves should therefore be appropriately marked with warning signs and chevron boards. (However, some studies examining the impact of such signage have shown that drivers appreciated the improved visibility of these locations and, paradoxically, responded by increasing their speed.)
- Horizontal road markings, whether in the form of centre lines or edge lines, also influence driving behaviour in curves. A Czech before-and-after study on the implementation of horizontal road markings (Havránek et al., 2020) showed a reduction in speed with both centre and edge lines. Both types of markings also affected the driving trajectory: while the

presence of a centre line shifted the driving line towards the road edge, edge lines caused the trajectory to move closer to the centre of the carriageway (Havránek et al., 2020; Vyskočilová et al., 2023).

Lane and road width

- Studies of road widening found reduced crash rates on rural roads. In urban areas, however, widening led to more crashes, partly because of longer pedestrian crossings, which increase exposure and risk.
- Wider roads are also associated with higher driving speeds among road users involved in crashes (Bucsuházy et al., 2018).
- Narrower lanes encourage drivers to position themselves closer to the centre and drive more slowly.

Shoulder width

- Generally, widening the shoulder reduces the number of crashes, as it allows for evasive manoeuvres in hazardous situations or for stopping the vehicle. However, a wider shoulder can also create the impression of a broader and more comfortable road, potentially leading to higher driving speeds and increased crash severity.
- Simulator studies have shown that the presence of a shoulder causes the driving line to shift closer to the edge of the road. This may reduce the number of head-on collisions with oncoming vehicles while increasing the number of roadway departures.

Traffic medians

- Physical separation of opposing lanes prevents head-on collisions. However, on curves, medians may contribute to an increased number of crashes.
- Naturalistic studies have confirmed that medians enhance drivers' subjective sense of safety, which in turn can lead to higher driving speeds.

At-grade intersections

- Intersections are generally high-risk locations because of their multiple potential conflict points. One key factor in driving is decision-making speed: the more complex the intersection, the higher the likelihood of a crash. Four-arm intersections are particularly hazardous, as they involve a greater number of conflict points. Reducing these points is a solution; for example, roundabouts can halve crash rates.
- A critical manoeuvre at intersections, especially for older drivers, is turning left, which requires yielding to oncoming traffic and accurately judging the size of gaps to complete the manoeuvre.
- Failure to give way is often influenced by the design of the traffic environment. Restricted visibility may prompt drivers to slow down when

approaching an intersection; conversely, good visibility can sometimes encourage higher speeds or overtaking. Intersections should ideally be placed on straight road segments, as visibility can be limited in horizontal or vertical curves.

- Visibility is also affected by the angle of an intersection: a right angle is ideal, while intersections with skewed angles are riskier.
- Alignment between the physical and perceived boundaries of an intersection is also important: intersections should be clearly defined in their layout and should not create a misleading sense of having the right of way ("psychological priority").

Railway crossings

- Railway level crossings are considered high-risk locations too. The majority of crashes occur at the so-called passive crossings, which are not equipped with active control devices and thus require drivers to rely on their own risk perception. The most significant risk factors include the intensity of road and rail traffic, limited visibility, and high driving speed (Vivek et al., 2021). Other contributing factors include the inconspicuousness of the crossing area or a short crossing length (Ambros et al., 2019).

Roadside safety barriers

- The most commonly used roadside safety barriers are guard rails. Their primary purpose is to reduce the severity of the consequences of a crash: they serve to contain or redirect vehicles (and to absorb the kinetic energy of the impact). They may also influence the likelihood of a crash by improving the road's visual guidance. At the same time, however, they constitute a fixed roadside obstacle.
- Simulator studies have shown that drivers tend to adopt a trajectory closer to the centre of the road (further away from the guard rails). As with traffic medians, an increased sense of safety was observed, leading to higher driving speeds.

Roadside vegetation

- Vegetation represents a fixed roadside obstacle: studies have shown that establishing "clear zones" free of fixed objects helps reduce the number of crashes. Factors that influence crash severity include vehicle impact speed and tree trunk diameter (Bucsuházy et al., 2022).
- Simulator studies have demonstrated that the presence of trees affects driver behaviour: when positioned at a greater distance, they were perceived as optical guidance; when positioned closer to the road, they were perceived as a hazard, resulting in reduced speed and a shift towards the centre of the road (Calvi, 2015).

Advertisements and billboards

- Numerous studies – typically employing driving simulators and eye-tracking rather than being conducted in real traffic conditions – have shown that advertising distracts drivers: it reduces attention to road signs, increases mental workload, prolongs reaction times, impairs vehicle control, and frequently results in lane deviations.
- Moreover, advertising structures constitute fixed obstacles.
- Dynamic billboards, featuring changing content and LED illumination (Oviedo-Trespalacios et al., 2019), are particularly distracting.

Lighting

- Lighting has a positive effect on safety, particularly in urban areas, especially in relation to pedestrian crossings.
- However, the installation of lighting may also lead to increased driving speeds (Elvik, 2017).

Infrastructure for cyclists (cycle lanes and cycle paths)

- The allocation of dedicated lanes for cyclists leads to a reduction in cycling-related crashes. However, at intersections, the number of cyclist crashes tends to increase, with the complexity of the intersection playing a key role. Multi-lane roundabouts and intersections without physically separated cycle lanes are particularly risky (Poudel & Singleton, 2021; Prati et al., 2018). The most common and/or severe crash scenarios involving cyclists include failure to give way, a vehicle cutting across the cyclist's path when turning, and a rear-end collision with a cyclist (Bíl et al., 2016; Thomas et al., 2019).
- Experience from the Netherlands, the country with the strongest tradition of cycling, shows that wider cycle paths, lighting, and lower speed limits at intersections contribute to greater cyclist safety. On the contrary, parked vehicles, tram traffic, and fixed roadside obstacles are factors that worsen cyclist safety (Niaki et al., 2021).

Pedestrian infrastructure

- A review by Moradi et al. (2019) identified the following risk factors as the most common contributors to greater injury severity in vehicle–pedestrian collisions: vehicle speed, lighting conditions or lack of lighting, and multi-lane roads.
- In terms of inadequate pedestrian crossing design, there are numerous potential negative factors: inappropriate placement or excessive length of the pedestrian crossing, missing, poorly designed, or undersized pedestrian refuge islands, insufficient sightlines, missing or improperly placed guidance elements for persons with reduced mobility or orientation, and inadequate, inappropriate, or absent lighting (Haque et al., 2025).

- A significant proportion of pedestrian injuries also result from falls (without the involvement of other road users), which are typically not recorded as they do not count as traffic crashes. Critical factors contributing to these falls include uneven or slippery surfaces, stairways, and busy intersections (Schepers et al., 2017; Šucha et al., 2024).

Traffic calming measures

- Roads should be designed in such a way as to prevent high driving speeds. This applies especially in urban areas, where motor vehicles and vulnerable road users (such as pedestrians and cyclists) are more likely to interact. Where such design is lacking, traffic calming must be implemented, whether in the form of site-specific or area-wide measures.
- The impact of site-specific interventions, such as speed humps or pedestrian refuge islands, on crash rates is difficult to determine. Their effectiveness is usually indicated by the reduction in speed (typically in the range of 30–50%; Jurewicz, 2009). However, it is often the case that speed reduction is localised to the intervention site only.
- Area-wide calming (30 km/h or 20 mph zones) is therefore a more suitable approach. It involves the systematic separation of different transport modes through measures such as the use of physical elements and adjusted speed limits. The positive effects include reduced speeds, lower crash rates, and improved environmental quality (Yannis & Michelaraki, 2025).

Summary

The aim of the second section was to describe how road infrastructure affects road users.

The characteristics of the traffic environment, along with other factors, define the driving tasks that road users must deal with. A major challenge in designing and developing traffic environments lies in finding a balance between central cognitive processing on the one hand and the external stimuli on the other hand. Infrastructure planning must therefore take into account the following psychological aspects (Risser et al., 2020):

- *Anticipation and orientation.* The optical guidance should be free from inconsistencies to make it easy for road users to perceive the direction of travel. The visual field, vehicle control elements, and the surrounding space should be arranged in such a way as to facilitate cognitive processing.
- *Driving smoothness.* Both horizontal (straight segments and curves) and vertical (ascents and descents) alignments should support a smooth driving flow without abrupt speed adjustments.
- *Perception and comprehension of elements.* The geometric layout of the road can lead users to adopt higher speeds in locations where this is inappropriate. The design of the traffic environment shapes road user behaviour, potentially encouraging risk-taking. Situational factors influence

both driver expectations and personal motivation in scenarios requiring strategic adaptation. The objective demands of the situation and the driver's subjective perception of it should not contradict one another.

- *Habituation and desensitisation.* It should be noted that continuous exposure to the same sensory stimuli leads to a gradual decrease in activation. For example, drivers may fail to respond to a warning sign after having encountered it several times before.
- *Mental workload.* If the road layout increases a driver's mental workload, information processing may deteriorate, and situational awareness may be reduced – for instance, when driving on a very narrow road. Conversely, when the workload is low (such as on long straight motorway sections), the level of activation may drop. This may lead to reduced attention and compensatory activities, such as interacting with in-vehicle devices or making phone calls, which further consume the remaining mental capacity needed for information processing.

If drivers do not receive enough timely information about their environment, or are overloaded with information, their responses may become slower and more error-prone. To prevent errors, the infrastructure should be designed in such a way as to reflect human limitations and expectations, increasing the likelihood of successful and rapid information processing and appropriate responses to traffic situations. In short, the goal is not to surprise the driver.

In view of the aforementioned types of errors, the key risk factors can be divided into two groups:

- environmental complexity (the more complex the environment, the more challenging decision-making in time and space becomes),
- inappropriate speed (speed is a universal risk factor – increased speed correlates with a higher likelihood and severity of crashes).

Strategy options therefore include one or more of the following (ideally in combination):

- simplification of the environment (e.g. by channelising traffic flows or separating transport modes),
- speed reduction (e.g. through route design changes, traffic calming measures, signing and marking).

Examples of infrastructure modifications based on these principles include:

- *Roundabouts* have fewer conflict points than standard four-arm intersections (see Figure 17.2), thereby reducing the risk of potential collisions. The deflection of the vehicle path around the central island also contributes to speed reduction. Movement through roundabouts is governed by simple rules that do not require additional signing.

- *2+1 roads*, a form of cross-sectional design with alternating lanes (see Figure 17.3), serve to increase capacity and safety by reducing the number of head-on collisions during overtaking manoeuvres.
- *Road narrowing*, a principle of traffic calming, leads to lower driving speeds (as narrower roads implicitly encourage slower driving) and facilitates pedestrian movement (by shortening the length of pedestrian crossings).

There are numerous risk characteristics of which the impact can be reduced or minimised through traffic infrastructure. Several examples have also been given where such measures do not have the intended effect. For instance, improving visibility or sight distance at a critical location may paradoxically increase risk, as drivers tend to choose higher speeds. Similarly, increasing the road width in an attempt to facilitate safer passage often leads to an increase in driving speed. This highlights the importance of the perception of subjective risk: road users behave and make decisions according to the level of risk they perceive. These are examples of *risk behaviour adaptation*. In the context of this chapter, the definition of compensation (Rudin-Brown & Jamson, 2013) can be formulated as behaviour that follows the implementation of a change (the introduction of a measure) in the traffic environment which was not intended by the initiator of the change.

An illustrative example is provided by a Norwegian study of the risk presented by horizontal curves with small radii (Elvik, 2022):

- A single sharp curve is risky, while sections with several sharp curves are associated with lower crash rates.
- This is explained by driver behavioural adaptation: with isolated (unexpected) curves, drivers do not adjust their speed accordingly; repeated curves, however, are anticipated by drivers, who therefore reduce their speed.

Behavioural adaptation thus influences the effectiveness of road safety measures. Evans (2004) notes that the resulting effectiveness of a measure is determined not only by its technical (intended) safety benefit but also by feedback, that is, the degree of adaptation. If no behavioural adaptation occurs, the resulting effectiveness equals the technical effectiveness. However, if adaptation does occur, the resulting effectiveness is lower than the technical effectiveness (it may even be zero or counterproductive, that is, leading to increased risk and crash rates). Concerning horizontal curves, Elvik (2022) therefore notes that a potential measure (straightening the alignment, i.e. increasing the curve radius) may lead to increased driving speeds and thus reduced effectiveness.

These examples demonstrate that in designing measures aimed at influencing risk factors and improving road safety, it is essential to consider not only the technical aspects but also road user behaviour and its potential

Figure 17.2 An intersection (left) versus roundabout diagram showing the number of collision points.

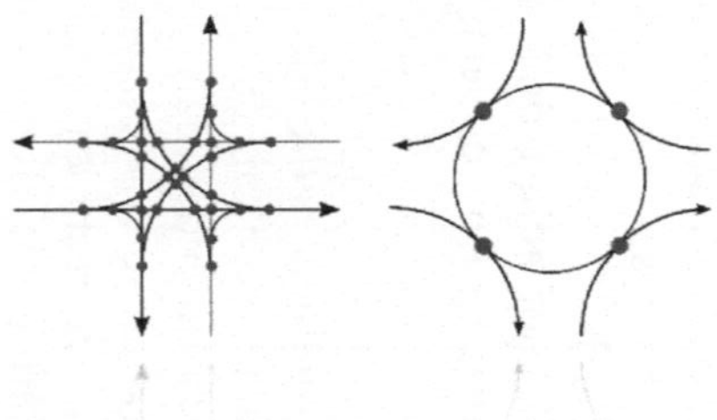

(Adapted from Wikimedia Commons)

Figure 17.3 Alternating passing lane configuration (2 + 1)

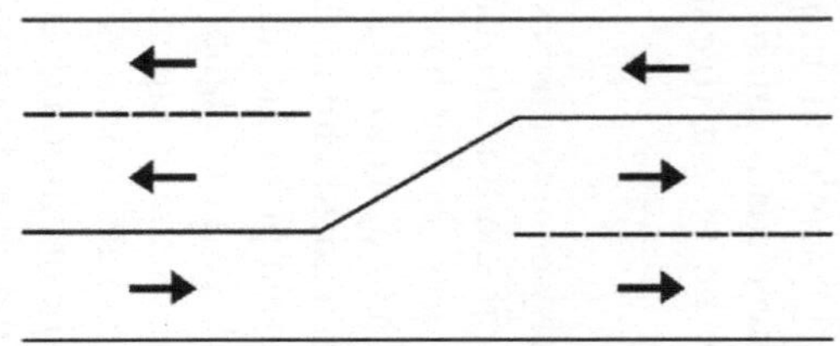

adaptations. In this context, the concepts of self-explaining and forgiving roads, which are discussed in more detail next, come to the fore.

How is transport infrastructure created?

The previous part described how transport infrastructure affects road users. This next section follows up by describing how the identification of risk factors has influenced the development of approaches to designing transport infrastructure.

This account is structured into three phases: the rise of automobilism, traffic calming, and Vision Zero.

Phase one: the rise of automobilism

The invention of the automobile at the end of the 19th century and the subsequent rise of automobilism in the 20th century had a significant impact on both society and transport. Šmíd (2004) notes that as early as the 1920s, during the advent of functionalism, there was an uncritical fascination with the automobile as a symbol of progress and prosperity, while its negative aspects – which became increasingly apparent with the expansion of car use – were overlooked. From that time, the automobile became a key reference point in urban planning: in order to fully exploit its technical potential, it was considered necessary to maximise space for it and to adapt roads accordingly.

The negative impacts of automobilism were especially evident in urban areas (Šmíd, 2004):

- *Transport function*: The transport function became the dominant role of street space. The shaping of streets was subordinated to the needs of motor traffic.
- *Environmental function*: As the number of cars on the streets increased, the environmental quality of human settlements deteriorated markedly.
- *Aesthetic function*: The dominance of the transport function inevitably affected the appearance of streets. Moving and especially parked cars now dominate the streetscape.
- *Compositional, structural, and spatial-forming function*: Instead of a space integrating multiple functions, the street became a single-purpose area, dedicated to traffic. Streets came to be viewed primarily as roads, and squares as car parks. Because of the significantly degraded environment, the threat to pedestrians, and the difficulty of crossing them, main traffic arteries became barriers.
- *Residential, social, and cultural function*: All of these changes were inevitably reflected in the lives of residents. The result was a profound shift in lifestyle: the street ceased to play the role of a residential space, and the

loss of the street's residential function also led to a breakdown of social ties and contacts.

- *Commercial function*: As busy streets lost their attractiveness, the appeal of shops located there – and shopping in the city centre in general – diminished. At the same time, commercial activity increasingly shifted to suburban shopping centres accessible only by car.

Phase two: traffic calming

In response to the aforementioned negative effects, new features and concepts began to emerge from the 1960s onwards, which gradually came to be known collectively as *traffic calming*. This term encompasses a set of measures and tools intended to increase the utility of streets, improve the environment, and enhance safety – particularly for pedestrians and cyclists – at the expense of the previously superior position of motor traffic. The primary goal of traffic calming is to align the character of the street space with the function of the road that runs through it.

The definition earlier suggests that *traffic calming* is a broad and evolving concept that has come to encompass both minor local modifications to residential streets and large-scale interventions or comprehensive redesigns of urban and municipal road networks. Traffic calming should not be regarded solely as a matter of traffic engineering; rather, it is closely related to integrated urban and transport planning, public engagement in decision-making processes, mobility management, and efforts to promote sustainable transport.

Traffic calming measures can be further divided as follows:

- Physical measures:
  - entrance treatments at the boundary of urban areas (refuge islands, chicanes, tight curves, etc.)
  - roadway width reductions
  - chicanes
  - kerb extensions
  - traffic islands (refuge, divider, or raised)
  - transverse elements (speed humps and cushions) and raised intersections
  - traffic medians and side strips, greenery
  - road surface treatment (at intersections)
  - minimal radii for horizontal and vertical curves
  - intersection corner design
- Organisational and psychological measures:
  - signing and marking
  - one-way street conversions
  - staggered parking layouts
  - lane width adjustments (travel lanes and auxiliary lanes)
  - uncontrolled intersections following the priority-to-the-right rule

It is also possible to distinguish between site-specific interventions and area-wide schemes (such as pedestrian zones, residential zones, 30 km/h zones, and shared spaces).

Phase three: Vision Zero

The aforementioned traffic calming elements can be regarded as a transitional phase, involving rather local modifications to infrastructure built during the rapid expansion of automobilism, when the needs of all road users and road safety were not the centre of attention.

However, to achieve a true shift in road safety, it is necessary to change the overall paradigm of how transport infrastructure is perceived. The current mode of thinking is referred to as Vision Zero, and is based on the following premises:

- It is unacceptable for road traffic to result in fatalities or serious injuries.
- The human factor is involved in every traffic crash. However, since driver errors can never be entirely eliminated, the road environment (i.e. the road and its surroundings) should be designed in such a way as to prevent road casualties.
- This entails a shift in the notion of responsibility: it is no longer solely the responsibility of the road user who caused the crash, but also the shared responsibility of those who shape the parameters of the transport system. As these include road designers and infrastructure administrators, such thinking influences the planning and operation of road infrastructure.

According to the Vision Zero approach, road infrastructure should be both self-explaining and forgiving:

- A *self-explaining road* can generally be understood as comprehensible. The assumption is that the traffic environment should generate appropriate expectations regarding driver behaviour. The aim is for the driver to understand intuitively what kind of driving behaviour is expected of them, including speed adjustment. Features known to affect driving speed can be used to enhance self-explaining road qualities. Wider and more comfortable roads generally lead to higher speeds, whereas the presence of vegetation, pavements, and pedestrian crossings typically encourages drivers to reduce their speed (Ambros et al., 2021).
- A *forgiving road* is one that can, to some extent, compensate for driver error or atypical vehicle behaviour resulting from mechanical failure – that is, it helps minimise health consequences. Possible measures include removing of roadside trees and roughening the road surface to reduce the risk of crashes in adverse weather conditions. Guard rails can also substantially mitigate the consequences of crashes. In areas with high levels of motorcycle traffic, underrun protection should be installed beneath the guard rails. The road edge should be clearly marked with a high-quality edge line,

ideally textured, so that drivers become more aware when crossing it. The vibrations caused by running over such a textured edge line may also help awaken a driver experiencing microsleep.

This approach also has implications for setting speed limits, which should take into account the vulnerability of the human body (PIARC, 2019):

- 30 km/h in areas with vulnerable road users
- 50 km/h at locations with potential side-impact collisions (at intersections)
- 70 km/h at locations with potential head-on collisions (where traffic moving in opposite directions is not physically separated)

It is desirable that road safety be considered not only during the designing of a road but throughout its entire life cycle (planning, operation, modification, etc.). This purpose is served by a range of procedures governed by the European Directive on road infrastructure safety management.

Summary

Road infrastructure comprises all the fixed segments of the transport system necessary for the operation and organisation of traffic. Each of these elements is involved in influencing road users and therefore represents one of the five components affecting the transport system. In terms of perception and behaviour (and consequently safety), elements of the roadside surroundings, such as vegetation and advertising structures, collectively referred to as the traffic environment, can also be considered influential factors.

Initially, methods for assessing these influences, including crash-based analyses, observations, and experiments, were introduced. From a research perspective, it is ideal when conclusions from studies employing multiple methods converge. However, divergent findings may also occur, reflecting the distinct characteristics (strengths and limitations) of each method.

Subsequently, selected factors influencing human behaviour in traffic environment (i.e. the road and its immediate surroundings) were presented, including:

- horizontal curves
- cross section (lane width, road width, shoulder width, and central median strip)
- at-grade intersections
- railway level crossings
- roadside safety barriers
- roadside elements (vegetation, advertisements, and billboards)
- lighting
- cycling infrastructure (cycle lanes and cycle paths)
- pedestrian infrastructure
- traffic calming measures

The following psychological aspects must be taken into account when planning infrastructure:

- anticipation and orientation
- smoothness of driving flow
- perception of elements and their comprehension
- habituation and desensitisation effects
- mental workload

In summary, human factor errors primarily arise because drivers either lack timely information about the environment or are overwhelmed by excessive information, which may delay their reactions and increase the likelihood of mistakes. This underscores the importance of perceived subjective risk: road users behave and make decisions according to the level of risk they sense.

To prevent errors, infrastructure can be designed or adapted to correspond with human limitations and expectations, thereby increasing the likelihood of successful and timely information processing and correct responses to traffic situations. Modifications to the traffic environment should therefore aim to simplify the environment – for example, through channelising traffic flows or separating transport modes – or to reduce vehicle speeds. Good examples include roundabouts, 2 + 1 roads, and road narrowing.

The description of how the identified risk factors have influenced the evolution of infrastructure design approaches is structured into three phases: the rise of automobilism, traffic calming, and Vision Zero.

The rise of automobilism led to negative impacts, with the traffic function dominating over other functions, particularly in urban areas.

Traffic calming was a response to these negative effects. It encompasses a set of measures and tools intended to increase the utility of the road, improve the environment, and enhance safety. Measures can be physical, organisational, or psychological, and can be distinguished as either site-specific or area-wide.

However, achieving a real improvement in safety requires a paradigm shift in how traffic infrastructure is perceived. The current philosophy is known as Vision Zero, the motto of which is that it is unacceptable for deaths and serious injuries to continue occurring in road traffic. Accordingly, road infrastructure should be both self-explaining and forgiving. This also relates to the choice of speed limits, which should respect the vulnerability of the human body.

References

Ambros, J., Perůtka, J., Skládaný, P., & Tučka, P. (2019). Enhancing the insight into Czech railway level crossings' safety performance. *International Journal of Rail Transportation*, *8*, 99–108. https://doi.org/10.1080/23248378.2019.1612791

Ambros, J., Turek, R., Šragová, E., Petr, K., Šucha, M., & Frič, J. (2021). How fast would you (or should you) drive here? Investigation of relationships between official speed limit, perceived speed limit, and preferred speed. *Transportation Research*

Part F: Traffic Psychology and Behaviour, *83*, 164–178. https://doi.org/10.1016/j.trf.2021.09.003

Babojelić, K., & Novačko, L. (2020). Modelling of driver and pedestrian behaviour – a historical review. *Promet – Traffic & Transportation*, *32*(5), 727–745. https://doi.org/10.7307/ptt.v32i5.3524

Bíl, M., Bílová, M., Dobiáš, M., & Andrášik, R. (2016). Circumstances and causes of fatal cycling crashes in the Czech Republic. *Traffic Injury Prevention*, *17*(4), 394–399. https://doi.org/10.1080/15389588.2015.1094183

Bobermin, M. P., Silva, M. M., & Ferreira, S. (2021). Driving simulators to evaluate road geometric design effects on driver behaviour: A systematic review. *Accident Analysis and Prevention*, *150*, 105923. https://doi.org/10.1016/j.aap.2020.105923

Bucsuházy, K., Zůvala, R., Doecke, S., & Ambros, J. (2018). *Investigation of pre-crash vehicle speed*. Fourth International Conference on Traffic and Transport Engineering, Belgrade.

Bucsuházy, K., Zůvala, R., Valentová, V., & Ambros, J. (2022). Factors related to severe single-vehicle tree crashes: In-depth crash study. *PLOS One*, *17*(1), e0248171. https://doi.org/10.1371/journal.pone.0248171

Calvi, A. (2015). Does roadside vegetation affect driving performance? Driving simulator study on the effects of trees on drivers' speed and lateral position. *Transportation Research Record*, *2518*(1), 1–8. https://doi.org/10.3141/2518-01

Carsten, O., & Hamish Jamson, A. (2011). Driving simulators as research tools in traffic psychology. In B. E. Porter (Ed.), *Handbook of traffic psychology* (pp. 87–96). Academic Press.

Centre for Automotive Safety Research. (2021, October 11). *Crash investigation*. The University of Adelaide. https://set.adelaide.edu.au/centre-for-automotive-safety-research/crash-investigation

Centrum dopravního výzkumu, v. v. i. (2022). Homepage | *Reconstruction of real fatal accident*. www.vyzkumnehod.cz/en/

Das, S., & Maurya, A. K. (2017). Modelling of motorised two-wheelers: A review of the literature. *Transport Reviews*, *38*(2), 209–231. https://doi.org/10.1080/01441647.2017.1317049

Elvik, R. (2017). *Road lighting: Effects on road safety, other benefits and cost-benefit analysis*. 3rd International Street Lighting & Smart Controls Conference, Brisbane.

Elvik, R. (2022). Which is the more important for road safety – road design or driver behavioural adaptation? *Traffic Safety Research*, 2, 000009. https://doi.org/10.55329/pvir7839

Elvik, R., Høye, A., Vaa, T., & Sørensen, M. (2009). *The handbook of road safety measures* (2nd ed.). Emerald. https://doi.org/10.1108/9781848552517

Evans, L. (2004). *Traffic safety*. Science Serving Society.

Fisher, D. L., Rizzo, M., Caird, J. K., & Lee, J. D. (2011). *Handbook of driving simulation for engineering, medicine, and psychology*. CRC Press.

German In-Depth Accident Study (GIDAS). (n.d.). *Home*. www.gidas.org/start-en.html

Haque, F., Kidwai, F. A., ul Haq, M. F. (2025). A systematic review of proactive approach for pedestrian safety at intersections. *Transportation Research Part F: Traffic Psychology and Behaviour*, *109*, 792–808. https://doi.org/10.1016/j.trf.2025.01.013

Havránek, P., Zůvala, R., Špaňhel, J., Herout, A., Valentová, V., & Ambros, J. (2020). How does road marking in horizontal curves influence driving behaviour? *European Transport Research Review*, *12*(33). https://doi.org/10.1186/s12544-020-00425-7

Jurewicz, C. (2009). Impact of LATM treatments on speed and safety. *Road & Transport Research*, *18*(4), 14–22.

Laureshyn, A., & Várhelyi, A. (2018). *The Swedish traffic conflict technique – observer's manual*. Lund University.

Moradi, A., Ameri, P., Rahmni, K., Najafi, M., Jamshidi, E., Fakhri, Y., Khazaei, S., Moeini, B., & Amjadian, M. (2019). Factors affecting the severity of pedestrian traffic crashes. *Archives of Trauma Research*, *8*(2), 46–56.

National Highway Traffic Safety Administration (n.d.). *Crash injury research*. www.nhtsa.gov/research-data/crash-injury-research

Niaki, M. N., Wijlhuizen, G. J., & Dijkstra, A. (2021). *Safety enhancing features of cycling infrastructure: Review of evidence from Dutch and international literature*. SWOV. https://swov.nl/system/files/publication-downloads/r-2021-20.pdf

OECD. (2018). *Speed and crash risk*. www.itf-oecd.org/sites/default/files/docs/speed-crash-risk.pdf

Oviedo-Trespalacios, O., Truelove, V., Watson, B., & Hinton, J. A. (2019). The impact of road advertising signs on driver behaviour and implications for road safety: A critical systematic review. *Transportation Research Part A: Policy and Practice*, *122*, 85–98. https://doi.org/10.1016/j.tra.2019.01.012

Papadimitriou, E., Filtness, A., Theofilatos, A., Ziakopoulos, A., Quigley, C., & Yannis, G. (2019). Review and ranking of crash risk factors related to the road infrastructure. *Accident Analysis and Prevention*, *125*, 85–97. https://doi.org/10.1016/j.aap.2019.01.002

PIARC. (2019). *Road safety manual: A guide for practitioners*. www.piarc.org/ressources/documents/a4cc6ab-34401-Road-Safety-Manual-Guid-World-Road-Association-EN.pdf?utm_source=chatgpt.com

PIARC. (2023). ***Term of the road dictionary: Transport infrastructure***. www.piarc.org/en/activities/Road-Dictionary-Terminology-Road-Transport/term-sheet/86714-en-transport+infrastructure

Poudel, N., & Singleton, P. A. (2021). Bicycle safety at roundabouts: A systematic literature review. *Transport Reviews*, *41*(5), 617–642. https://doi.org/10.1080/01441647.2021.1877207

Prati, G., Puchades, V. M., De Angelis, M., Fraboni, F., & Pietrantoni, L. (2018). Factors contributing to bicycle-motorised vehicle collisions: A systematic literature review. *Transport Reviews*, *38*(2), 184–208. https://doi.org/10.1080/01441647.2017.1314391

Risser, R. (2004). Philosophy of traffic calming. *Asian Journal of Transport and Infrastructure*, *11*(1), 1–9.

Risser, R., Šucha, M., Chaloupka-Risser, C., & Wagner, T. (2020). Infrastructure. In M. Šucha, L. Kluppels, R. Risser, & D. deVol (Eds.), *Selected topics on psychology and traffic safety* (pp. 57–59). Kirschbaum Verlag.

Roshandel, S., Zheng, Z., & Washington, S. (2015). Impact of real-time traffic characteristics on freeway crash occurrence: Systematic review and meta-analysis. *Accident Analysis and Prevention*, *79*, 198–211. https://doi.org/10.1016/j.aap.2015.03.013

Rudin-Brown, C. M., & Jamson, S. L. (2013). *Behavioural adaptation and road safety: Theory, evidence and action*. CRC Press.

Schepers, P., den Brinker, B., Methorst, R., & Helbich, M. (2017). Pedestrian falls: A review of the literature and future research directions. *Journal of Safety Research*, *62*, 227–234. https://doi.org/10.1016/j.jsr.2017.06.020

Singh, H., & Kathuria, A. (2021). Analyzing driver behavior under naturalistic driving conditions: A review. *Accident Analysis and Prevention*, *150*, 105908. https://doi.org/10.1016/j.aap.2020.105908

Šmíd, P. (2004). *Jak zklidnit dopravu v obcích: Příručka pro zástupce místní samosprávy* [How to calm traffic in municipalities: A handbook for local government representatives]. Partnership Foundation.

Sohel Mahmud, S. M., Ferreira, L., Hoque, M. S., & Tavassoli, A. (2019). Micro-simulation modelling for traffic safety: A review and potential application to heterogeneous traffic environment. *IATSS Research*, *43*(1), 27–36. https://doi.org/10.1016/j.iatssr.2018.07.002

Šucha, M., Šragová, E., Suriaková, B., Risser, R., Mazalová, R., Oviedo-Trespalacios, O., Włodarczyk, A., Aquino, S. D., Rusli, R., Useche, S. A., Martínez-Buelvas, L., Silva, M., Çelik, A. K., Park, J., Bastos, J. T., Enea, V., Argandar, G. D., Ramezani, S., Barboza-Palomino, M., . . . Wang, Y. (2024). Self-reported pedestrian falls in 15 countries worldwide. *Transportation Research Part F: Traffic Psychology and Behaviour*, *105*, 267–283. https://doi.org/10.1016/j.trf.2024.07.009

Theofilatos, A., & Yannis, G. (2014). A review of the effect of traffic and weather characteristics on road safety. *Accident Analysis and Prevention*, *72*, 244–256. https://doi.org/10.1016/j.aap.2014.06.017

Thomas, L., Nordback, K., & Sanders, R. (2019). Bicyclist crash types on national, state, and local levels: A new look. *Transportation Research Record*, *2673*(6), 664–676. https://doi.org/10.1177/0361198119849056

TRL (2012, May 29). *Road accident in-depth study (RAIDS)*. www.trl.co.uk/projects/road-accident-in-depth-study – raids-

Useche, S. A., Faus, M., & Alonso, F. (2022). Is safety in the eye of the beholder? Discrepancies between self-reported and proxied data on road safety behaviors – a systematic review. *Frontiers in Psychology*, *13*, 964387. https://doi.org/10.3389/fpsyg.2022.964387

van Haperen, W., Riaz, M. S., Daniels, S., Saunier, N., Brijs, T., & Wets, G. (2019). Observing the observation of (vulnerable) road user behaviour and traffic safety: A scoping review. *Accident Analysis and Prevention*, *123*, 211–221. https://doi.org/10.1016/j.aap.2018.11.021

van der Horst, R. (2006). *Finding a position of naturalistic driving studies in the methodological spectrum*. Workshop on Naturalistic Driving Studies, Munich.

Vivek, A. K., Khan, T., & Mohapatra, S. S. (2021). Safety and associated parameters influencing performance of rail road grade crossings: A critical review of state of the art. *Journal of Safety Research*, *79*, 257–272. https://doi.org/10.1016/j.jsr.2021.09.007Get rights and content

Vyskočilová, L., Bucsuházy, K., Zůvala, R., Striegler, R., & Ambros, J. (2023). What influences lateral position in horizontal curves? *Results in Engineering*, *18*, 101146. https://doi.org/10.1016/j.rineng.2023.101146

Wang, C., Quddus, M. A., & Ison, S. G. (2013). The effect of traffic and road characteristics on road safety: A review and future research direction. *Safety Science*, *57*, 264–275. https://doi.org/10.1016/j.ssci.2013.02.012

Yannis, G., & Michelaraki, E. (2025). Effectiveness of 30 km/h speed limit – a literature review. *Journal of Safety Research*, *92*, 490–503. https://doi.org/10.1016/j.jsr.2024.11.003

Young, W., Sobhani, A., Lenné, M. G., & Sarvi, M. (2014). Simulation of safety: A review of the state of the art in road safety simulation modelling. *Accident Analysis and Prevention*, *66*, 89–103. https://doi.org/10.1016/j.aap.2014.01.008

18 Safety measures based on psychological principles

Kristýna Josrová

Introduction

Traffic infrastructure enables transport, but it is not merely about reaching a destination. While it improves the efficiency of travel, the focus must also be on making it safer. Many governments and agencies have made this a priority by adopting Vision Zero, which aims to eliminate road traffic injuries and fatalities entirely (Kim et al., 2017).

Still, to err is human. Most traffic accidents are caused by human error. People are not flawlessly functioning beings who consistently apply what they have learned. Relying solely on adherence to rules and recommendations therefore means accepting a certain level of accidents and injuries. Rather than ignoring human fallibility, we should seek ways to prevent mistakes and mitigate their consequences. The Safe System approach is designed to anticipate human shortcomings and shape the traffic system in such a way that accidents are averted – and when they do occur, their impact is minimised (Khan & Das, 2024).

In addition to reducing the number of injuries, our efforts to improve traffic safety should also aim to promote public health more broadly by supporting diverse forms of movement. After all, transport infrastructure includes the creation of spaces where people can walk, cycle, or interact socially. Enabling opportunities for active mobility benefits both physical and mental well-being (Sundling & Jakobsson, 2023).

When designing infrastructure, it is essential to consider who the end users are and how we want them to behave and experience their journey. On this basis, we can choose appropriate measures that encourage the desired behaviour. Sometimes the aim is a fast and direct car journey; at other times, the goal is to create spaces where pedestrians can move about freely and socialise near moving vehicles. In most cases, there is no one-size-fits-all solution – rather, it is about finding the optimal balance. That is why, we must keep core values in mind and continue to search for new and better approaches. The implementation of new measures and the redesign of traffic infrastructure should go hand in hand with rigorous evaluation of their effectiveness and limitations. Careful research can uncover counter-intuitive

DOI: 10.4324/9781003608394-18

effects of various interventions: some well-intentioned measures may end up doing more harm than good, while others that appear risky at first may lead to safer outcomes.

This chapter presents a range of approaches and measures aimed at making traffic safer, more intuitive, easier to navigate, and more welcoming to multiple modes of transport. Speed management, traffic calming measures, self-explaining roads, and shared spaces will be discussed next.

Speed management and traffic calming

Speed is a key factor in road traffic, often determining whether a crash will occur – and, if it does, influencing the severity of the outcome, which frequently results in injury or death. Driving at an appropriate speed also provides a buffer, giving drivers time to register what is happening around them and respond accordingly. As motorised vehicles have become faster and thus more dangerous, the need to address the pressing issue of speed has grown increasingly urgent. This chapter introduces measures that encourage drivers to adopt appropriate speeds according to the area and environment. It includes traffic calming measures, typically used in urban and residential areas, as well as speed management strategies more commonly applied in non-residential zones. By understanding how drivers perceive and react to different stimuli, we can apply this knowledge to road geometry – designing roads and their surroundings in ways that promote the desired driving behaviour.

Traffic calming measures

Traffic calming measures (TCMs) are infrastructure elements primarily used in urban areas to reduce vehicle speeds and thus enhance the safety of vulnerable road users. In some cases, their aim is also to reduce traffic volume by discouraging drivers from using a particular route (Hydén, 2021). Most TCMs fall into two main categories: road narrowing and vertical deflection. Examples include speed humps, raised crosswalks, rumble strips, curb extensions, chicanes, traffic islands, coloured pavements, landscaping, mini-roundabouts, and speed displays (Fitzgerald & Halliday Inc., 2008). This section briefly introduces several of these measures to illustrate the principles used to deter speeding. Speed humps (or speed breakers) are small vertical obstacles designed to create discomfort for drivers who fail to reduce their speed. A well-designed speed hump, with an appropriate height and shape, does not cause discomfort if the driver is travelling at the intended speed (Hydén, 2021). These measures have proven effective at lowering driving speeds and, as a result, reducing the number and severity of crashes (Monyatsi, 2021).

Lane and road width have a significant influence on speed choice; wider lanes (those exceeding 3.1 metres) tend to encourage higher speeds (Karim, 2015). Narrowing the roadway prompts drivers to slow down in order to safely navigate what is perceived as a more challenging environment. Such segments are often viewed as riskier, which encourages drivers to adapt their speed accordingly (Distefano & Leonardi, 2019). Curb extensions are one example of road narrowing; they alert drivers, shorten pedestrian crossing distances, and improve mutual visibility between drivers and pedestrians (Fitzgerald & Halliday Inc., 2008).

A key aspect of TCMs is that they provide immediate feedback to drivers about whether their behaviour is appropriate. Ignoring a speed hump results in physical discomfort, failing to respond to narrowing features may risk vehicle damage, and speed displays offer direct visual feedback, prompting drivers to adjust their speed. Over time, as drivers become familiar with specific TCMs, they learn to modify their behaviour in order to avoid these negative consequences.

Speed management

The easier a road feels to drive on, the more riskily drivers tend to behave. Straight segments and long sight distances create a sense of comfort, often prompting drivers to increase their speed. In such cases, perceived safety is offset by riskier behaviour. Conversely, when the road environment appears more challenging, drivers become more cautious – an effect particularly noticeable on curves. If the radius of a curve is sufficient, it is effective at prompting drivers to reduce their speed. The same applies to sufficiently steep uphill or downhill gradients, which can have a similar moderating effect on driving speed (Sheykhfard et al., 2024).

Paradoxically, roads that appear safer can become more dangerous. In efforts to provide greater safety margins, roads and lanes were often widened. However, research shows that there is an optimal lane width that encourages lower speeds and leads to fewer crashes. Wider lanes tend to invite higher speeds: the wider the lane, the faster drivers tend to travel (Azin et al., 2025). Road markings also influence perception: wider longitudinal lane markings have been shown to increase perceived speed (Garach et al., 2022).

Beyond road design, the surrounding landscape, roadside elements, and overall environment play an important role in how drivers perceive and regulate their speed. A wide, empty road can create a psychological invitation to speed. Without visual reference points, it becomes difficult for drivers to gauge their speed accurately. To identify an appropriate speed for a particular location and situation, drivers need such reference points. Naturally occurring or deliberately placed elements alongside the road help drivers modulate their pace. These objects, even if only perceived peripherally, serve as cues that assist in speed estimation.

Monotonous or empty environments, such as flat open areas, offer little sensory input, making it easy for drivers to misjudge their speed. This effect is common on rural roads and highways. In suburban or residential settings, vegetation and roadside trees help drivers make more accurate speed estimates. Close solid structures, such as guard rails, buildings, or vertical elements such as trees and lamp posts, can create a sense of spatial restriction. When placed near the roadside, these elements blur as drivers pass, creating a tunnel effect that enhances the sensation of speed and encourages slowing down. In this way, the environment provides drivers with valuable feedback (Rak & Bilík, 2025).

Self-explaining roads

A self-explaining road (SER) is one that is intuitively understandable to the driver. Upon encountering it, the driver immediately recognises how to behave. The environment is designed to be easily interpreted, subtly guiding the driver towards the desired actions. The road and its surroundings are structured to support safe driving behaviour – clear, intuitive, and undemanding – so that drivers can promptly grasp what is expected of them. The design and appearance of a location prepare drivers for specific conditions and risks inherent to that area (Theeuwes, 2021).

Human behaviour is strongly influenced by context. Therefore, encouraging appropriate driver behaviour begins with making it clear where the driver is. Throughout life, people develop subconscious schemas that help them categorise different environments and adapt their behaviour accordingly. Just as one instinctively lowers one's voice in a library or looks both ways at a railway crossing, drivers also distinguish between types of roads on the basis of their experience. Without conscious thought, most drivers can differentiate between motorways, local roads, rural byways, and city streets. Each of these environments calls for a different approach: different speeds, different expectations, and different hazards. For example, one might expect to encounter a tram in an urban setting, but a deer on a rural road. Roads vary in appearance; some are wide, with guard rails and unidirectional traffic, while others have traffic lights, pavements, and cyclists (Theeuwes & Godthelp, 1995).

Visual cues are essential in helping drivers orient themselves. The appearance of a road, including surface texture, lane width, visibility, lighting, geometry, and the presence of other traffic elements, plays a key role in informing drivers about where they are and how they should behave. Drivers perceive their environment holistically and can often identify their location at a glance on the basis of characteristic visual features (Theeuwes & Godthelp, 1995). Since drivers rely heavily on appearance, suitable infrastructure design can make driving both easier and safer.

Drivers' expectations, shaped by accumulated experience, are central to their perception and decision-making. SERs work on a subconscious level, using design elements that match drivers' mental models of specific road types. This allows drivers to anticipate not only how they should behave but also how other road users are likely to behave. SERs are effective because they create expectations that are consistent with the actual conditions at a given location (Theeuwes, 2021). Experience plays a decisive role in how incoming information is perceived. Drivers tend to notice familiar elements placed in expected locations. Experience has taught them what to look for and where to find potential hazards. If the layout or placement of traffic elements deviates from expectations, it can lead to misinterpretation or missed cues. Therefore, the positioning of traffic infrastructure must align with subconscious expectations formed through prior experience. When something is unexpected, it may go unnoticed, failing to alert the driver. For this reason, road design must reflect drivers' expectations, and, when necessary, guide their attention towards unfamiliar or critical features (Theeuwes, 2021). To design an effective SER, three core principles must be applied: the road must be easily recognisable, distinguishable, and interpretable.

Interpretability means that the environment naturally prompts appropriate actions – slowing down, checking for pedestrians, or preparing to give way. It should be immediately obvious how all road users are expected to behave. The environment itself communicates the rules of engagement for safe travel (Theeuwes, 2021).

Recognisability refers to the presence of clear, typical attributes associated with specific road types. Drivers develop mental representations of different road categories, so roads that look like a motorway or priority road will be treated as such. An easily recognisable motorway, for example, has all the expected visual cues, and avoids features that do not belong on that type of road (Theeuwes, 2021).

Distinguishability ensures that road types are visually distinct from one another. When each type of road has a radically different appearance, it triggers unconscious categorisation, leading drivers to adopt behaviour that fits the environment (Theeuwes, 2021).

These three principles are interlinked. Each road type should look and feel different so that misinterpretation is virtually impossible. Roads of the same category should have a consistent appearance, allowing drivers to recognise them quickly. The environment should guide the driver towards behaviour that is appropriate for the location. Road infrastructure and layout not only signal the type of road but also encourage the intended behaviour, indicating, for example, the appropriate speed or which hazards to be aware of (Theeuwes, 2021).

When these principles are properly implemented, SERs become highly effective tools for enhancing safety for both drivers and other road users. Through thoughtful design and modification of the traffic environment, SERs help to reduce vehicle speeds and improve behaviour. Moreover, they are generally well received by local road users (Charlton & Baas, 2006).

Shared space

Transport infrastructure is typically designed to separate different modes of transport, with the aim of ensuring both efficiency and safety. This means that cars are expected to use the road, pedestrians the pavement, and, ideally, bicycles are confined to dedicated cycle lanes. Each group has its own designated area and interacts with others only when necessary – for example, when crossing a street. The concept of shared space, however, is grounded in an opposite approach: multiple modes of transport coexisting within the same physical environment. Rather than separating users, it seeks to integrate them. In shared spaces, there is an absence of physical segregation: no elevated curbs, no barriers, and no clearly defined carriageway for motorised vehicles. Traffic signs are minimised or removed entirely, which means road users cannot rely on the standard traffic rules and conventions. It is unclear how others will behave or who has priority. This deliberate uncertainty prompts greater caution. As a result, shared space requires communication between all road users; people must make eye contact, observe one another, and negotiate priority. This uncertainty is purposeful: it encourages everyone to consider the presence of others, anticipate their actions, and adjust accordingly, fostering more considerate behaviour (Hamilton-Baillie, 2008).

Shared space is intended to enhance the presence of pedestrians, support their freedom of movement, and promote social interaction, thereby contributing to a more liveable environment. Drivers are expected to adapt to this setting, and the physical space reflects this expectation. Street furniture, such as benches, trees, or fountains, is deliberately placed in such a way as to emphasise the pedestrian nature of the environment. Together with other infrastructural features, such elements nudge drivers to reduce speed and accommodate vulnerable road users, ultimately improving safety. The need to communicate and adjust behaviour fosters mutual respect between different groups of road users (Karndacharuk et al., 2014).

Shared spaces are effective at reducing vehicle speeds, which contributes to increased safety and a lower incidence of collisions (Tzouras et al., 2024). Nevertheless, cars still pose a potential threat to vulnerable road users, and not all pedestrians may feel comfortable sharing space with motor vehicles. Some may be reluctant to engage with traffic and instead choose to avoid the area altogether. While shared space often enhances the experience of its intended user groups, it may not work equally well for everyone. Likewise, drivers may feel uneasy navigating a space where they must interact more directly with pedestrians and cyclists – although this unease is central to the concept, as it promotes more cautious and attentive driving. When implementing elements of shared space, it is essential not to ask only whether it will work, but rather how it will work – and for whom. Designing such spaces is not merely a matter of combining the right elements to produce a functioning layout. It requires a carefully considered design process, followed by evaluation and refinement over time (Ruiz-Apilánez et al., 2017). Ultimately, a space can only be truly shared if it is developed in collaboration with the people who use it.

Summary

Traffic infrastructure is primarily focused on ensuring efficient transport, but it must also aim to promote safety. As human error accounts for the majority of traffic accidents, it is essential to create a traffic system that anticipates human limitations and mitigates the consequences of collisions. In designing such systems and interventions, it is important to consider the needs of different groups of road users and the broader functions that the space can serve. Moreover, it is essential to test and evaluate the effectiveness of the measures that are implemented.

Speed plays a key role in road safety. To encourage drivers to adopt safe speeds, a range of speed management and traffic calming measures are implemented, many of which lead to a reduction in vehicle speed. In urban environments, measures such as road narrowing and vertical deflections (e.g. speed humps), which can cause discomfort when traversed at high speed, are commonly used. Drivers' behaviour is influenced by factors such as visibility, lane width, road geometry, and the surrounding environment. Traffic design utilises modifications to these elements to encourage appropriate speeds.

Self-explaining roads (SERs) are designed to be intuitively comprehended by road users. Their layout and visual features elicit appropriate driver behaviour by clearly communicating the type of environment and what is expected in that context – purely through design. As the different types of roads are made easily recognisable, drivers are able to adjust their speed, attention, and expectations accordingly, contributing to safer driving. SERs rely particularly on visual cues that communicate the intended behaviour, allowing drivers to navigate without needing to interpret rules consciously. For this to work effectively, roads of the same type should have a consistent appearance, while different road types should be visually distinct, making misinterpretation unlikely.

Shared space refers to areas where multiple modes of transport coexist without physical segregation. It relies on mutual communication and caution, as traditional demarcations, such as curbs or road markings, are removed. Instead, features such as street furniture, trees, and benches subtly indicate the presence and priority of pedestrians, who are the primary focus of these environments. Shared space promotes liveability, encourages slower driving, and enhances social interaction. The uncertainty experienced by drivers – because of the lack of formal rules – leads to more cautious and attentive behaviour, contributing to overall safety.

References

Azin, B., Ewing, R., Yang, W., Promy, N. S., Kalantari, H. A., & Tabassum, N. (2025). Urban arterial lane width versus speed and crash rates: A comprehensive study of road safety. *Sustainability*, *17*(2), 628. https://doi.org/10.3390/su17020628

Charlton, S. G., & Baas, P. H. (2006). *Speed change management for New Zealand roads* (No. 300). Land Transport New Zealand.

Distefano, N., & Leonardi, S. (2019). Evaluation of the benefits of traffic calming on vehicle speed reduction. *Civil Engineering and Architecture*, 7(4), 200–214. https://doi.org/10.13189/cea.2019.070403

Fitzgerald & Halliday, Inc. (2008). *Traffic calming resource guide*. https://scrcog.org/wp-content/uploads/upwp/studies/TrafficCalming_ResourceGuide_Final.pdf

Garach, L., Calvo, F., & De Oña, J. (2022). The effect of widening longitudinal road markings on driving speed perception. *Transportation Research Part F: Traffic Psychology and Behaviour*, *88*, 141–154. https://doi.org/10.1016/j.trf.2022.05.021

Hamilton-Baillie, B. (2008). Shared space: Reconciling people, places and traffic. *Built Environment*, *34*(2), 161–181. https://doi.org/10.2148/benv.34.2.161
Hydén, C. (2021). Traffic calming: The way ahead in mixed traffic. In G. Tiwari & D. Mohan (Eds.), *Transport and safety: Springer transactions in civil and environmental engineering* (pp. 129–146). Springer. https://doi.org/10.1007/978-981-16-1115-5_7
Karim, D. M. (2015, June 6). *Narrower lanes, safer streets* [Conference paper presentation]. Conference of Canadian Institute of Transportation Engineers, Regina, Canada.
Karndacharuk, A., Wilson, D. J., & Dunn, R. (2014). A review of the evolution of shared (street) space concepts in urban environments. *Transport Reviews*, *34*(2), 190–220. https://doi.org/10.1080/01441647.2014.893038
Khan, M. N., & Das, S. (2024). Advancing traffic safety through the safe system approach: A systematic review. *Accident Analysis & Prevention*, *199*, 107518. https://doi.org/10.1016/j.aap.2024.107518
Kim, E., Muennig, P., & Rosen, Z. (2017). Vision zero: A toolkit for road safety in the modern era. *Injury Epidemiology*, *4*(1). https://doi.org/10.1186/s40621-016-0098-z
Monyatsi, L. (2021, 5. July). *The effectiveness of speed humps as traffic calming for accident reduction* [Conference paper presentation]. 39th International Southern African Transport Conference.
Rak, M., & Bilík, M. (2025). The effect of roadside on speed perception. *Acta Polytechnica CTU Proceedings*, *52*, 116–121. https://doi.org/10.14311/APP.2025.52.0116
Ruiz-Apilánez, B., Karimi, K., García-Camacha, I., & Martín, R. (2017). Shared space streets: Design, user perception and performance. *Urban Design International*, 22, 267–284. https://doi.org/10.1057/s41289-016-0036-2
Sheykhfard, A., Haghighi, F., Saeidi, S., SafariTaherkhani, M., & Das, S. (2024). Understanding the influence of environmental factors on driver speed: A structural equation modeling analysis. *IATSS Research*, *48*(3), 427–439. https://doi.org/10.1016/j.iatssr.2024.08.001
Sundling, C., & Jakobsson, M. (2023). How do urban walking environments impact pedestrians' experience and psychological health? A systematic review. *Sustainability*, *15*(14), 10817. https://doi.org/10.3390/su151410817
Theeuwes, J. (2021). Self-explaining roads: What does visual cognition tell us about designing safer roads? *Cognitive Research: Principles and Implications*, 6(1), 1–15. https://doi.org/10.1186/s41235-021-00281-6
Theeuwes, J., & Godthelp, H. (1995). Self-explaining roads. *Safety Science*, *19*(2–3), 217–225. https://doi.org/10.1016/0925-7535(94)00022-U
Tzouras, P. G., Kepaptsoglou, K., & Vlahogianni, E. I. (2024). Empirical investigation of shared space traffic: A comparison to conventional urban road environment. *International Journal of Transportation Science and Technology*, *15*, 122–135. https://doi.org/10.1016/j.ijtst.2023.08.001

19 Driver rehabilitation programmes

Petr Zámečník and Darina Havlíčková

Introduction

The traffic safety management framework incorporates a range of structural tools designed to educate drivers in safe road behaviour and prevent risky driving practices. However, structural measures, including public awareness campaigns and demerit point systems, tend to have limited effectiveness among specific populations such as drivers who habitually violate traffic regulations. For these groups, more targeted or individualised interventions are required. Although the proportion of drivers who operate a vehicle under the influence or commit serious traffic violations is relatively small, they are responsible for a high number of road accidents with severe consequences (Bingham & Shope, 2004; Simpson et al., 1996; Peck, 1993; Williams, 2003). Specific rehabilitation programmes have been developed for such drivers with the aim of modifying their behaviour to improve road safety outcomes (Zámečník et al., 2018). Neither the literature nor practice offers a consistent definition of what constitutes rehabilitation programmes or, in other words, corrective interventions for drivers who repeatedly violate traffic laws. Some jurisdictions use the term rehabilitation programmes, while others adopt terminology derived from medical or educational contexts, such as driver improvement courses, remedial training courses, or educational measures (e.g. the Dutch "educatieve maatregelen" or the German "Nachschulung"). In the Czech Republic, the term therapeutic programmes is also employed. Nevertheless, rehabilitation programmes remains the most widely used designation, typically referring to therapeutic and educational courses grounded in psychological interventions. The most frequently cited definition describes the programmes as "systematic measures for traffic offenders (in particular drunk drivers and speed offenders) aiming at a change of their behaviour in order to prevent further offences" (Bartl et al., 2002, p. 3). These programmes focus on the development of greater consideration for others by encouraging a change in attitude and enhancing social responsibility. The primary deficit of this high-risk driver target group lies in their deliberate non-compliance with traffic rules, rendering skill-based training or mere education largely ineffective or even counterproductive (Bartl et al., 2002).

DOI: 10.4324/9781003608394-19

Across Europe, these programmes take various forms (the only exception being the Vatican, where no such measures are in place). Approaches range from purely therapeutic interventions (e.g. in Germany, Austria, and the Czech Republic), through combined models (e.g. in Belgium and the Netherlands), to those with a stronger educational focus (e.g. in Switzerland and Spain). Programmes typically integrate both educational and therapeutic components, although the psychological element is the most prominent by principle. They are usually implemented in a group setting, although some countries, such as Germany, also provide individualised formats. Recidivism reduction rates for group courses generally stand at around 50% (Bartl et al., 2002), while individual modalities can reduce recidivism by as much as 90%. Drivers displaying marked aggression or otherwise unable to engage effectively in group sessions have been found to benefit more from individualised approaches (Bartl et al., 2002; Davies et al., 1999). Overall, rehabilitation programmes appear to be an appropriate and effective means of improving the behaviour of high-risk drivers. However, national and cultural particularities must be taken into account to ensure both optimal outcomes and the efficient use of resources.

The development of rehabilitation programmes

Evidence collected since the 1950s has shown that some drivers commit significantly more traffic-related violations and criminal offences than others (e.g. Chapman, 1954) and that road accidents are a major social issue. However, in the mid-20th century, road safety systems were still in their infancy, primarily focusing on the personal responsibility of individual road users rather than on systemic responsibility, an approach that would only become fully established with the adoption of the Vision Zero strategy. Early findings were thus only based on the observation that fines and threats of punishment had little deterrent effect on the most dangerous drivers. Initial remediation efforts therefore relied on education and the reinforcement of individual responsibility for accidents. These proved to be only marginally effective, as the issue for these drivers was neither ignorance nor inexperience. In response to the continued rise in road accidents, new measures were sought to counter this trend. In the 1960s, the United States developed the concept of corrective programmes, which was later introduced in German-speaking countries in Europe during the 1970s. Although these initiatives were rooted in a discourse centred on individual responsibility for traffic accidents, they represent a valid intervention tool. Today, rehabilitation programmes are an established component of road safety management systems, and their role, place, and purpose within the Safe System paradigm are actively discussed.

Following their transfer from the United States to Europe, the programmes evolved into a variety of forms, and there is currently no unified system of rehabilitation measures. Nevertheless, EU Member States are encouraged to implement these programmes (Schulze et al., 2012). In the initial wave of

European implementation during the 1970s, Germany and Austria were the early adopters. In Austria, the programmes began as pilot schemes within the prison system and were gradually extended to other populations, including novice drivers. However, a more recent extension of rehabilitation programmes for drink-driving offenders to include the use of alcohol interlocks proved unsuccessful (because of a very low uptake on the part of drivers). In Germany, too, the programmes have undergone several rounds of expansion since the 1970s and remain among the most diverse in Europe, offering voluntary and mandatory options as well as group-based and individual formats. A second wave of implementation took place in the 1990s, during which roughly half of the EU countries introduced such measures: the United Kingdom in 1991, France in 1992, Belgium in 1995, the Netherlands in 1996, and Finland, Switzerland, and Italy in 1997. The most recent implementation wave occurred between 2005 and 2015, with programmes introduced in Spain (2005), Slovakia and Slovenia (2011), and Estonia (2015). The last remaining EU country to adopt rehabilitation programmes was the Czech Republic, which introduced them in 2024.

In general, developments in most countries have followed a similar pattern – programmes initially targeted drink-driving offenders and were subsequently expanded to address other groups. The first pilot projects were conducted either within academic settings (e.g. Belgium and Spain) or in prisons (e.g. Austria and the Czech Republic). In later phases of implementation, countries often adopted existing models (e.g. Italy based its approach on the German model). Research activity largely mirrored these waves of implementation, with the second half of the 1990s marking a "golden age" of research into this field, followed by another notable surge of scholarly interest a decade later. Over the past ten years, research has become more sporadic and typically focuses on specific aspects, as the topic is now generally considered to be relatively well covered in terms of effectiveness, determinants of effectiveness, and methods, as well as from a practical perspective. As regards legislation, the past decade has seen minor updates across the European Union, the United States of America, and Australia, particularly in relation to the scaled-up use of alcohol interlocks.

Context and scope of the programmes

As outlined in the previous section, the structure and targeting of programmes vary between countries. The most comprehensive summaries of their design and effectiveness are provided by --Bartl et al. (2002) and Hufnagl (2007), who identify two general target groups: individuals who drive under the influence of alcohol or drugs and drivers who commit other serious traffic offences. In the case of drink-driving, the procedural differences between countries are relatively minor, typically being limited to variations in blood alcohol concentration thresholds that determine eligibility or mandatory participation in a programme. The situation is more complex for serious traffic offenders, as there is a wide range of possible violations. In such cases, the applicable procedures and the requirement to participate in a programme are triggered by either a demerit point system or a list of specified offences.

Countries typically take into account two key contextual factors when determining the scope of their rehabilitation programmes: the technological and knowledge base in the field of rehabilitation, and practical considerations (such as the structure of the demerit point system, long-term trends in road traffic accidents, and the availability of qualified professionals to deliver the programmes). As a result, programmes are generally designed to target the groups of drivers that pose the greatest risk to road safety in a given country. Data-driven, evidence-based decision-making considers the influence of various personal characteristics and lifestyle factors on driving behaviour. Accordingly, research typically begins by identifying a set of risk factors that define the target group. These insights first serve to guide the structural definition of the measures (i.e. which drivers should be required to attend a programme) and subsequently inform the specific approach to working with these drivers (i.e. the types of intervention shown to be effective).

The initial considerations are statistics on traffic accidents and offences, along with related sociodemographic factors, including drink-driving, speeding, and disregarding traffic rules. From a psychological perspective, only age and driving experience tend to be taken into account, as it is difficult to reflect other personal characteristics in policy design. In terms of age, young novice drivers represent the highest-risk group, as they cause a relatively high number of accidents and consistently exhibit risky behaviour, regardless of their cultural background or nationality (see, for example, Gulliver & Begg, 2007; Jonah, 1990; Parry, 1968; Rhodes & Pivik, 2011). In terms of driving experience, novice drivers again stand out: inexperience is a significant factor in traffic accidents (Feng & Donmez, 2013; McCartt et al., 2003). According to statistical and sociodemographic indicators, the most frequently targeted groups for rehabilitation programmes are therefore individuals who drive under the influence of alcohol, young and inexperienced drivers, and speeders. Unfortunately, the system does not make it possible to use gender as a criterion, despite it being a strong predictor of risky behaviour. Indeed, men are more likely to speed, drive under the influence, and engage in other forms of risky traffic-related behaviour (Lawton et al., 1997; Reason, 1990; Wilsnack & Wilsnack, 1991). While women are approximately 10% more likely to be involved in an accident, the relative fatality rate for men is three times higher (Schwartz, 2008).

As for the next level, the second one, common attributes or characteristics of individuals committing specific offences were previously used to determine the appropriate types of interventions. According to Schulze et al. (2012), for example, the most common characteristics of drink drivers include young age, a history of multiple driving offences, a risky lifestyle, low levels of knowledge and risk perception, and higher levels of aggression and sensation-seeking. High levels of sensation-seeking are typical for young, risk-prone drivers (e.g. Marengo et al., 2012), while speeders tend to exhibit high levels of indifference to rules (Iversen & Rundmo, 2002; Machin & Sankey, 2008). Aggressive drivers often display high impulsivity (Dahlen et al., 2005; Underwood et al., 1999), and risk-taking behaviour is typically associated with low levels of altruism (Machin & Sankey, 2008; Ulleberg & Rundmo, 2003) and high levels of extraversion (Moncel et al., 2025; Twenge et al., 2001). However, it should be noted at this point that risky drivers comprise a heterogeneous group, and no single characteristic on its own necessarily implies a risky driver (Hubicka et al., 2008; Nochajski & Stasiewicz, 2006). In general, the notion of determining a degree of risk on the basis of specific characteristics is debatable, as a certain trait may be associated with risk in one context but act as a protective factor in another. Research into the relationships between certain characteristics or their combinations with accident rates and risk levels has thus not resulted in improved safety outcomes (Hollnagel, 2004; Sagberg & Bjørnskau, 2006), and "profiling" is generally regarded as ethically problematic (Jones & Lacey, 2000).[1]

In terms of both their content and focus, the programmes have evolved from a narrower approach centred purely on traffic-related topics and efforts to reduce the riskiness of specific traits to a more comprehensive perspective that incorporates lifestyle, health, family, and personal development. A recent trend involves long-term support, including follow-up consultations and sessions where participating drivers can revisit the strategies they have learnt (Robertson et al., 2013). Other developments focus on combining group and individual approaches (Nochajski & Stasiewicz, 2006; Wells-Parker et al., 1995), the use of motivational interviewing techniques (Stein & Lebeau-Craven, 2002), and online interventions (Zhao et al., 2023).

From a psychological perspective, mild- to moderate-risk individuals are the most suitable candidates for inclusion in the programmes. These are participants without severe pathologies, such as substance use and attention disorders. Importantly, the programmes are designed as tools for changing or regulating traffic-related behaviour, rather than as treatment for substance use disorders or other psychiatric conditions. Such tools can only be effective if the driver is capable of exercising control over their own behaviour. Otherwise, the programmes are likely to be of limited or no effectiveness. Drivers in this mild- to moderate-risk category, that is, those who typically participate in the programmes, consider themselves conventional drivers; they tend to downplay their risky behaviour and are convinced that they are not breaking any rules, despite being aware of the risks associated with such behaviour (Nallet et al., 2008). Motivated by short-term rewards, these drivers are generally unresponsive to punitive measures as a means of behavioural change. By contrast, psychological interventions, such as raising awareness of risk and the negative consequences of one's behaviour, may help reduce the perceived advantages of delinquent behaviour and therefore appear as effective measures for reducing recidivism.

The Safe System approach offers a broader perspective, from which rehabilitation programmes have the potential only to reduce recidivism, but not to prevent accidents entirely. Even if these programmes were to run indefinitely, there would still be road fatalities (caused by first-time offenders, non-recidivists, or erring drivers). According to the current Safe System paradigm, knowledge gained from rehabilitation activities should also inform system-level adjustments, so that "risky" drivers cannot cause death or serious injury. In other words, it should be leveraged for broader preventive measures, rather than endlessly correcting recidivists and "putting out fires" caused by a poorly designed system.

Theoretical background

Currently, the main theoretical foundation for rehabilitation programmes is the GDE (Goals for Driver Education) model, also referred to as Guidelines for Driver Education (Hatakka & Keskinen, 1999; Peräaho et al., 2003). Expanding and better conceptualising previous models introduced by Rasmussen (1983) and Michon (1985), the GDE model makes it possible to address driver behaviour from a hierarchical perspective. GDE was developed and refined in the late 1990s (Hatakka & Keskinen, 1999) and comprehensively published three years later (Hatakka et al., 2002). It divides driver behaviour into four levels, with the core assumption that higher levels control and direct behaviours at lower levels. The levels are interrelated and interdependent, with changes at one level resulting in changes at others (particularly downwards). The higher levels (Level 4 in the model) provide the foundation for lifestyle in general as well as behaviour in a specific traffic context. In terms of effecting change, Level 4 is therefore more stable and significant in comparison to the lower levels. To simplify, the first two levels (the lower ones) can be associated with abilities or skills, while the higher two levels relate to will and motivation. This represents the difference between "skill" and "will", that is, between what the driver is capable of doing and what they are willing to do. A driver's abilities and knowledge are thus of relatively little importance; what matters is how they apply them. This difference is evident with the elderly, for example, whose skills may be lower, but their will to drive within their capabilities is stronger. For rehabilitation programmes, recognising the distinction between ability and will is crucial, as is understanding the interconnectedness and hierarchical nature of the four levels in the GDE model.

The first level (vehicle control) represents a necessary but not sufficient capacity for safe driving. It encompasses a range of skills needed to operate a vehicle, including steering, accelerating, braking, shifting gear, evasive manoeuvres, and mirror use. Safe driving requires full control over the vehicle. The second level concerns the mastering of traffic situations, which, again, is a necessary but not sufficient precondition of safe driving. It involves a set of skills – such as anticipation and adaptation, interaction, risk perception, and knowledge of traffic rules – that help drivers navigate road traffic and manage specific traffic situations. Like the first level, this is also strongly shaped by experience. The third level covers the goals and context of driving.

It includes the purpose and motivation for a journey and the environment and circumstances of the drive (e.g. time of day, passengers, cargo, route, and the driver's physical and mental condition). The choice of circumstances under which driving takes place may result from a conscious decision or a habit. Finally, the fourth level refers to motives and life goals and skills. The vehicle is viewed as a means of achieving one's broader goals for life. The way it is used reflects the individual's lifestyle and is in accord with it. According to GDE, lifestyle and individuality (including self-control, social background, gender, age, group affiliation, and social factors which all influence one's attitudes, motives, and behaviour) are the key factors for safe driving. This level thus occupies a central position in relation to the others. Abilities, skills, impairments, or external constraints are not as crucial in themselves as the awareness and acknowledgement of the limitations they imply. The principles of GDE explain why rehabilitation programmes place so much emphasis on addressing lifestyle and planning the driving context, rather than merely focusing on the skills needed to control the car.

GDE thus serves as an important general framework for rehabilitation programmes, although it is not their sole theoretical foundation. In addition to this theory, the programmes also draw on insights from other psychological approaches and research. One of the most prominent among these is the Transtheoretical Model (TTM) (Prochaska & DiClemente, 2005), also known as the stages of change. TTM works with individuals' readiness to change their behaviour towards a new or healthier approach, describes the processes of change, and provides strategies to guide clients through its defined stages (Prochaska & DiClemente, 2005; Prochaska et al., 2008). The five core stages are: pre-contemplation (the client is not ready to change), contemplation (awareness of the problematic nature of one's behaviour and consideration of change), preparation (readiness for change, laying the groundwork for change), action (behaviour change), and maintenance (stabilising the change and preventing relapse). General evaluations of psychotherapy programmes based on TTM (particularly in the area of addiction) show considerable variation in effectiveness, as do evaluations of specific programmes for drivers (see, for example, Friman et al., 2017). As with conventional rehabilitation programmes, it remains unclear whether the final effect reflects the guiding principles of the programme or its specific implementation. Other theoretical concepts used in rehabilitation programmes include the health action process approach (HAPA) (Schwarzer, 2008) and cognitive dissonance theory (Festinger, 1957). Nevertheless, the programmes derive their theoretical grounding and effectiveness primarily from GDE and TTM.

Effectiveness of rehabilitation programmes

In theory (as discussed in the previous section), rehabilitation programmes should be effective. Is this the case in practice? When we exclude drivers who lack the potential to control their behaviour, the effectiveness of rehabilitation programmes is relatively high. This refers to effectiveness in reducing recidivism rates and preventing reoffending (see Watson et al., 2015). Attitudinal changes, which carry some evaluative potential, can also be observed (see, e.g. Parker et al., 1995). Changes in knowledge have sometimes been demonstrated, which is particularly relevant in areas such as promoting the separation of driving from alcohol and drug use. In educational courses, this is the primary goal of the intervention. Finally, changes in emotional experiences or prosocial behaviour – which have a strong influence on driving behaviour – can be observed (see, e.g. Trick et al., 2012).

Evaluation research (e.g. --Bartl et al., 2002; Robertson et al., 2013) suggests that rehabilitation programmes hold significantly greater potential for achieving desirable behavioural change in high-risk drivers than punitive measures. For example, custodial sentences have limited effectiveness, as a large proportion of those who are imprisoned reoffend and are reincarcerated within three years (Vito et al., 2015) – sanctions alone offer little corrective potential for individuals with a risky lifestyle. Group-based rehabilitation programmes have been reported to reduce recidivism by an average of a "magic" 50% (Bartl et al., 2002). Zámečník et al. (2018) note that the rehabilitative potential of the overall high-risk driver population depends heavily on external factors: the likelihood of being caught, the intensity of the programme, the number of drivers referred to it, and the curriculum of the course, including its length, the instructors, and other aspects. Moreover, the role of what is measured to assess effectiveness is important, whether it is offences or accidents, attitudinal change or increased knowledge, and the length of follow-up time frames over which these outcomes are monitored. For these reasons, the figures reported across studies often vary.

Regarding programmes aimed at drink-driving, most studies evaluating psychotherapeutically oriented courses report very positive outcomes, in terms of both recidivism and alcohol-related accidents (DeYoung, 1997; Pratt et al., 2000). In particular, this applies to courses accompanied by additional measures, such as licence suspension (Freeman & Liossis, 2002) or the use of alcohol interlocks. In such cases, recidivism among the target group has been reduced by an average of 60%, and responsibility for traffic accidents by as much as 80% (Bjerre & Thorsson, 2008). Importantly, these must be courses that are grounded in therapeutic intervention (Bartl et al., 2002). The high effectiveness of such approaches is further supported by Schützenhöfer and Krainz (1999), Davies et al. (1999), Jones et al. (1996), and Winkler et al. (1990). The most extensive evaluation studies were carried out by Bartl et al. (2002), who reported an average reduction of recidivism by 50%, and as part of the European DRUID project, which found an average reduction of 45.5% (ranging from 15% to 71%).

Courses focused on non-alcohol-related offences have also shown high effectiveness. In these cases, the reduction in recidivism is more challenging to measure, as the drivers involved are not associated with a single predominant offence (e.g. they may become disqualified after reaching the demerit point limit or committing multiple serious offences). In their meta-analysis, Masten and Peck (2004) report an average reduction in recidivism of 8%. Similarly, low effectiveness was found by Elvik et al. (2009). However, these low figures may result from the inclusion of interventions without psychological or therapeutic components (such as warning letters or discussions with professionals) among rehabilitation programmes. This is also how the findings of these studies were interpreted. Although some analyses show no significant positive effect as a result of their confounding psychological and non-psychological programmes, later research that has addressed this limitation indicates a major improvement in effectiveness (Hufnagl, 2007). For instance, a positive effect can be observed in compliance with speed limits, including changes in attitudes towards speeding (Delhomme et al., 2008; Fylan & Stradling, 2014; Elliott et al., 2003). One evaluation refers to as many as 90% of drivers who reported having changed their behaviour as a result of the course (Fylan & Stradling, 2014).

If set up correctly, rehabilitation programmes can be considered highly successful in terms of their effectiveness (i.e. lower recidivism rates). If not, their effectiveness is limited. These are, of course, the two extremes. In practice, a continuum of more or less effective approaches exists, reflected in the wide variability of results in evaluation studies (see, for example, the European projects ANDREA, BESTPOINT, DRUID, and SUPREME). It can be concluded that if they adhere to psychotherapeutic approaches, are delivered by psychologists, and allow for individualised content and engagement, the programmes can lead to a 50% reduction in recidivism. The specific percentage figure, however, should be interpreted with caution, as it represents an average – an inadequate metric for capturing the effects of such a variable and comprehensive intervention. Evaluations always concern specific applications of the programme, and there is a significant distinction between the method and the way it is implemented. The same method, when applied to different participants, by different facilitators, and within varying sociocultural contexts, yields different results. The motivation of the facilitators is a crucial factor, for example, as it is through them that the programme is realised. The final effectiveness is also critically influenced by the nature of the evaluation system – for example, whether an offence different from the one that led to the driver's referral to the programme also counts as recidivism. The length of the follow-up period is also important, as it dramatically affects exposure to risk (in terms of both offending and being caught). Some programme graduates do not immediately resume driving after completing rehabilitation. The effectiveness may thus appear high in statistical terms, while the reality may differ. However, bias can also occur in the opposite direction: the reoffending rate may appear higher because one of the aims of rehabilitation programmes is to promote personal responsibility. This leads to less defiance towards the system, which affects how guilt or shared responsibility is determined. These factors, in turn, influence the technical calculation of reoffending rates.

In addition to effectiveness in reducing recidivism, it is also important to focus on other benefits. Rehabilitation programmes have a number of significant secondary effects that greatly enhance their potential. These include improvements in knowledge and attitudes related to road safety, a reduction in external attribution, better ability to manage stressful situations, a positive feedback experience, the development of healthier aspirations and attitudes, more realistic self-assessment, greater awareness of the problem, and the adoption of positive coping strategies (Zámečník et al., 2018). Additionally, changes in problem denial, readiness for change, reductions in impulsivity and risk-taking scores, alleviation of sleep problems, and lower rates of offending have been observed. Last but not least, there is improvement both in separating drinking and driving and in quality of life. All of these aspects also have a positive impact on society and public health. Psychotherapy-centred courses, in particular, offer broader benefits, as they target fundamental life beliefs and lifestyles. Drivers participating in the programmes learn coping strategies to manage emotionally charged or challenging situations, draw on support from family and friends, and recognise the benefits of following rules. This is subsequently reflected not only in specific offences but also in driving behaviour and life in general.

When considering rehabilitation programmes as part of road safety management, it is important not to apply this label to measures that lack corrective potential. These include warning letters or safe driving courses incorporated into driver education and targeting only the first two levels of the GDE matrix. Such measures may even be counterproductive (OECD/ECMT, 2006) and provide no added value to road safety.

Note

1 That does not mean, however, that individual characteristics are completely disregarded. For example, learning style and motivation are taken into account under the Risk-Need-Responsivity (RNR) principle. The point is simply that the "trait riskiness" is not placed at the core of the intervention.

References

Barjonet, P. E., & Tortosa, F. (2001). Transport psychology in Europe: A historical approach. In P. E. Barjonet (Ed.), *Traffic psychology today* (pp. 13–29). Springer. https://doi.org/10.1007/978-1-4757-6867-1_2

Bartl, G., Assailly, J. P., Chatenet, F., Hatakka, M., Keskinen, E., & Willmes-Lenz, G. (2002). EU-project "Andrea": Analysis of driver rehabilitation programmes. *Final report*. Kuratorium für Verkehrssicherheit. https://road-safety.transport.ec.europa.eu/system/files/2021-07/andrea_final_report.pdf

Bingham, C. R., & Shope, J. T. (2004). Adolescent developmental antecedents of risky driving among young adults. *Journal of Studies on Alcohol*, *65*(1), 84–94. https://doi.org/10.15288/jsa.2004.65.84

Bjerre, B., & Thorsson, U. (2008). Is an alcohol ignition interlock programme a useful tool for changing the alcohol and driving habits of drink-drivers? *Accident Analysis & Prevention*, *40*(1), 267–273. https://doi.org/10.1016/j.aap.2007.06.008

Chapman, A. L. (1954). An epidemiological approach to traffic safety. *Public Health Reports (1896)*, *69*(8), 773–775. www.ncbi.nlm.nih.gov/pmc/articles/PMC2024269/

Dahlen, E. R., Martin, R. C., Ragan, K., & Kuhlman, M. M. (2005). Driving anger, sensation seeking, impulsiveness, and boredom proneness in the prediction of unsafe driving. *Accident Analysis & Prevention*, *37*(2), 341–348. https://doi.org/10.1016/j.aap.2004.10.006

Davies, G. P., Harland, G., & Broughton, J. (1999). *Drink/driver rehabilitation courses in England and Wales* (TRL Report 426). Transport Research Laboratory.

Delhomme, P., Kreel, V., & Ragot, I. (2008). The effect of the commitment to observe speed limits during rehabilitation training courses for traffic regulation offenders in France. European Review of Applied Psychology, 58(1), 31–42. https://doi.org/10.1016/j.erap.2005.07.005

DeYoung, D. J. (1997). An evaluation of the effectiveness of alcohol treatment, driver license actions and jail terms in reducing drunk driving recidivism in California. *Addiction*, *92*(8), 989–997.

Elliott, M. A., Armitage, C. J., & Baughan, C. J. (2003). Drivers' compliance with speed limits: An application of the theory of planned behavior. *Journal of Applied Psychology*, *88*(5), 964–972. https://doi.org/10.1037/0021-9010.88.5.964

Elvik, R., Høye, A., Vaa, T., & Sørensen, M. (2009). *The handbook of road safety measures* (2nd ed.). Emerald Insight.

Feng, J., & Donmez, B. (2013). Design of effective feedback: Understanding driver, feedback, and their interaction. Driving Assessment Conference, 7, 404–410. https://doi.org/10.17077/drivingassessment.1519

Festinger, L. (1957). *A theory of cognitive dissonance*. Stanford University Press.

Freeman, J., & Liossis, P. (2002). Drink driving rehabilitation programs and alcohol ignition interlocks: Is there a need for more research? *Road and Transport Research*, *11*(4), 3–13.

Friman, M., Huck, J., & Olsson, L. E. (2017). Transtheoretical model of change during travel behavior interventions: An integrative review. International Journal of Environmental Research and Public Health, 14(6), 581. https://doi.org/10.3390/ijerph14060581

Fylan, F., & Stradling, S. (2014). Behavioural change techniques used in road safety interventions for young people. *European Review of Applied Psychology*, *64*(3), 123–129. https://doi.org/10.1016/j.erap.2014.02.003

Gulliver, P., & Begg, D. (2007). Personality factors as predictors of persistent risky driving behavior and crash involvement among young adults. *Injury Prevention*, *13*(6), 376–381. https://doi.org/10.1136/ip.2007.015925

Hatakka, M., & Keskinen, E. (1999). Analysis of matrix. In G. Bartl, R. Cummins, C. Evers, M. Hatakka, E. Keskinen, M. Pannacci, & G. Stummvoll (Eds.), *Final report of European Commission project DAN*. Kuratorium für Verkehrssicherheit.

Hatakka, M., Keskinen, E., Gregersen, N. P., Glad, A., & Hernetkoski, K. (2002). From control of the vehicle to personal self-control: Broadening the perspectives to driver education. *Transportation Research Part F: Traffic Psychology and Behaviour*, *5*(3), 201–215. https://doi.org/10.1016/S1369-8478(02)00018-9

Hollnagel, E. (2004). *Barriers and accident prevention*. Routledge.

Hubicka, B., Laurell, H., & Bergman, H. (2008). Criminal and alcohol problems among Swedish drunk drivers – predictors of DUI relapse. International Journal of Law and Psychiatry, 31(6), 471–478. https://doi.org/10.1016/j.ijlp.2008.09.003

Hufnagl, G. (2007). *SUPREME: Summary and publication of best practices in road safety in the member states*. Proceedings of the European Transport Conference 2007, Leiden, The Netherlands.

Iversen, H., & Rundmo, T. (2002). Personality, risky driving and accident involvement among Norwegian drivers. *Personality and Individual Differences*, *33*(8), 1251–1263. https://doi.org/10.1016/S0191-8869(02)00010-7

Jonah, B. A. (1990). Age differences in risky driving. *Health Education Research*, *5*(2), 139–149. https://doi.org/10.1093/her/5.2.139

Jones, R. K., & Lacey, J. H. (2000). *State of knowledge of alcohol-impaired driving: Research on repeat DWI offenders*. National Highway Traffic Safety Administration.

Jones, R. K., Lacey, J. H., & Wiliszowski, C. H. (1996). *Evaluation of alternative programs for repeat DWI offenders*. Mid-America Research Institute. https://doi.org/10.21949/1524810

Lawton, R., Parker, D., Manstead, A. S. R., & Stradling, S. G. (1997). The role of affect in predicting social behaviors: The case of road traffic violations. *Journal of Applied Social Psychology*, *27*, 1258–1276. https://doi.org/10.1111/j.1559-1816.1997.tb01805.x

Machin, M. A., & Sankey, K. S. (2008). Relationships between young drivers' personality characteristics, risk perceptions, and driving behaviour. *Accident Analysis & Prevention*, *40*(2), 541–547. https://doi.org/10.1016/j.aap.2007.08.010

Marengo, D., Settanni, M., & Vidotto, G. (2012). Drivers' subtypes in a sample of Italian adolescents: Relationship between personality measures and driving behaviors. Transportation Research Part F: Traffic Psychology and Behaviour, 15(5), 480–490. https://doi.org/10.1016/j.trf.2012.04.001

Masten, S. V., & Peck, R. C. (2004). Problem driver remediation: A meta-analysis of the driver improvement literature. *Journal of Safety Research*, *35*(4), 403–425. https://doi.org/10.1016/j.jsr.2004.06.002

McCartt, A. T., Shabanova, V. I., & Leaf, W. A. (2003). Driving experience, crashes and traffic citations of teenage beginning drivers. Accident Analysis & Prevention, 35(3), 311–320. https://doi.org/10.1016/S0001-4575(02)00006-4

Michon, J. A. (1985). A critical view of driver behavior models: What do we know, what should we do? In L. Evans & R. C. Schwing (Eds.), Human behavior and traffic safety (pp. 485–524). Plenum Press. https://doi.org/10.1007/978-1-4613-2173-6_19

Moncel, C., Osmont, A., & Dauvier, B. (2025). Associations between the big five personality traits and everyday and experimental risk-taking: A literature review in adolescence and adulthood. *Personality and Individual Differences*, *236*. https://doi.org/10.1016/j.paid.2024.112982

Nallet, N., Bernard, M., & Chiron, M. (2008). Individuals taking a French driving licence points recovery course: Their attitudes towards violations. *Accident Analysis & Prevention*, *40*(6), 1836–1843. https://doi.org/10.1016/j.aap.2008.08.014

Nochajski, T. H., & Stasiewicz, P. R. (2006). Relapse to driving under the influence (DUI): A review. Clinical Psychology Review, 26(2), 179–195. https://doi.org/10.1016/j.cpr.2005.11.006

Organisation for Economic Co-operation and Development, & European Conference of Ministers of Transport (OECD/ECMT). (2006). *Young drivers: The road to safety*. OECD Publishing. https://doi.org/10.1787/9789282113356-en

Parker, D., Reason, J. T., Manstead, A. S., & Stradling, S. G. (1995). Driving errors, driving violations and accident involvement. *Ergonomics*, *38*(5), 1036–1048. https://doi.org/10.1080/00140139508925170

Parry, M. H. (1968). *Aggression on the road: A pilot study of behaviour in the driving situation*. Tavistock Publications.

Peck, R. C. (1993, May 6). The identification of multiple accident correlates in high-risk drivers with specific emphasis on the role of age, experience & prior traffic violation frequency: In international symposium on alcohol, drugs and driving, Santa Monica, California. *Alcohol, Drugs and Driving*, *9*(3–4), 145–166.

Peräaho, M., Keskinen, E., & Hatakka, M. (2003). *Driver competence in a hierarchical perspective: Implications for driver education*. University of Turku, Traffic Research.

Pratt, T. C., Holsinger, A. M., & Latessa, E. J. (2000). Treating the chronic DUI offender: "Turning point" ten years later. *Journal of Criminal Justice*, *28*(4), 271–281. https://doi.org/10.1016/S0047-2352(00)00046-5

Prochaska, J. O., & DiClemente, C. C. (2005). The transtheoretical approach. In J. C. Norcross & M. R. Goldfried (Eds.), *Handbook of psychotherapy integration* (pp. 147–171). Oxford University Press. https://doi.org/10.1093/med:psych/9780195165791.003.0007

Prochaska, J. O., Redding, C. A., & Evers, K. E. (2008). The transtheoretical model and stages of change. In K. Glanz, B. K. Rimer, & K. Viswanath (Eds.), *Health behavior and health education: Theory, research, and practice* (pp. 97–121). Jossey-Bass Inc Pub.

Rasmussen, J. (1983). Skills, rules, and knowledge; signals, signs, and symbols, and other distinctions in human performance models. *IEEE Transactions on Systems, Man, and Cybernetics*, *13*(3), 257–266. https://doi.org/10.1109/TSMC.1983.6313160

Reason, J. (1990). *Accident proneness: The history of an idea*. Cambridge University Press.

Rhodes, N., & Pivik, K. (2011). Age and gender differences in risky driving: The roles of positive affect and risk perception. *Accident Analysis & Prevention*, *43*(3), 923–931. https://doi.org/10.1016/j.aap.2010.11.015

Robertson, A. A., Gardner, S., Xu, X., Chi, G., & McCluskey, D. L. (2013). Mississippi's DUI offender intervention: 40 years of programming and research. *Journal of Offender Rehabilitation*, *52*(2), 138–155.

Sagberg, F., & Bjørnskau, T. (2006). Hazard perception and driving experience among novice drivers. Accident Analysis & Prevention, 38(2), 407–414. https://doi.org/10.1016/j.aap.2005.10.014

Schützenhöfer, A., & Krainz, D. (1999). Auswirkung von driver improvement-Massnahmen auf die Legalbewaehrung. *Zeitschrift für Verkehrsrecht*, *44*(4), 138–143.

Schulze, H., Schumacher, M., Urmeew, R., & Auerbach, K. (2012). *DRUID final report: Work performed, main results and recommendations*. European Commission

Schwartz, J. (2008). Gender differences in drunk driving prevalence rates and trends: A 20-year assessment using multiple sources of evidence. *Addictive Behaviors*, *33*(9), 1217–1222. https://doi.org/10.1016/j.addbeh.2008.03.014

Schwarzer, R. (2008). Modeling health behavior change: How to predict and modify the adoption and maintenance of health behaviors. *Applied Psychology*, *57*(1), 1–29. https://doi.org/10.1111/j.1464-0597.2007.00325.x

Simpson, H. M., Mayhew, D. R., & Beirness, D. J. (1996). *Dealing with the hard core drinking driver*. Traffic Injury Research Foundation.

Stein, L. A. R., & Lebeau-Craven, R. (2002). Motivational interviewing and relapse prevention for DWI: A pilot study. *Journal of Drug Issues*, *32*(4), 1051–1070.

Trick, L. M., Brandigampola, S., & Enns, J. T. (2012). How fleeting emotions affect hazard perception and steering while driving: The impact of image arousal and valence. Accident Analysis & Prevention, 45, 222–229. https://doi.org/10.1016/j.aap.2011.07.006

Twenge, J. M., Baumeister, R. F., Tice, D. M., & Stucke, T. S. (2001). If you can't join them, beat them: Effects of social exclusion on aggressive behavior. Journal of Personality and Social Psychology, 81(6), 1058–1069. https://doi.org/10.1037/0022-3514.81.6.1058

Ulleberg, P., & Rundmo, T. (2003). Personality, attitudes, and risk perception as predictors of risky driving behaviour among young drivers. *Safety Science*, *41*(5), 427–443. https://doi.org/10.1016/S0925-7535(01)00077-7

Underwood, G., Chapman, P., Wright, S., & Crundall, D. (1999). Anger while driving. Transportation Research Part F: Traffic Psychology and Behaviour, 2(1), 55–68. https://doi.org/10.1016/S1369-8478(99)00006-6

Vito, G. F., Higgins, G. E., & Tewksbury, R. (2015). The effectiveness of parole supervision: Use of propensity score matching to analyze reincarceration rates in Kentucky. Criminal Justice Policy Review, 28(7), 627–640. https://doi.org/10.1177/0887403415609717

Watson, B., Siskind, V., Fleiter, J. J., Watson, A., & Soole, D. W. (2015). Assessing specific deterrence effects of increased speeding penalties using four measures of recidivism. Accident Analysis & Prevention, 84, 27–37. https://doi.org/10.1016/j.aap.2015.08.006

Wells-Parker, E., Bangert-Drowns, R., McMillen, R. C., & Williams, M. (1995). Final results from a meta-analysis of remedial interventions with drink/drive offenders. *Addiction*, *90*(7), 907–926.

Williams, A. F. (2003). Views of U.S. drivers about driving safety. *Journal of Safety Research*, *34*(5), 491–494. https://doi.org/10.1016/j.jsr.2003.05.002

Wilsnack, S. C., & Wilsnack, R. W. (1991). Epidemiology of women's drinking. *Journal of Substance Abuse*, *3*(2), 133–157. https://doi.org/10.1016/S0899-3289(05)80033-1

Winkler, W., Jacobshagen, W., & Nickel, W. (1990). Zur Langzeitwirkung von Kursen für wiederholt alkoholauffällige Kraftfahrer: Untersuchung nach 60 Monaten Bewährungszeit. *Blutalkohol*, 27(3), 154–174.

Zámečník, P., Kurečková, V., Řezáč, P., Havlíčková, D., Shaw, D., Czekóová, K., Špiláková, B., & Salazar, M. (2018). *Rehabilitační programy pro řidiče*. Centrum dopravního výzkumu, v.v.i.

Zhao, L., Zhao, Y., Bu, L., Sun, H., Tang, W., Li, K., Zhang, W., Tang, W., & Zhang, Y. (2023). Design method of a smart rehabilitation product service system based on virtual scenarios: A case study. *IEEE Transactions on Neural Systems and Rehabilitation Engineering*, *31*, 4570–4579. https://doi.org/10.1109/TNSRE.2023.3333049

20 Post-accident psychological care

Martin Kořán

Introduction

This chapter provides a brief outline of the history of psychological care for survivors of road accidents suffering from post-accident stress reactions, post-accident driving phobia, or post-traumatic stress disorder (PTSD). It describes the psychological support provided by various institutions, including the Police of the Czech Republic, the Fire Rescue Service, the Czech Psychosocial Intervention Team, and the Psychosocial Intervention Service System of the emergency medical services, as well as by traffic and clinical psychologists. While short-term acute psychological reactions following an accident may subside spontaneously within days or weeks, some individuals may develop subsequent post-traumatic stress disorders accompanied by various psychosomatic conditions. In the immediate aftermath of an accident, victims should receive acute psychological care in the form of crisis intervention. To address the long-term psychological consequences of a road crash (such as post-accident driving phobia or PTSD), it is advisable to apply extended psychotherapeutic approaches, such as cognitive behavioural therapy (CBT), prolonged exposure therapy (PE), or psychedelic-assisted psychotherapy, all of which are described in this chapter.

Psychological impacts of traffic accidents

According to Červinka et al. (2023), serious road traffic accidents are the most common type of traumatic event in the context of Euro-American society. In the Czech Republic, serious road accidents (alongside rape) represent the most frequent source of post-traumatic stress disorders (Höschl et al., 2004). The scholarly literature on the psychological effects of traffic accidents on drivers and other road users, approached as major stress-inducing situations, is relatively extensive (e.g. Beck & Coffey, 2007; Brown et al., 1992, Chapter 11; Butler et al., 1999; Červinka, 2022; Kořán, 1993, 2013, 2016a; Meyer & Steil, 1998; Neusarová, 2009). A road accident occurs unexpectedly and not only results in significant damage to property and physical health but also poses a threat to the mental health of

DOI: 10.4324/9781003608394-20

those involved. Experts, particularly traffic psychologists, have studied the circumstances leading up to serious road accidents, as well as the immediate and long-term responses to accidents and the occurrence of both short-term and long-term post-accident reactive disorders. Survivors of serious road accidents often suffer from acute stress reactions, post-traumatic stress disorder (PTSD), or post-traumatic fear of driving (post-accident vehophobia), generalised anxiety, depression, fear of blame and reproach, feelings of guilt, concerns about punishment, and other related conditions. These states may be accompanied by intrusive thoughts and images, harrowing dreams, nightmares, flashbacks (recurrent or persistent intrusive mental images of the event with a compulsive nature), or various psychosomatic issues, such as cardiovascular disorders, gastrointestinal problems, eating disorders, and sleep disturbances (Ehring et al., 2006; Vágnerová, 2004). The potential consequences of a road accident can be either short-term or long-term. A short-term acute stress reaction, as a transitional mental disorder, may subside on its own within a few days. Many people can recover and cope with the consequences of an accident on their own within approximately four to six weeks, largely resuming their lives as they were before the accident. It is important to recognise that expressions of grief following a traumatic event are normal reactions to an abnormal situation, that is, explain to a client that *behaving normally in an abnormal situation is not normal*. Long-term post-traumatic stress disorder following a road accident can become entrenched, potentially leading to incapacity for work, avoidance of driving, social withdrawal, and the chronic persistence of physical symptoms. Červinka (2022) provides a comprehensive review of research studies indicating that the average prevalence of PTSD developed as a result of serious traffic accidents is 22% (ranging from 6.3% to 58.3%). These disorders affect both the at-fault parties and the victims of accidents, including the parents of children injured or killed in such events. Professional mental health assistance is typically sought by individuals who struggle to cope with the aftermath of such life events or who remain unable to come to terms with the consequences even years later. Mental health professionals recommend treating PTSD by intensive psychotherapy, including early crisis intervention and supportive psychotherapy, which can help prevent a more severe course of the disorder. Specific approaches to brief dynamic and cognitive-behavioural psychotherapy have been found to be effective (Raboch & Pavlovský, 2003).

Specific phobias

Individuals who have been involved in traffic collisions may develop specific phobias, primarily characterised by intense experiences of terror and fear. These driving-related phobias include accident phobia (dystychiphobia) and post-accident vehophobia (Neuman, 2021). *Vehophobia* refers to a severe and long-term post-accident fear of driving to an extent that affects an individual's quality of life and significantly impedes daily activities. Individuals suffering from this phobia or driving-related anxiety may experience the following conditions (Patterson, 2024; Stephens, 2021):

- *Intense fear of driving*
- *Feelings of distress near or at the accident site*
- *Even thinking of driving is paralysing*
- *Overwhelming fear of causing another traffic accident*
- *Fear of having a panic attack while driving*
- *Fear of hurting other road users while driving*
- *These thoughts may be impossible to dispel, despite conscious effort.*

These experiences are comparable to a panic attack or a severe anxiety disorder. There are driving schools that specialise in training individuals with this phobia (Curtis, 2022).

Post-accident care in Europe

According to a 2011 European Parliament resolution, post-accident care is considered one of the cornerstones of road safety in Europe. While the document primarily mentions medical care and trauma management provided by first responders at the accident scene and during transport and subsequent treatment, effective psychological care is also an integral component (European Parliament Resolution, 2011).

In 1997, the European Federation of Psychologists' Associations (EFPA) established a *Working Group on Crisis, Disaster, and Trauma Psychology, which became a Standing Committee in 2005* (Malikova & Saari, 2009). This group has been concerned with intervention and prevention efforts in relation to crisis and disaster situations, focusing particularly on managing the post-traumatic effects of various natural and human-caused disasters. The Czech Republic has been actively involved in this group and has, in recent years, collaborated closely with members of the EFPA *Standing Committee on Traffic Psychology (SCTP)*, a cooperation that was particularly significant in the early 21st century, when several terrorist attacks in Europe involved the use of transport systems (Kořán, 2011).

Since 2007, *The European Network for Traumatic Stress* (*TENTS* – www.tentsproject.eu) has been available in Europe to deal with the consequences of major road accidents and terrorist attacks involving means of transport, such as the 2004 Madrid train station bombings, the July 2005 London Underground and bus explosions, and recent vehicle-ramming attacks targeting unsuspecting crowds. One of the TENTS task forces, in which the Czech Republic also participated through the psychological team of the Ministry of the Interior, worked to develop a broad knowledge base for post-traumatic stress management to help victims of catastrophic events, including large-scale road accidents.

The Czech Republic is also involved in the *European Federation of Road Traffic Victims – Fédération Européenne des Victimes de la Route (FEVR)*, a non-governmental organisation comprising national associations of victims of road crashes and their families. Being a part of a network of similar organisations across Europe (fevr.org/post-crash-response), the FEVR seeks to provide free moral, *psychological, emotional*, legal, and social *support* to road accident survivors and their families, mainly through its member organisations (Kořán et al., 2007b).

Psychological first aid

A valuable resource for post-accident psychological support is the *Psychological First Aid, Field Operations Guide, 2nd Edition*, published by the National Center for PTSD in 2006. This handbook outlines a supportive intervention designed for use in the immediate aftermath of disasters or terrorist attacks. It provides step-by-step instructions for providers of psychological first aid, guiding them in conducting rapid assessments of survivors' immediate needs and implementing appropriate supportive measures. The objective of psychological first aid is to reduce the distress caused by a traumatic event and to foster both short-term and long-term adaptive coping mechanisms. It is primarily intended for working with individuals or families who have experienced a tragic event (such as the loss of a loved one, a road accident, or another catastrophic incident). The Guide articulates the following core actions that effective early psychological assistance should involve:

1. *Contact and engagement – approach survivors in a non-intrusive, compassionate, and helpful manner*
2. *Ensuring safety and comfort*
3. *Stabilisation – calm and orient emotionally overwhelmed or disoriented survivors*
4. *Gathering information about current needs and concerns that facilitates tailored psychological first aid interventions*
5. *Practical assistance – offer practical help to survivors in response to their immediate needs*
6. *Connection with social supports – help establish contacts with primary support networks, such as family, friends, or community resources*
7. *Information on stress reactions and coping strategies*
8. *Referral to collaborative services needed both at the time and in the future.*

Psychotherapy offers a range of effective approaches to coping with the psychological consequences of road accidents. These include *crisis intervention, trauma-focused cognitive behavioural therapy (TF-CBT)*, which helps people change their behavioural patterns, *eye movement desensitisation and reprocessing (EMDR)*, and *prolonged exposure therapy* (Foa et al., 2019; Popiel et al., 2015). In recent years, psychotherapy incorporating the use of psychedelics, such as MDMA (MethylenDioxyMethylAmphetamine), ketamine, or psilocybin, has been successfully trialled in the United States of America, the United Kingdom, and Australia for the treatment of PTSD, particularly in individuals who have experienced an extremely stressful event that they continuously relive in persistent intrusive memories, recurring thoughts, nightmares, and fantasies. While no studies specifically examine the effects of psychedelic-assisted psychotherapy on drivers or road accident survivors, numerous studies describe its effect on neuroplasticity, behaviour, and experiencing, for instance, in veterans of the US and British armed forces, firefighters, and first responders. Since spring 2025, psychedelic-assisted psychotherapy has been widely administered to Ukrainian soldiers with post-traumatic stress disorder (*psychedelicseurope.org*). This approach facilitates a more rapid transformation of psychological processes and responses to internal and external stimuli (Biscoe et al., 2023; Choi et al., 2024).

Crisis intervention

The general principles for the psychological management of the consequences of road accidents (whether for those responsible for the accident, victims, or their relatives) are outlined in most crisis intervention handbooks and have been successfully applied to post-accident care (Baštecká, 2005; Špatenková, 2011; Vodáčková et al., 2002). The purpose of crisis intervention is to mitigate the immediate situation and stabilise the individual (Špatenková, 2011). Špatenková refers to this form of psychological support as a time-limited therapeutic contact focused on the problem that triggered the distress. The key aspects of such interventions for road accident victims should include creating a friendly atmosphere, providing emotional support, being available without imposing, listening when the individual wishes to talk but not forcing them to, ensuring psychological stabilisation, reducing negative symptoms, and facilitating the resumption of adaptive functioning.

According to Špatenková (2011), the *specific characteristics of crisis intervention* are as follows:

- *Immediate assistance* (help should be provided as quickly as possible);
- *Risk reduction* (establishing emotional support and ensuring a sense of safety);
- *A focus on the problem "here and now"* (addressing the current situation and issue);
- *Time limitation* (up to six to ten sessions held regularly at least once per week; in the case of short-term hospitalisation, up to six to ten days or nights);
- *Intensive contact between the crisis interventionist* and the client (the frequency of interactions can be relatively high, sometimes even daily);
- *A structured, active, and sometimes even directive approach* on the part of the crisis interventionist (ranging from patient, empathetic listening to directive intervention in situations posing a threat to health or life);
- *An individualised approach* (as crisis is a subjective experience, different individuals may differ in their responses to the same crisis situations and, therefore, support must be tailored to attend to their individual needs).

A thorough account of crisis intervention procedures exceeds the scope of this chapter. This topic is explored in greater detail by Baštecká (2005), Špatenková (2011), and Vodáčková et al. (2002).

Cognitive behavioural therapy (CBT)

The CBT process is based on a learning model. According to Praško et al. (2007), the therapist's role in treating PTSD is to help the individual revisit memories of a traumatic event (e.g. using the Trauma Memory Questionnaire or a flashback record) while expressing the affective experience involved. In addition, the patient needs to be desensitised to traumatising memories of the traffic accident and allowed to regain a sense of control over their experiences.

At the beginning, the therapist and client establish a contract defining the baseline situation and specifying the goal to be achieved. In addition to cognitive work, CBT involves skills training and gradual exposure to feared stimuli (systematic desensitisation), either in the client's thoughts or in real life. Another technique is *flooding*, in which the client is exposed to the feared situation (e.g. experiences associated with the road accident) at its full intensity and remains in that state until their anxiety begins to diminish. CBT therapists help clients unlearn negative thought patterns and reactions to challenging situations in their lives (such as experiences associated with a road accident) and develop new, positive responses. Following a structured, step-by-step approach (e.g. orientation and emotional support, abreaction, education on PTSD, etc.), CBT helps break stress-inducing instances into smaller manageable components, setting and achieving short-term goals by working to modify the client's thoughts, emotions, and responses in distressing situations that trigger and maintain PTSD. The cognitive component (restructuring) focuses on modifying thought patterns that influence emotional experiences, while the behavioural component targets changes in the client's behaviour and lifestyle (Praško & Novák, 2007).

A comprehensive discussion of cognitive behavioural psychotherapy (CBT) procedures exceeds the scope of this chapter. Those seeking more detailed coverage of the topic are encouraged to refer to Becková (2018), Praško et al. (2007), Pešek et al. (2013), or Šlepecký et al. (2018).

Prolonged exposure therapy (PE)

Prolonged exposure therapy (PE) is a form of behavioural therapy and cognitive behavioural therapy intended for the treatment of PTSD. *PE* involves guiding the patient to revisit the traumatic event. Recalling the memory and reliving the fear it entails ultimately leads to the reprocessing of the experience (McLean & Foa, 2011). Two types of PE are distinguished: *imaginal* exposure and *in vivo* exposure. In imaginal exposure, the patient, guided by the therapist, reconstructs the road accident in a detailed, present-tense narration, while the therapist provides psychoeducation about the nature of trauma and trauma reactions. The patient's imaginal exposure account is recorded, and the emotions provoked by the imaginal exposure are then discussed by the patient and therapist during therapy sessions. Assisted by the therapist, the patient processes these emotions and practises controlled breathing.

During *in vivo exposure* (repeated confrontation with anxiety-provoking situations and objects), the patient faces trauma cues outside the therapist's office (in their home environment). The therapist and patient together identify potential cues and situations associated with trauma-related fear (compiling a list of stressful scenarious ranked by their expected level of distress) and agree on which cues to confront as part of *in vivo* exposure between sessions. The patient is encouraged to proceed stepwise to experience success in facing less fear-provoking stimuli and managing the associated emotions before moving to more challenging situations. The exposure must last long enough for corrective learning to occur. The goal of this treatment is to facilitate the processing of traumatic memories and reduce anxiety and avoidance by exposure to trauma-related reminders. For example, *in vivo* exposure can be effective in helping a road accident survivor to overcome fear and return to driving. Prolonged exposure therapy is a flexible treatment approach that can be tailored to individual needs and facilitates the psychological processing of traumatic events and the mitigation of trauma-induced mental disorders. A typical PE programme for PTSD treatment consists of between 8 and 15 individual 90-minute sessions. Prolonged exposure therapy leads to clinical improvement in 40–75% of patients with chronic PTSD (Foa et al., 2019; McLean & Foa, 2011). A controlled Polish trial evaluating the treatment of PTSD following a motor vehicle accident showed that prolonged exposure therapy was more effective in achieving PTSD remission than pharmacological treatment (Popiel et al., 2015).

McLean and Foa (2011) expected the future development of additional treatment methods benefiting from technological advancements, such as the use of virtual reality to deliver PE. Realistic virtual recreations of the client's traumatic experiences may enhance outcomes by facilitating the activation of emotional and cognitive structures, known to be an important factor for successful emotional processing. The use of virtual reality in psychological therapy for driving phobia (vehophobia) was explored by Elphinston et al. (2022). Having reviewed the existing evidence, they concluded that virtual reality (VR) is a technological system integrating three-dimensional computer graphics, auditory, and other sensory inputs to create a realistic interactive experience. VR-based procedures offer a number of advantages over traditional cognitive behavioural approaches and have the potential to improve access to psychological treatment. A key factor is the sense of realism. The latest AI-driven technology creates the illusion that the user is genuinely present in the simulated world, incorporating visual, auditory, and haptic sensations. For example, simulation technology eliminates the risks associated with real-life driving; simulation allows the therapist to control environmental factors such as weather conditions or traffic density. VR-assisted exposure therapy can be conducted remotely and outside therapy sessions, potentially improving access to treatment and reducing its cost. According to Elphinston et al. (2022), this line of research has the potential to increase the accessibility of psychotherapy for patients who have developed driving phobia (vehophobia) following motor vehicle accidents.

EMDR-assisted therapy

Guided by the adaptive information processing model, **eye movement desensitisation and reprocessing (EMDR)** was developed for the treatment of PTSD (Shapiro, 2007). EMDR is a controversial comprehensive individual psychotherapy typically conducted once or twice per week over a total of 6 to 12 sessions. While other therapeutic techniques focus on directly changing the emotions, thoughts, and reactions stemming from traumatic experiences, EMDR therapy targets memory itself, aiming to modify how traumatic memories are stored in the brain, thereby reducing or eliminating trauma symptoms. Clinical observations during EMDR therapy suggest that an accelerated learning process is stimulated by standardised EMDR procedures, which involve the use of eye movements and other forms of rhythmic bilateral stimulation (such as tones or finger taps by the therapist). When clients briefly focus on a traumatic memory while simultaneously attending to bilateral stimulation (BLS), the vividness and emotional intensity of the memory become reduced. In comparison to CBT, EMDR has shown comparable or even faster results. Ališahevič-Gelo and Hasanovič (2021) also confirmed the effectiveness of EMDR in individuals with traffic accident-related trauma who had not responded to pharmacological treatment.

In their meta-analysis, Davidson and Parker (2001) concluded that EMDR did not appear to be more effective than other exposure techniques in treating patients with PTSD.

The scope of this work does not allow for a more detailed description of the method. Specific manuals for EMDR-assisted treatment can be found on the American Psychological Association (APA) website: www.apa.org/ptsd-guideline/treatments/eye-movement-reprocessing.

Post-traumatic growth

Post-traumatic growth (PTG) is the ability to make use of a tragic situation in achieving a positive life change (e.g. improvements in interpersonal relationships, shifts in the value system, and a greater appreciation for life) or developing stronger emotional resilience (Agin-Liebes et al., 2024). Many individuals with such experience have re-evaluated their lives, initiating a process of psychological and relational healing. According to Tedeschi and Calhoun (2004), PTG encompasses the following major aspects:

- improvement in interpersonal relationships
- an increased sense of personal strength
- identification of new possibilities

- positive spiritual development
- greater appreciation of life

Some individuals who have lost a loved one in a road accident may channel their experience into active participation in helping the community (e.g. through organisations such as the Czech Association of Road Accident Victims), as supporting others in similar situations becomes a significant consequence of a personal tragedy (socialised non-reconciliation with a traumatic event). Group work with relatives of road accident victims can take place in the form of *debriefing* sessions (a crisis intervention method involving guided conversation aimed at coming to terms with a traumatic event) or self-help and support groups. Meeting other people who have personally experienced a road accident or lost a loved one in such an event can help survivors overcome their onerous distrust of the rest of the world, often voiced as "those who haven't been through it can't understand". This sentiment is frequently expressed by new members of group meetings of the Czech Association of Road Accident Victims which serve as a source of empathetic moral and psychological support, provided not only by fellow-victims and their relatives but also by attending psychologists.

Recommendation for helping clinicians

Individuals who work with traumatised victims of road accidents should be aware that they themselves are at risk of burnout, secondary post-traumatic reactions or post-traumatic stress disorders, or symptoms similar to those experienced by their clients. Therefore, they should engage in ongoing self-care focusing on both body and mind (e.g. supervision or consultations with experienced professionals, maintaining a balance between personal and professional life, stress compensation, pursuing hobbies and recreational activities, and finding the positive aspects of their work) (for more details, see Špatenková, 2011, p. 46).

References

Agin-Liebes, G., Ekman, E., Anderson, B., Malloy, M., Haas, A., & Woolley, J. (2024). Participant reports of mindfulness, posttraumatic growth, and social connectedness in psilocybin-assisted group therapy: An interpretive phenomenological analysis. *Journal of Humanistic Psychology*, *64*(4), 564–591. https://doi.org/10.1177/00221678211022949

Ališahevič-Gelo, E., & Hasanovič, M. (2021). The effectiveness of treating a traffic accident trauma with EMDR after failed psychofarmacological treatment: A case report. *Psychiatria Danubina*, *33*(1), 88–90.

Baštecká, B. (2005). *Terénní krizová práce: Psychosociální intervenční týmy* [Field crisis work: Psychosocial intervention teams]. Grada Publishing.

Beck, J. G., & Coffey, S. F. (2007). Assessment and treatment of posttraumatic stress disorder after a motor vehicle collision: Empirical findings and clinical observations. *Professional Psychology: Research and Practice*, *38*(6), 629–639. https://doi.org/10.1037/0735-7028.38.6.629

Becková, J. S. (2018). *Kognitivně behaviorální terapie* [Cognitive behavioural therapy]. TRITON.
Biscoe, N., Bonson, A., Slavin, M., Busuttil, W., Mac Manus, D., Cox, A., & Murphy, D. (2023). Psilocybin-assisted psychotherapy for treatment of PTSD in UK armed forces veterans: A feasibility study protocol. *European Journal of Trauma and Dissociation*, *7*(4), https://doi.org/10.1016/j.ejtd.2023.100359
Brown, T. M., Pullen, I. M., & Scott, A. I. (1992). *Emergentní psychiatrie* [Handbook of emergency psychiatry]. J. Kocourek-Psychoanalytické nakladatelství.
Butler, D. J., Moffic, H. S., & Turkal, N. W. (1999). Post-traumatic stress reactions following motor vehicle accidents. *American Family Physician*, *60*(2), 524–531.
Červinka, D. (2022). *Klíčové faktory rozvoje posttraumatické stresové poruchy u obětí vážných dopravních nehod* [Main aspects of PTSD development in serious road accident victims] [Master's thesis, Palacký University Olomouc]. https://theses.cz/id/611ymr/
Červinka, D., Vondráčková, L., Sulíková, B., & Adamovská, E. (2023). Posttraumatic stress disorder after a car accident in the Czech population: Prevalence and risk factors for its development. *Transactions on Transport Sciences*, *14*(3), 16–24. https://doi.org/10.5507/tots.2023.014
Choi, C., Johnson, D. E., Chen-Li, D., & Rosenblat, J. (2024). Mechanism of psilocybin on the treatment of posttraumatic stress disorder. *Journal of Psychopharmacology*, 1–5. https://doi.org/10.1177/02698811241286771
Curtis, D. (2022). *About me*. www.thedrivinganxietycoach.co.uk/about-me/
Davidson, P. R., & Parker, K. C. (2001). Eye movement desensitization and reprocessing (EMDR): A meta-analysis. *Journal of Consulting and Clinical Psychology*, *69*(2), 305–316. https://doi.org/10.1037//0022-006X.69.2.305
Ehring, T., Ehlers, A., & Glucksman, E. (2006). Contribution of cognitive factors to the predictions of post-traumatic stress disorder, phobia and depression after motor vehicle accidents. *Behaviour Research and Therapy*, *44*(12), 1699–1716. https://doi.org/10.1016/j.brat.2005.11.013
Elphinston, R. A., Vaezipour, A., Fowler, J. A., Russell, T. G., & Sterling, M. (2022). Psychological therapy using virtual reality for treatment of driving phobia: A systematic review. *Disability and Rehabilitation*, *45*(10), 1582–1594. https://doi.org/10.1080/09638288.2022.2069293
European Parliament Resolution. (2011). *European road safety: European Parliament resolution of 27 September 2011 on European road safety 2011–2020 (2010/2235(INI))*. https://eur-lex.europa.eu/legal-content/CS/TXT/?uri=CELEX%3A52011IP0408
Foa, E., Hembree, E. A., Rothbaum, B. O., & Rauch, S. (2019). *Prolonged exposure therapy for PTSD: Emotional processing of traumatic experiences – therapist guide (treatments that work)*. Oxford University Press.
Höschl, C., Libiger, J., & Švestka, J. (2004). *Psychiatrie* [Psychiatry] (2nd ed.). Tigris.
Kořán, M. (1993). *Úloha psychologa při zvládání následků dopravních nehod* [Final thesis of the postgraduate course in traffic psychology, Palacký University Olomouc].
Kořán, M. (2011). Stálá komise pro dopravní psychologii při Evropské federaci psychologických asociací – činnost a výhledy [Standing committee traffic psychology EFPA – activities and perspectives]. *E-Psychologie*, *5*(1), 78–80.
Kořán, M. (2013). Péče o účastníky nehod a jejich blízké [Care for survivors of road accidents and their loved ones]. In M. Šucha, V. Rehnová, M. Kořán, & D. Černochová (Eds.), *Dopravní psychologie pro praxi*. Grada.
Kořán, M. (2016). Následky dopravních nehod a ponehodová péče [Traffic accident consequences]. In M. Šucha, I. Vlčková, D. Černochová, P. Zámečník, & V. Rehnová (Eds.), *Dictionary of traffic psychology terminology* (pp. 182–183). Palacký University Olomouc.

Kořán, M., Vondruška, S., Vondrušková, M., Tisovská, D., & Civínová, E. (2007b). *Přehled systému péče o oběti dopravních nehod v evropských zemích* [Overview of the system of care for road accident victims in European countries]. Czech Association of Road Accident Victims.
Malikova, J., & Saari, S. (2009). (I98) standing committee on crisis, disaster, and trauma psychology: Goals within the European and worldwide network. *Prehospital and Disaster Medicine*, *24*(Suppl. 1), 137.
McLean, C. P., & Foa, E. B. (2011). Prolonged exposure therapy for post-traumatic stress disorder: A review of evidence and dissemination. *Expert Review of Neurotherapeutics*, *11*(8), 1151–1163. https://doi.org/10.1586/ern.11.94
Meyer, C., & Steil, R. (1998). Posttraumatic stress disorder after motor vehicle accidents. *Der Unfallchirurg*, *101*, 878–893. https://doi.org/10.1007/s001130050354
National Child Traumatic Stress Network. (2006). *Psychological first aid: Field operations guide* (2nd ed.). National Center for PTSD. www.nctsn.org/resources/psychological-first-aid-pfa-field-operations-guide-2nd-edition
Neuman, F. (2021). Driving phobia: An ideal treatment. *Psychology Today*. www.psychologytoday.com/us/blog/fighting-fear/201207/driving-phobia-ideal-treatment
Neusarová, J. (2009). Dopravní nehodovost a její následky na životy zasažených lidí. *Prevence úrazů, otrav a násilí*, 2, 179–184.
Patterson, E. (2024, 23. August). Fear of driving (vehophobia): Symptoms, causes and treatment. *Choosing Therapy*. www.choosingtherapy.com/fear-of-driving/
Pešek, R., Praško, J., & Štípek, P. (2013). *Kognitivně-behaviorální terapie v praxi: Pro terapeuty, studenty a poučené laiky* [Cognitive-behavioral therapy in practice: For therapists, students, and informed laypersons]. Portál.
Popiel, A., Zawadzki, B., Pragłowska, E., & Teichman, Y. (2015). Prolonged exposure, paroxetine and the combination in the treatment of PTSD following a motor vehicle accident: A randomized clinical trial – the "TRAKT" study. *Journal of Behavior Therapy and Experimental Psychiatry*, *48*, 17–26. https://doi.org/10.1016/j.jbtep.2015.01.002
Praško, J., Možný, P., & Šlepecký, M. (2007). *Kognitivně behaviorální terapie psychických poruch* [Cognitive-behavioral therapy for mental disorders]. Triton.
Praško, J., & Novák, T. (2007). Kognitivně behaviorální terapie posttraumatické stresové poruchy [Cognitive-behavioural therapy for post-traumatic stress disorder]. In J. Praško, P. Možný, & M. Šlepecký (Eds.), *Kognitivně behaviorální terapie psychických poruch* [Cognitive-behavioural therapy for mental disorders]. Triton.
Raboch, J., & Pavlovský, P. (2003). *Psychiatrie* [Psychiatry]. Triton.
Shapiro, F. (2007). EMDR, adaptive information processing, and case conceptualization. *Journal of EMDR Practice and Research*, *1*(2), 68–87. https://doi.org/10.1891/1933-3196.1.2.68
Šlepecký, M., Praško, J., Kotianová, A., & Vyskočilová, J. (2018). *Třetí vlna v kognitivně-behaviorální terapii: Nové směry*. Portál.
Špatenková, N. (2011). *Krizová intervence pro praxi* [Crisis intervention in practice]. Grada.
Stephens, J. (2021). Treating post-accident vehophobia – the fear of driving. *Stephens Law*. www.stephenslaw.com/blog/treating-post-accident-vehophobia-the-fear-of-driving/
Tedeschi, R. G., & Calhoun, L. G. (2004). Posttraumatic growth: Conceptual foundations and empirical evidence. *Psychological Inquiry*, *15*(1), 1–18. https://doi.org/10.1207/s15327965pli1501_01
Vágnerová, M. (2004). *Psychopatologie pro pomáhající profese* [Psychopathology for helping professions] (3rd ed.). Portál.
Vodáčková, D., Špatenková, N., Ševčík, O., & Kol. (2002). *Krizová intervence* [*Crisis intervention*]. Portál.

21 Autonomous mobility and psychology issues

Vít Gabrhel and Jakub Štefec

Introduction

Autonomous vehicles (AVs) – that is, vehicles which do not require the presence of a human driver – are not yet a common feature of everyday traffic. Nevertheless, the very notion that it might be the case in the future is already arousing both interest and concern among the public. It is therefore no surprise that key stakeholders from both the public and private sectors are seeking to understand perceptions of AVs and the degrees of awareness of the topic across various population groups, ranging from the general public (Zmud et al., 2016) to older adults (Casley et al., 2013) and transport experts (Begg, 2014). Attitudes towards new technologies and willingness to adopt them clearly involve psychological dimensions that must not be overlooked during the implementation of AV technology. One such dimension is trust (Becker & Axhausen, 2017), especially in technologies such as autonomous vehicles, which are expected to have significant implications for an aspect of daily life that concerns everyone: mobility.

However, the relevance of AVs to the field of traffic psychology extends well beyond public attitudes or willingness to adopt the technology. Automated driving systems also pose new challenges for vehicle ergonomics and driver cognition and motor control (Ko et al., 2019), for example, in terms of the reaction time required for a human operator to resume control ("fallback"). Even the fundamental elements of current transport systems need to be re-examined as a result of the changes introduced by AVs. For example, how will AVs interact with vulnerable road users such as pedestrians and cyclists? How will this interaction reshape traffic infrastructure? Might a shift away from private car ownership towards shared mobility services be anticipated? Will the car, which is currently perceived as a private possession, come to be regarded instead as a shared service or app-based platform?

Although the above questions might seem somewhat distant from present-day realities, it already makes sense to begin considering possible answers. While projections for the widespread deployment of AVs range from several years to a few decades, technologies enabling the automation

DOI: 10.4324/9781003608394-21

of certain driving tasks are already being implemented in both road and rail transport (Orlovska et al., 2020). These include advanced driver assistance systems (ADAS), such as cruise control, automated parking assistance, and Tesla Autopilot. As regards rail transport, passengers can already travel on driverless metro or underground lines in a number of cities worldwide, including Copenhagen, Shanghai, Singapore, Barcelona, and Vancouver.

Before discussing the domains relevant to traffic psychology in which the influence of AVs can be expected, it may be useful first to define what constitutes an autonomous vehicle.

Conceptualisation of autonomous vehicles

The Society of Automotive Engineers (*SAE*) has standardised the *Levels of Driving Automation* to clarify the distinct stages of automated vehicle control (Smith, 2013). In fact, this classification has become the framework through which legislators, researchers, and other stakeholders – in both the public and private sectors – approach the topic. The notion of AVs adopted in the present chapter is based on this conceptual framework.

The SAE taxonomy (SAE J3016; Smith, 2013) defines a total of six levels of driving automation. These levels represent qualitatively distinct categories rather than a continuous spectrum. At one end, *Level 0* (i.e. *no automation*) assumes full human responsibility for executing the dynamic driving task (DDT), which includes steering, acceleration, and braking. At the other extreme, *Level 5* (i.e. *full automation*) refers to a system capable of performing all aspects of the DDT under all conditions and in all driving scenarios (whether on a motorway or in urban environments).

Between these end points, there are several levels involving various degrees of automation. For instance, *Level 4* (*high automation*) allows human intervention under specific circumstances, such as adverse weather conditions that impair AV sensor performance. However, at this level, the automated driving system must be capable of performing a fallback manoeuvre to safely assume control of the DDT in the event that the human operator fails to respond to an emergency takeover request (Chaloupka & Risser, 2020; Emzivat et al., 2017; Smith, 2013), even if this entails bringing the vehicle to a complete stop (Yu & Luo, 2019). The Levels of Driving Automation taxonomy is summarised in Figure 21.1.

Specific areas of relevance to traffic psychology where the impact of AVs can be anticipated will now be considered. For the purposes of this chapter, attention will be focused on two broadly defined domains: road safety (including issues related to a human operator's *cognitive* and *motor* functioning and *vulnerable road users*) and the socio-psychological aspects of AVs (including *attitudes* and *willingness to use* AVs, as well as the *identity* of the driver).

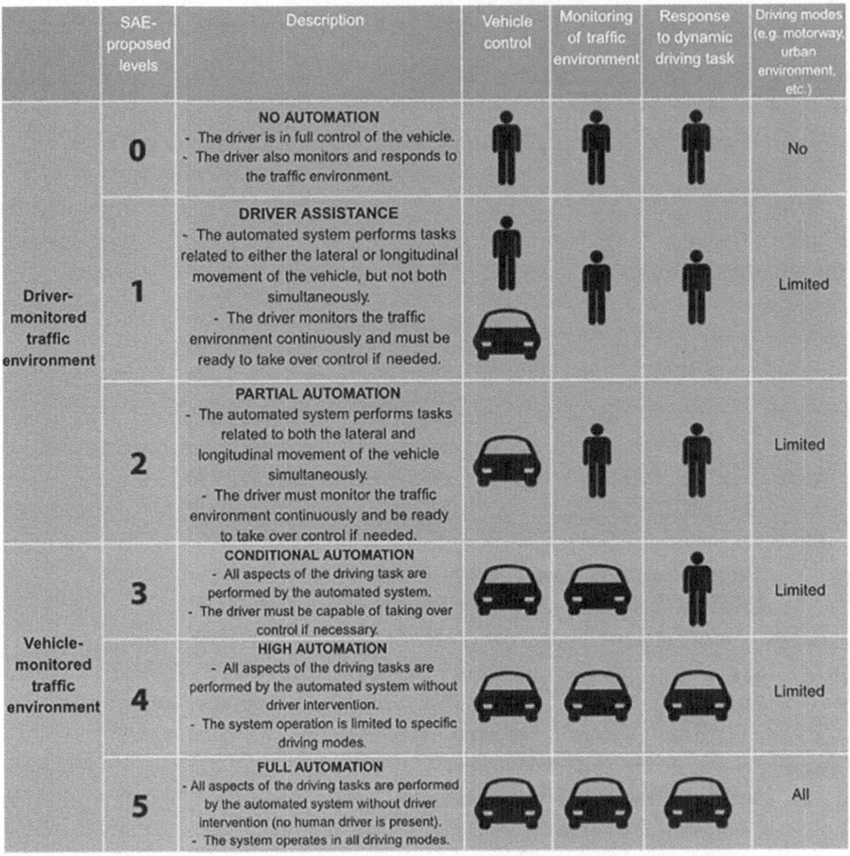

	SAE-proposed levels	Description	Vehicle control	Monitoring of traffic environment	Response to dynamic driving task	Driving modes (e.g. motorway, urban environment, etc.)
Driver-monitored traffic environment	0	**NO AUTOMATION** - The driver is in full control of the vehicle. - The driver also monitors and responds to the traffic environment.				No
	1	**DRIVER ASSISTANCE** - The automated system performs tasks related to either the lateral or longitudinal movement of the vehicle, but not both simultaneously. - The driver monitors the traffic environment continuously and must be ready to take over control if needed.				Limited
	2	**PARTIAL AUTOMATION** - The automated system performs tasks related to both the lateral and longitudinal movement of the vehicle simultaneously. - The driver must monitor the traffic environment continuously and be ready to take over control if needed.				Limited
Vehicle-monitored traffic environment	3	**CONDITIONAL AUTOMATION** - All aspects of the driving task are performed by the automated system. - The driver must be capable of taking over control if necessary.				Limited
	4	**HIGH AUTOMATION** - All aspects of the driving tasks are performed by the automated system without driver intervention. - The system operation is limited to specific driving modes.				Limited
	5	**FULL AUTOMATION** - All aspects of the driving tasks are performed by the automated system without driver intervention (no human driver is present). - The system operates in all driving modes.				All

Figure 21.1 Vehicle autonomy levels according to the SAE International.

Source: adapted from Havlíčková et al. (2021).

Road safety

Background

Existing research into road safety finds human error to be the main cause of traffic accidents (e.g. Gouy et al., 2012; Sabey & Taylor, 1980; Treat, 1979). According to the U.S. National Highway Traffic Safety Administration (NHTSA), the federal agency responsible for road safety, as many as 94% of serious crashes are attributed to human error (Allahyar et al., 2016). For instance, human error may involve speeding because of overconfidence in one's own driving abilities (Delhomme et al., 2009).

Experts in road safety therefore assume that by reducing or eliminating human error through the widespread adoption of autonomous vehicles, the number of traffic accidents will decline. It is worth noting that similar expectations are shared by a substantial proportion of people across the globe (Schoettle & Sivak, 2014).

Cognition, motor skills, and ergonomics

The anticipated benefits of AVs are accompanied by expected challenges arising from the expansion of this technology. One such challenge specifically concerns road safety. As shown in Figure 21.1, automation levels 0 to 3 assume the human driver's ability to intervene manually (e.g. in reaction to the failure of an autonomous driving system). Will drivers who become accustomed to relying on the autonomous system be able to respond flexibly and take control of the vehicle when the situation so requires? Is there a risk that driving skills may deteriorate as a result of automated driving use, which would further impair the driver's ability to regain control when necessary (Havlíčková et al., 2020)?

These questions are relevant because one of the presumed advantages of driving automation from the human operator's perspective lies in the ability to engage freely in other activities (secondary to driving), such as working, sleeping, making phone calls, or playing video games, that it provides them with. And while there are drivers who continue to monitor traffic despite this opportunity, it is generally observed that the tendency to engage in non-driving-related tasks rises with increasing levels of vehicle automation (Carsten et al., 2012; Jamson et al., 2013). The low error rate and high level of reliability of the autonomous driving system may thus reduce the driver's readiness to react and assume manual control if needed.

Simulator studies (Young & Stanton, 2007) and naturalistic driving studies (Rudin-Brown & Parker, 2004) have identified reduced situational awareness in drivers of vehicles equipped with ADAS features such as automated parking, adaptive cruise control, or lane centring. Significantly, lower situational awareness can lead to delayed reaction times when braking (Young & Stanton, 2007) or when needing to change lanes because of a failure of lane-keeping systems (Desmond et al., 1998). Inappropriate trust in the autonomous driving system may therefore become a source of error and heightened risk in certain situations (Gouy et al., 2012; Johansson, 2014).

On the other hand, the degree of reduced situational awareness, that is, the extent to which drivers stop monitoring road traffic, varies. It depends partly on the situational demands (e.g. motorway driving vs. city centre driving), but also on the nature of the secondary activities (reading a book, using a smartphone, or sleeping), or the length of time spent driving in automated mode (Funkhouser & Drews, 2016). In the study by Jamson et al. (2013), drivers tended to pay more attention to traffic when traffic intensity increased, despite being in a highly automated vehicle. This finding is supported by earlier research (e.g. Bainbridge, 1983), which links attention to traffic with driver activation levels, suggesting that merely observing an autonomous system is less engaging and satisfying than active driving.

A potential solution lies in vehicle ergonomics designed in such a way as to support the driver's optimal level of activation. In-vehicle sensors and communication with the infrastructure (*vehicle-to-infrastructure* or V2I) or with other autonomous vehicles (*vehicle-to-vehicle* or V2V) may provide and integrate information otherwise unavailable to the human driver, thereby enhancing the driver's level of activation – for example, when driving in a queue with limited forward visibility (Aria et al., 2016). However, in order for the human operator to work with this information, it must be presented in a comprehensible format, sufficiently salient to attract the operator's attention, and updated regularly (Norman, 1990; Seppelt & Lee, 2007). Previous research has, for instance, identified differences in reaction time depending on whether a take-over request (TOR) was delivered through speech or non-speech audio cues (Ko et al., 2019).

Nevertheless, the above issues also have a legislative and technological solution, involving the implementation of vehicles with high (Level 4) or full (Level 5) automation.

Vulnerable road users

AVs present a major challenge to the current concept of road safety and mobility in a broader sense. While AVs offer prospects for enhancing the safety of vulnerable road users (Wang et al., 2020), the dynamics of mutual interactions may generate unintended systemic consequences. At present, pedestrians and cyclists, as vulnerable road users, are typically more motivated than drivers to comply with traffic regulations, particularly in situations that are critical for safety (Peters, 2006). Unlike human drivers, AVs are designed to drive more cautiously and to follow traffic rules strictly, even in contexts where human drivers may not. A general awareness of the principles guiding AV performance may thus lead pedestrians or cyclists to act less predictably, regardless of whether such behaviour involves a violation of the law. For instance, if cyclists were to ignore a "Stop, give way!" sign, or if pedestrians were to enter the carriageway outside designated crossing areas, AVs – trained on data from such environments – would be incentivised to drive more cautiously (Deb et al., 2017). This is because, in the process of *machine learning*, AV behaviour is reinforced on the basis of what is coded as desirable, such as avoiding collisions with pedestrians. It is therefore expected that AVs will adapt their behaviour to changing conditions in accordance with the set goal(s). As a result, cyclists may feel entitled to "claim the lane" and make use of road infrastructure that lacks dedicated cycling elements (e.g. cycle lanes) to an even greater extent than at present.

Socio-psychological aspects of autonomous vehicles

Autonomous vehicles (AVs) eliminate the need for a human driver. In addition, they reduce the need for private car ownership. These changes ultimately pose a challenge to entrenched understanding of attributes associated with driving, such as individualism, social status, personal ownership, and the perceived freedom that comes with it (Peters, 2006). Some drivers view their cars as extensions of their self-identity and regard the act of driving as an expression of self-identity (Wang et al., 2020). For such drivers, driving is not merely about getting from point A to point B, but also something adventurous and thrilling (Steg, 2005). Driving may thus contribute to a sense of control and self-efficacy, or simply serve as a source of enjoyment.

Driving is a practice accompanied by emotional attachment and may thus become part of a driver's *self-concept* (Gössling, 2017; Schwanen & Lucas, 2011; Šucha, 2014). Consequently, the process of a human operator being immersed in a secondary activity rather than remaining the driver responsible for the dynamic driving task (DDT) represents a substantial shift in their *self-perception* (Shabanpour et al., 2018). The proliferation of AVs may thus require individuals to re-evaluate their *identity*. For this reason, some people who currently drive may reject the use of AVs, as they might perceive it as incompatible with their *self-identity* (Craig et al., 2019).

However, the perceived enjoyment of driving does not necessarily imply a threat to one's *identity* when using AVs (Craig et al., 2019; Sanders et al., 2012). Individuals who find driving stimulating may still be willing to use an AV in situations involving driver impairment, whether it is due to medication, alcohol, or fatigue. This apparent paradox reveals that the skill-based component of driving constitutes only one aspect of a driver's identity. It also clarifies why, in certain contexts, AV use may be the preferred option for drivers who generally enjoy driving – for example, during motorway travel, where manual driving may not be particularly engaging (Thapa et al., 2021).

Willingness to use autonomous vehicles

The introduction to this chapter highlighted the importance of examining AV technology in terms of public attitudes and willingness to adopt it, which are crucial preconditions for its successful implementation. So how does the public (or rather, specific segments of it) perceive AVs? To what extent are people willing to use them? And what factors come into play?

Although AVs are not yet widespread, the majority of respondents in several surveys reported having heard of them (Gabrhel et al., 2019; Schoettle & Sivak, 2014), and they generally expected positive changes from this technology, whether in terms of improved road safety, increased fuel efficiency, or enhanced performance of the transport system as a whole. Nevertheless, most respondents did not consider it likely that they would use AVs as part of their everyday mobility (Gabrhel et al., 2019; Schoettle & Sivak, 2014; Zmud et al., 2016). Commonly reported reasons include the high cost of AVs, as well as concerns about safety and a general lack of trust in the technology (Gabrhel et al., 2019; Lang et al., 2016). On the contrary, prior experience with advanced driver assistance systems (ADAS) appears to predict a greater willingness to use AVs. This finding is in line with the *innovation diffusion theory* (IDT), which suggests that personal experience, or at least the opportunity to observe a technology in use, increases the likelihood of adopting it (Meuter et al., 2005; Strömberg et al., 2016; Yuen et al., 2018). Across age groups, having the opportunity to see or test AVs in person is frequently mentioned as the preferred way to learn more about the technology. This is particularly relevant for subpopulations such as older adults, who could benefit significantly from AVs in terms of maintaining mobility and a healthy lifestyle, yet often show low levels of willingness to adopt the technology. This reluctance may result from their limited access to information, which may be due to their use of different media channels (e.g. television) than those through which AV-related information is typically disseminated (primarily the Internet).

Cultural differences also play a role in shaping willingness to use AVs. The willingness of Czech respondents to use AVs is comparable to that of respondents in Germany and the Netherlands, with around 40% expressing an intention to use AVs for daily mobility. By comparison, this figure was 85% in India, 75% in China, and 36% in Japan (Lang et al., 2016). These differences may reflect varying expectations regarding the benefits (i.e. the instrumental value) of AV use in comparison with current vehicles. The different levels of intention to use AVs may also mirror respondents' evaluations of their countries' current transport systems, particularly with respect to issues such as traffic congestion or road safety (Lang et al., 2016).

The role of emerging technologies

The ever-present process of modernisation is driving a fundamental transformation in the automotive industry. The intelligent transportation system (ITS) employs various automation technologies, computers, control mechanisms, and communication components to enhance the safety and efficiency of transport. Autonomous vehicles (AVs) are an integral part of this system. Their widespread adoption, however, depends on numerous factors, including safety and reliability (Biswas & Wang, 2023). Emerging technologies such as the Internet of Things (IoT) (Biswas & Wang, 2023), artificial intelligence (AI) (Nallaperuma et al., 2019), and blockchain (Sharma et al., 2020) may play a key role in overcoming these challenges.

The IoT is a network of interconnected devices that collect, transmit, and process data from the surrounding environment using embedded sensors (Shashank et al., 2021). In the future, the IoT could serve various purposes, such as monitoring traffic in real time and planning optimal routes on the basis of variables that would include speed and traffic density (Sarrab et al., 2020). In addition to monitoring traffic situations, IoT in-vehicle sensing devices could also track a person's biometric indicators such as blood pressure or heart rate (Biswas & Wang, 2023).

To process the enormous amount of input from the traffic system effectively, the IoT requires substantial computing power. The role of AI in analysing the massive datasets generated by IoT is thus vital. Artificial intelligence and machine learning (ML) algorithms are used to analyse traffic data, making possible not only traffic optimisation but also the prediction of traffic scenarios (Nallaperuma et al., 2019).

In addition to the challenge of processing large volumes of data, the issue of personal data protection also arises. One possible solution lies in the utilisation of blockchain technology. Blockchain is a digital, decentralised system that can be employed effectively to create reliable, anonymous, immutable, and autonomous records based on predefined consensus algorithms. This significantly increases the transparency and security of data, as stored records cannot be retroactively modified or falsified (Sharma et al., 2020).

Beyond technological issues, ITS also raise important ethical and social concerns. These include data protection, intrusions into citizens' privacy, a sense of constant surveillance, and mistrust of new technologies. Another problem may be the involvement of older populations, for whom such technologies may have unintended consequences. Increased reliance on smart technology can lead to feelings of helplessness in the event of system failure, particularly if users have become overly dependent on technology for everyday tasks (Machidon, 2015). Although smart transport infrastructure has the potential to improve traffic conditions significantly, gaining public trust in the system remains crucial to its successful implementation (Jaramillo-Alcazar et al., 2023).

Conclusion

Autonomous vehicles (AVs) represent a transformative technology in the field of everyday mobility. The taxonomy known as the "Levels of Driving Automation" defines six stages in total. "Level 0" (i.e. "no automation") assumes full human control of the vehicle, whereas "Level 5" (i.e. "full automation") implies a fully autonomous driving system requiring no human intervention. Existing research into road safety identifies human error as the primary cause of traffic crashes. Experts in traffic safety therefore assume that eliminating the human factor (as a source of error) through the widespread adoption of AVs could significantly reduce crash rates.

However, the expected benefits of AVs are accompanied by challenges associated with the implementation of the technology. One such challenge specifically concerns traffic safety. Will drivers who become accustomed to relying on autonomous systems be able to respond quickly and resume control of the vehicle when circumstances demand it? Could reliance on automation impair driving skills, further reducing the ability to re-engage in the driving task when necessary?

Low error rates and the high level of reliability of autonomous systems may lead to a decline in drivers' readiness to respond and take manual control. In other words, undue trust in an autonomous driving system may, in certain situations, become a source of error and heightened danger.

On the other hand, the extent of reduced situational awareness, that is, the degree to which drivers stop monitoring the traffic environment, varies according to the context (e.g. motorway driving versus navigating urban centres). One possible solution lies in vehicle ergonomics designed to support optimal driver activation – for example, by communicating in-vehicle sensor data that would otherwise be inaccessible to the driver. To be useful for a human operator, such information must be presented in a comprehensible format, be salient enough to capture the driver's attention, and be continuously updated.

AVs represent a major challenge for the current concept of road safety and mobility in more general terms. On the one hand, they hold out promise for improving the safety of vulnerable road users (Wang et al., 2020); on the other, the dynamics of mutual interactions may produce unintended systemic consequences.

AVs eliminate the need for a human driver. In addition, they reduce the need for private car ownership. These developments challenge deeply embedded understandings of driving-related attributes such as individualism, social status, personal ownership, and associated perceptions of freedom. However, the perceived pleasure of driving does not necessarily imply a threat to self-identity when using AVs.

Although AVs are not yet widespread, the majority of the respondents across various surveys reported some level of exposure to them. They also expressed a generally positive view of the potential changes AVs might bring, whether in terms of road safety, more efficient fuel consumption, or the transport system as a whole. Nevertheless, only a few consider the use of AVs in their everyday mobility to be likely.

Commonly indicated reasons include the high cost of the technology and concerns about safety or a general mistrust of automated driving. Conversely, prior experience with ADAS has been shown to predict a greater willingness to use AVs. This finding aligns with the innovation diffusion theory (IDT), which holds that personal experience, or even the opportunity to observe a technology in use, increases the likelihood of adoption. Finally, the opportunity to encounter or test AVs in person is frequently cited as a preferred way of learning more about the technology across age groups.

Unsurprisingly, cultural differences play a role in people's willingness to adopt AVs. These may be linked to the perceived benefits (i.e. instrumental value) of AV use in a given country in comparison to conventional vehicles. Differences in the intention to use AVs may also reflect the perceived performance of the existing transport system – particularly with regard to traffic congestion and road safety – in the countries where survey respondents live.

References

Allahyar, M., Becic, E., Chappell, S., Fisher, D., Lohrenz, M., Monk, C., & Philips, B. (2016). The evolving role of automation in transportation: Human factors lessons learned from the different modes. *Proceedings of the Human Factors and Ergonomics Society Annual Meeting*, *60*(1), 1971–1975. https://doi.org/10.1177/1541931213601449

Aria, E., Olstam, J., & Schwietering, C. (2016). Investigation of automated vehicle effects on driver's behavior and traffic performance. *Transportation Research Procedia*, *15*, 761–770. https://doi.org/10.1016/j.trpro.2016.06.063

Bainbridge, L. (1983). Ironies of automation. In J. Rasmussen, K. Duncan, & J. Leplat (Eds.), *Analysis, design and evaluation of man – machine systems* (pp. 129–135). Elsevier. https://doi.org/10.1016/B978-0-08-029348-6.50026-9

Becker, F., & Axhausen, K. W. (2017). Literature review on surveys investigating the acceptance of automated vehicles. *Transportation*, *44*(6), 1293–1306. https://doi.org/10.1007/s11116-017-9808-9

Begg, D. (2014). A 2050 vision for London: What are the implications of driverless transport? *Transport Times*. www.transporttimes.co.uk/Admin/uploads/64165-transport-times_a-2050-vision-for-london_aw-web-ready.pdf

Biswas, A., & Wang, H.-C. (2023). Autonomous vehicles enabled by the integration of IoT, edge intelligence, 5G, and blockchain. *Sensors*, *23*(4), 1963. https://doi.org/10.3390/s23041963

Carsten, O., Lai, F. C. H., Barnard, Y., Jamson, A. H., & Merat, N. (2012). Control task substitution in semiautomated driving. *Human Factors*, *54*(5), 747–761. https://doi.org/10.1177/0018720812460246

Casley, S., Jardim, A., & Quartulli, A. (2013). *A study of public acceptance of autonomous cars*. Worcester Polytechnic Institute. www.semanticscholar.org/paper/A-Study-of-Public-Acceptance-of-Autonomous-Cars-Jardim-Quartulli/92786721c34f411d347b1bbb3592b890a1af7a52

Chaloupka, C., & Risser, R. (2020). Communication between road users and the influence of increased car automation. *Transactions on Transport Sciences*, *10*(2), 5–17. https://doi.org/10.5507/tots.2019.014

Craig, K., Thatcher, J. B., & Grover, V. (2019). The IT identity threat: A conceptual definition and operational measure. *Journal of Management Information Systems*, *36*(1), 259–288. https://doi.org/10.1080/07421222.2018.1550561

Deb, S., Strawderman, L., Carruth, D. W., DuBien, J., Smith, B., & Garrison, T. M. (2017). Development and validation of a questionnaire to assess pedestrian receptivity toward fully autonomous vehicles. *Transportation Research Part C: Emerging Technologies*, *84*, 178–195. https://doi.org/10.1016/j.trc.2017.08.029

Delhomme, P., Verlhiac, J.-F., & Martha, C. (2009). Are drivers' comparative risk judgments about speeding realistic? *Journal of Safety Research*, *40*(5), 333–339. https://doi.org/10.1016/j.jsr.2009.09.003

Desmond, P. A., Hancock, P. A., & Monette, J. L. (1998). Fatigue and automation-induced impairments in simulated driving performance. *Transportation Research Board*, *1628*(1), 8–14. https://doi.org/10.3141/1628-02

Emzivat, Y., Ibanez-Guzman, J., Martinet, P., & Roux, O. H. (2017). *Dynamic driving task fallback for an automated driving system whose ability to monitor the driving environment has been compromised*. 2017 IEEE Intelligent Vehicles Symposium (IV), Los Angeles. https://doi.org/10.1109/IVS.2017.7995973

Funkhouser, K., & Drews, F. (2016). Reaction times when switching from autonomous to manual driving control: A pilot investigation. *Proceedings of the Human Factors and Ergonomics Society Annual Meeting*, *60*(1), 1854–1858. https://doi.org/10.1177/1541931213601423

Gabrhel, V., Ježek, S., & Havlíčková, D. (2019). Public opinion on connected and automated vehicles: The Czech context. *Transactions on Transport Sciences*, *10*(2), 42–52. https://doi.org/10.5507/tots.2019.011

Gössling, S. (2017). *The psychology of the car: Automobile admiration, attachment, and addiction*. Elsevier.

Gouy, M., Diels, C., Reed, N., Stevens, A., & Burnett, G. (2012). *Behavioural adaptation of the unequipped driver to small time headways held by automated vehicles in the traffic* [Conference presentation]. International Conference on Traffic and Transport Psychology (ICTTP), Groningen, The Netherlands.

Havlíčková, D., Gabrhel, V., Linkov, V., & Vondráčková, L. (2021). *Lidský faktor v éře autonomní mobility* [The human factor in the era of autonomous mobility]. Centrum dopravního výzkumu v. v. i.

Havlíčková, D., Zámečník, P., Adamovská, E., Gregorovič, A., Linkov, V., & Zaoral, A. (2020). Model of driving skills decrease in the context of autonomous vehicles. In N. Stanton (Ed.), *Advances in human factors of transportation: Proceedings of the AHFE 2019 international conference on human factors in transportation* (pp. 158–167). Springer. https://doi.org/10.1007/978-3-030-20503-4_16

Jamson, A. H., Merat, N., Carsten, O. M., & Lai, F. C. (2013). Behavioural changes in drivers experiencing highly-automated vehicle control in varying traffic conditions. *Transportation Research Part C: Emerging Technologies*, *30*, 116–125. https://doi.org/10.1016/j.trc.2013.02.008

Jaramillo-Alcazar, A., Govea, J., & Villegas-Ch, W. (2023). Advances in the optimization of vehicular traffic in smart cities: Integration of blockchain and computer vision for sustainable mobility. *Sustainability*, *15*(22). https://doi.org/10.3390/su152215736

Johansson, E. (2014). *Human factors in vehicle automation-activities in the European project adaptive* (Webinar Slides). VRA Webinar.

Ko, S., Zhang, Y., & Jeon, M. (2019). Modeling the effects of auditory display takeover requests on drivers' behavior in autonomous vehicles. *Proceedings of the 11th International Conference on Automotive User Interfaces and Interactive Vehicular Applications: Adjunct Proceedings*, 392–398. https://doi.org/10.1145/3349263.3351508

Lang, N., Rüsmann, M., Mei-Pochtler, A., Dauner, T., Komiya, S., Mosquet, X., & Doubara, X. (2016). *Self-driving vehicles, Robo-taxis, and the urban mobility revolution*. The Boston Consulting Group. www.bcg.com/publications/2016/automotive-public-sector-self-driving-vehicles-robo-taxis-urban-mobility-revolution

Machidon, O. M. (2015). Societal implications of current and emerging "smart" technologies. *International Journal of Technoethics*, *6*(1). https://doi.org/10.4018/ijt.2015010105

Meuter, M. L., Bitner, M. J., Ostrom, A. L., & Brown, S. W. (2005). Choosing among alternative service delivery modes: An investigation of customer trial of self-service technologies. *Journal of Marketing*, *69*(2), 61–83. https://doi.org/10.1509/jmkg.69.2.61.60759

Nallaperuma, D., Nawaratne, R., Bandaragoda, T., Adikari, A., Nguyen, S., Kempitiya, T., De Silva, D., Alahakoon, D., & Pothuhera, D. (2019). Online incremental machine learning platform for big data-driven smart traffic management. *IEEE Transactions on Intelligent Transportation Systems*, *20*(12), 4679–4690. https://doi.org/10.1109/TITS.2019.2924883

Norman, D. A. (1990). The problem of automation: Inappropriate feedback and interaction, not over-automation. In D. E. Broadbent, A. D. Baddeley, & J. T. Reason (Eds.), *Human factors in hazardous situations*. Clarendon Press.

Orlovska, J., Novakazi, F., Lars-Ola, B., Karlsson, M., Wickman, C., & Söderberg, R. (2020). Effects of the driving context on the usage of automated driver assistance systems (ADAS) – naturalistic driving study for ADAS evaluation. *Transportation Research Interdisciplinary Perspectives*, *4*, 100093. https://doi.org/10.1016/j.trip.2020.100093

Peters, P. F. (2006). *Time, innovation and mobilities: Travel in technological cultures*. Routledge.

Rudin-Brown, C., & Parker, H. (2004). Behavioural adaptation to lane departure warning. *Transportation Research Record*, *1803*, 30–37.

Sabey, B. E., & Taylor, H. (1980). The known risks we run: The highway. In R. C. Schwing & W. A. Albers (Eds.), *Societal risk assessment* (pp. 43–70). Springer. https://doi.org/10.1007/978-1-4899-0445-4_3

Sanders, C., Rogers, A., Bowen, R., Bower, P., Hirani, S., Cartwright, M., Fitzpatrick, R., Knapp, M., Barlow, J., Hendy, J., Chrysanthaki, T., Bardsley, M., & Newman, S. P. (2012). Exploring barriers to participation and adoption of telehealth and telecare within the whole system demonstrator trial: A qualitative study. *BMC Health Services Research*, *12*(1), 220. https://doi.org/10.1186/1472-6963-12-220

Sarrab, M., Pulparambil, S., & Awadalla, M. (2020). Development of an IoT based real-time traffic monitoring system for city governance. *Global Transitions*, *2*, 230–245. https://doi.org/10.1016/j.glt.2020.09.004

Schoettle, B., & Sivak, M. (2014). *A survey of public opinion about connected vehicles in the U.S., the U.K., and Australia*. 2014 International Conference on Connected Vehicles and Expo (ICCVE), Vienna, Austria. https://doi.org/10.1109/ICCVE.2014.7297637

Schwanen, T., & Lucas, K. (2011). Understanding auto motives. In K. Lucas, E. Blumenberg, & R. Weinberger (Eds.), *Auto motives* (pp. 3–38). Emerald Group Publishing. https://doi.org/10.1108/9780857242341-001

Seppelt, B. D., & Lee, J. D. (2007). Making adaptive cruise control (ACC) limits visible. *International Journal of Human Computer Studies*, *65*(3), 192–205. https://doi.org/10.1016/j.ijhcs.2006.10.001

Shabanpour, R., Golshani, N., Shamshiripour, A., & Mohammadian, A. (2018). Eliciting preferences for adoption of fully automated vehicles using best-worst analysis. *Transportation Research Part C: Emerging Technologies*, *93*, 463–478. https://doi.org/10.1016/j.trc.2018.06.014

Sharma, A., Awasthi, Y., & Kumar, S. (2020). *The role of blockchain, AI and IoT for smart road traffic management system*. 2020 IEEE India Council International Subsections Conference (INDISCON), Visakhapatnam, India. https://doi.org/10.1109/INDISCON50162.2020.00065

Shashank, S., Nischay, D., & Kiran, P. (2021). IOT based traffic management system. *International Journal of Scientific Research in Computer Science, Engineering and Information Technology*, *8*(1), 184–187. https://doi.org/10.32628/CSEIT1217451

Smith, B. W. (2013). *SAE levels of driving automation*. Center for Internet and Society, Stanford Law School. http://cyberlaw.stanford.edu/blog/2013/12/sae-levels-driving-automation

Steg, L. (2005). Car use: Lust and must. Instrumental, symbolic and affective motives for car use. *Transportation Research Part A: Policy and Practice*, *39*(2–3), 147–162. https://doi.org/10.1016/j.tra.2004.07.001

Strömberg, H., Rexfelt, O., Karlsson, I. M., & Sochor, J. (2016). Trying on change – trialability as a change moderator for sustainable travel behaviour. *Travel Behaviour and Society*, *4*, 60–68. https://doi.org/10.1016/j.tbs.2016.01.002

Šucha, M. (2014). What does it mean to be a driver and what does it mean to own a car? *Journal of Traffic and Transportation Engineering*, *3*, 216–221.

Thapa, D., Gabrhel, V., & Mishra, S. (2021). What are the factors determining user intentions to use AV while impaired? *Transportation Research Part F: Traffic Psychology and Behaviour*, *82*, 238–255. https://doi.org/10.1016/j.trf.2021.08.008

Treat, J. R. (1979). *Tri-level study of the causes of traffic accidents: Final report*. National Highway Traffic Safety Administration. https://trid.trb.org/view/491431

Wang, X., Wong, Y. D., Li, K. X., & Yuen, K. F. (2020). This is not me! Technology-identity concerns in consumers' acceptance of autonomous vehicle technology. *Transportation Research Part F: Traffic Psychology and Behaviour*, *74*, 345–360. https://doi.org/10.1016/j.trf.2020.06.005

Young, M. S., & Stanton, N. A. (2007). Back to the future: Brake reaction times for manual and automated vehicles. *Ergonomics*, *50*(1), 46–58. https://doi.org/10.1080/00140130600980789

Yu, J., & Luo, F. (2019). *Fallback strategy for level 4+ automated driving system.* 2019 IEEE Intelligent Transportation Systems Conference (ITSC), Auckland, NZ. https://doi.org/10.1109/ITSC.2019.8917404

Yuen, K. F., Wang, X., Ng, L. T. W., & Wong, Y. D. (2018). An investigation of customers' intention to use self-collection services for last-mile delivery. *Transport Policy*, *66*, 1–8. https://doi.org/10.1016/j.tranpol.2018.03.001

Zmud, J., Sener, I. N., & Wagner, J. (2016). *Consumer acceptance and travel behavior: Impacts of automated vehicles* (No. PRC 15-49 F). Texas A&M Transportation Institute.

22 Prospects for traffic psychology in the decades to come

Petr Zámečník and Darina Havlíčková

Introduction

When discussing the future *prospects for traffic psychology*, it is important to provide context as well as to consider the present situation. That is, what we know and what remains unknown, what conceptual and theoretical foundations we stand on, and what has turned out to be a dead end. The first point worth pausing on is the context and framework of the field. The current literature associated with traffic psychology encompasses a very wide range of topics, theories, and applications, resulting in an increasingly unclear disciplinary identity. The boundaries and scope of traffic psychology continue to expand, making it difficult to determine what the field currently addresses – or should address – and what lies beyond its remit. This ambiguity arises primarily from the fact that traffic psychology is not a classical psychological discipline, but rather an application of the entirety of psychological knowledge and methods across all the subfields of psychology to the domain of transport (since the early 20th century) and, more recently, to mobility (since the 1990s). Therefore, traffic psychology does not possess a distinct set of theories of its own, but rather draws upon theoretical frameworks from other disciplines and fields of study. In addition, traffic psychology reflects the transport systems and traffic culture of the country in which it is applied, and so the approaches, priorities, and areas of focus differ to some extent between countries. Delineating the contours of the field has also been complicated by the fact that, for a long time, there were no specialised journals (particularly academic ones) dedicated to traffic psychology. As a result, the breadth of the field was reflected in publications from related disciplines, such as general psychology, occupational safety and organisation, medicine, and ergonomics. Many of these works appeared in national languages or took the form of grey literature (e.g. institutional reports, project documentation, and the findings of expert groups). This limited the sharing of information within the field and, in turn, reinforced the distinct national approaches to traffic psychology, contributing to variability in focus and topics (Hagenzieker et al., 2014; Tortosa et al., 2003).

DOI: 10.4324/9781003608394-22

From what was, in the first third of the 20th century, a narrowly focused practical and pragmatic branch of psychology, traffic psychology has evolved into a distinct academic discipline (Barjonet & Tortosa, 2001). By the mid-20th century at the latest, it had come to represent an intersection of *transport psychology*, economics, ergonomics, medicine, and other disciplines. What began as narrowly defined *psychotechnics* gradually developed into traffic psychology, or, alternatively, *transport psychology*. Its scope is often associated with Rothengatter's (1997) definition, which describes traffic psychology as a discipline concerned with the study of road user behaviour, the psychological processes underlying this behaviour, and the relationship between behaviour and traffic accidents. However, the discussion of the future directions of traffic psychology and the understanding of the logic and added value of new perspectives and approaches requires more than just a definition and framework of this psychological field. To understand why we are where we are, and why we are heading in a particular direction, it is essential to clarify the logic behind the field's development, along with the paradigms and discourses that have shaped it.

For most of its existence, traffic psychology has explored various aspects of traffic safety, a topic which the field is associated with very closely. Like many other professionals concerned with traffic safety, traffic psychologists, too, generally subscribed to the myth of personal responsibility: the belief that traffic accidents are primarily caused by individual road users being flawed, making mistakes, or taking conscious risks. One of the key tasks for traffic psychology in the coming decades is thus to reflect upon, or even redefine, this and other foundational assumptions upon which the field is built and the nature of the activities it carries out. To this end, a brief summary of the evolution of thinking and practice in traffic psychology is presented next. It will serve as a basis for considering what lies ahead for the discipline.

Paradigms of traffic psychology and traffic safety

The first two decades of the 20th century, which can be considered the formative period of traffic psychology, saw it emerge as a practical discipline elaborating on Münsterberg's (1912) line of thought. Although traffic accidents were generally regarded as random or chance occurrences, the performance of certain occupations was to be professionalised. Münsterberg argued that different professions required different psychophysical qualities and differing kinds and degrees of these qualities, with some occupations and activities demanding professional conduct. It was then the task of psychologists to identify the qualities required for a given profession using the methods at their disposal and subsequently examine whether an individual wishing to take up that profession actually possessed such qualities. In response to the transport sector's demand for such a professionalisation framework for various categories of drivers, psychology gradually established the field of psychotechnics or technopsychology in Europe and applied psychology in the United States of America (Hagenzieker et al., 2014; Mira, 1924; van Drunen, 1997).

A major shift came with the First World War, which saw the application of psychotechnics to the military domain in the form of driving psychology (selection of drivers, pilots, ship captains, etc.). After the war, psychotechnics and driving psychology found their way into civilian life. In principle, traffic psychologists regarded accidents and road behaviour as outcomes determined by abilities and skills, which could be measured and used for differentiation. Baumgarten (1928) vividly illustrates the post-war period by describing a widespread hunger for psychotechnics in general, and in the field of transport specifically, with particular emphasis on public transport, especially trains and trams. This period also saw the establishment of the International Association of Applied Psychology (in 1921), which laid the foundations for the field as a scientific discipline.

The 1920s brought growing pressure on transport companies and public authorities to introduce psychometric methods, to rationalise driver selection on the basis of abilities needed for effective (safe) task performance, and to predict drivers' "accident-proneness". Some countries, however, remain entrenched in the intellectual tradition of this developmental stage to this day (Hagenzieker et al., 2014). Compared to the early 20th century, a significant change was the professionalisation of assessment and a conceptual shift in the analysis of accident causation – moving away from the notion of crashes as random events and towards a focus on the human factor. While this redirected attention to human error and provided traffic psychology with an opportunity for substantial development, it did not prevent the building of the myth of individual responsibility, a belief system that would cost Europe hundreds of thousands of lives over the course of the 20th century.

Despite advances in assessment, its widespread implementation, and the professionalisation of driving, the number of accidents, including fatal ones, continued to rise. Gradually, the focus shifted from practical skills to more general and complex abilities, particularly assessments of intelligence and personality (Baumgarten, 1949). While the second wave of traffic assessment efforts yielded many insights, the diagnostic orientation did not seem to lead to any substantial improvement in road safety. The emphasis on the human factor as the primary cause of accidents was thus increasingly challenged, and the search for other causal factors began. This shift marked the beginning of a paradigm change, with attention turning towards the "Three E's" – *E*ngineering, *E*ducation, and *E*nforcement.

From the 1950s onwards, rising motorisation and low levels of safety led to alarming crash rates, prompting governments to take the issue of road safety more seriously. This strengthened the research arm of the field, as the new paradigm involved cooperation with other disciplines. However, despite this shift, the situation did not improve; in fact, the number of fatal road crashes in most European countries doubled between 1955 and 1970. Between the 1950s and 1970s, both traffic safety and traffic psychology were dominated by an approach focused on identifying the causes of crashes. These causes were categorised under human, vehicle, and environmental factors (Haddon, 1968), with crash analyses typically seeking one primary cause. Measures to address these causes were selected from the Three E's framework. This mindset – that road safety will be improved through precise understanding and description of traffic crashes – unfortunately persists in many places today, even though it has been evident since at least the 1990s that there is no need to wait for crashes to occur before improving road safety (Shi et al., 2023).

From the late 1960s onwards, a multicausal approach that proposed that crashes result from a combination of factors gradually gained prominence (Haddon, 1972). Traffic psychology thus turned to the integration of external, internal, social, mental, cognitive, and motivational determinants. In practice, this included examination of risk perception, improvements in driver selection and training informed by learning theory, social persuasion (e.g. safety marketing), ergonomics, driver rehabilitation, and the assessment of attitudes (towards risk, safety, speed, seat belt use, alcohol consumption, etc.). Several important theories, including the *threat-avoidance model*, the *zero risk theory*, and the *risk homeostasis theory*, emerged during this period (Hagenzieker et al., 2014).

From the original view of crashes as accidental events and through the later emphasis on psychotechnics and the search for simple causes, this period saw the emergence of a multicausal approach, interdisciplinary cooperation, and the first signs of institutional responsibility. Although the human factor was no longer regarded as the sole or predominant cause of crashes, the notion of personal responsibility continued to prevail. Since the 1960s, however, there has been a general shift away from an enforcement-driven administrative approach, allowing psychology to introduce new interventions. It was proposed that people cannot be constantly monitored, and that internal motivation to follow rules must be strengthened through persuasion. This paved the way for the development of road safety campaigns and rehabilitation programmes (Tortosa et al., 2003).

The focus on crash events and the development of multicausal perspectives peaked in the 1980s, along with efforts to introduce technical solutions. From the 1950s onwards, growing numbers of road fatalities gave rise to a new paradigm – shifting the focus away from the causes of individual crashes (which remained largely unchanged) towards adapting the road system to its users instead. Traffic psychology began to develop this approach in the 1990s, when it became increasingly clear that human limitations are permanent: people will always make mistakes, break rules, and remain physically vulnerable to impacts. This brought about a new view of road safety based on the principle that people are fallible and that the road transport system should be designed to anticipate and accommodate their errors and physiological and psychological limitations (Belin et al., 2012). Countries that implemented this "Safe System" approach achieved significant success, with traffic fatalities reduced by 50 to 70% (WRI, 2018). The role of traffic psychology thus became to incorporate research on the human factor (i.e. the application of psychological and physiological principles) into the design and engineering of a safe system.

The OECD (1997) provides a useful summary of the shifts in thinking about the causes of accidents. It characterises the period up to 1920 as one in which accidents were viewed as random events (the myth of chance). The years 1920–1950 are described as the era of the human factor, focusing on risk or accident-proneness (the myth of individual responsibility). The period from 1950 to 1970 is viewed as being specifically marked by the search for accident causation (the myth of individual/shared responsibility). The following two decades featured multifactorial approaches, with a focus on passive safety and technical measures. Since the 1990s, accidents have increasingly been seen as the outcome of an integral road system, with measures aimed at adapting the system to its users (system responsibility).

However, the era of the Safe System is also the era of the end of traffic psychology as a discipline focused primarily on road safety. Since the 1990s, the subject and scope of traffic psychology have expanded rapidly, with this expansion being driven by technological developments (the emergence of entirely new subfields and areas of study) rather than a shift in safety paradigms. As a result, interaction with information and communication technologies (especially ADAS and ITS) and with other users who also engage with such systems has increased steadily since the 1990s. The goal of this *cognitive ergonomics* is to optimise the driver's communication with the environment (signalling and information systems) and with the vehicle itself (instrument panels) (Tortosa et al., 2003). This trend continued into the first decade of the 21st century, with the development of further technical measures (such as alcohol interlocks), human–machine interaction, and a broader focus on many additional aspects, particularly mobility. Traffic psychology began to address topics such as quality of life, environmental concerns, sustainability, and travel mode choice. In the second decade of the 21st century, its scope expanded further to include new mobility opportunities and requirements (micromobility, electromobility, inclusivity, transport for all), with much greater emphasis on the climate impacts of transport, and the discipline also plays a key role in the development of automation in transport (including automated vehicles and driver assistance systems). While the 20th century was defined, in both research and practice, by a single core theme – safety – the early 21st century is characterised by a broader focus on multiple topics approached in a more comprehensive and multidisciplinary manner.

Challenges for the 21st century

What lies ahead for traffic psychology in the third decade of the 21st century and beyond? As an applied discipline, traffic psychology is closely linked to the development of transport systems and transport policies, which are in turn strongly shaped by global trends. Currently, four global phenomena are particularly relevant for both transport systems and traffic psychology: population growth and urbanisation, demographic and social development, climate change, and technological advancement (Zámečník et al., 2022). While demand for transport is increasing, expanding infrastructure is neither economically nor spatially feasible. Each generation has – and will continue to have – distinct mobility needs. Compatibility – especially between the technology-oriented generations born after 2000 and the growing number of older adults resulting from population ageing – will be a major issue. The urgent need to prevent global warming from exceeding the critical threshold of 1.5°C above pre-industrial levels is putting pressure on the transport sector to undergo transformation in terms of technology, user behaviour, and system principles. Technological developments are bringing new mobility models and options, such as Mobility as a Service (MaaS), vehicle sharing, and automated vehicles, although their effects on individual and collective behaviour remain uncertain. In the light of these trends and their implications, traffic-related psychological research will increasingly focus on areas such as communication and social marketing, intercultural and ethical aspects, inclusivity and mobility for all, sustainability, public health impacts, and the broader relationship between mobility and quality of life.

Contemporary traffic psychology is increasingly embracing technological and experimental approaches, including the use of virtual reality, augmented reality, and other immersive environments. In response to global phenomena, the paradigm shift in road safety, and specific developments within the transport sector, the future of traffic psychology is expected to evolve across four main dimensions: technological (adaptation to emerging technologies), societal (particularly the consequences of the climate crisis), safety-oriented (road safety and responsibility for crashes), and disciplinary (fostering collaboration and sharing of expertise).

Disciplinary challenges

Traffic psychology has long faced – and continues to face – a certain degree of fragmentation and inconsistency in approach. Tortosa's call for traffic psychologists to work towards the development of a single European training programme and career model remains relevant to this day. European countries show major differences in core areas such as the psychological assessment of drivers and rehabilitation programmes. Similarly, there is significant variation in both undergraduate and postgraduate education. Although several pan-European associations and institutional networks (such as EFPA SCTP and TPI) exist, the level of cooperation needs to be substantially deepened in terms of both practice and research. In fact, traffic psychology is marked by a high degree of interdisciplinarity and national specificities, which makes the sharing of scientific knowledge all the more vital. This calls for the development of more specialised academic journals and an increase in their visibility and impact. While journals such as *Transportation Research Part F (TRF)* and *Traffic and Transport Theory and Safety (TOTS)* are available, there remains considerable scope for additional periodicals that focus more specifically on discipline-relevant topics and the dissemination of underutilised materials, such as grey literature. This would also help improve the public relations profile of the field, which, despite its critical relevance to transport and safety, still struggles to "sell itself" effectively to public and political authorities, as well as to the general public. In the near future, the promotion of the field will probably depend on its stronger integration and institutionalisation.

An important aspect of closer cooperation and institutionalisation at the European level is the field's underlying philosophy, which is likely to become a subject of continued discussion in the coming decades. Through their commitment and focus, traffic psychologists contribute to development and progress in transport, and yet it is not totally clear who all the beneficiaries of their findings are. As a science, traffic psychology must remain neutral and evidence-based; as an applied discipline, it must uphold professionalism and expertise. At the same time, it is a field with a considerable impact on human lives, quality of life, and society in general, which lends it a strong ethical dimension. In principle, traffic psychology is oriented towards the public as its primary client, and its findings should be applied in the service of this role. The objective is not merely to neutrally examine everything that emerges, but also to advocate action and help shape the direction of transport developments and policies. In practical terms, this could mean prioritising efforts to help society reduce its dependence on automobiles, rather than supporting marketing strategies that reinforce such dependence. In relation to safety, for example, traffic psychology should aim to define human limitations in terms of capabilities and behaviour as a pivotal knowledge base for systemic change, rather than as a basis for placing greater pressure on individuals.

Safety-related challenges

Two core areas of traffic psychology – assessment and therapy – developed within a paradigm associated with the myth of individual responsibility. However, the adoption of the system approach has brought about a shift in the purpose of these key activities, and of traffic psychology as a whole. This calls for disciplinary reflection aimed at aligning our practices with current knowledge, including its ethical dimension. This involves, on one hand, a principled rejection of the myth of individual responsibility as the supposed primary factor in road safety, and on the other, a redefinition of the purpose and nature of key professional activities. Regarding individual responsibility, although many EU member states, the United States, and others are formally committed to the systemic approach and Vision Zero, in practice they often fall short of implementing them. This is despite the fact that, especially over the last decade, numerous studies have clearly demonstrated the benefits of a safe system and specific strategies for its implementation (e.g. Belin et al.,

2012; Elvik et al., 2009; Guttman, 2014; Hyder et al., 2022; McAndrews, 2013; WHO, 2017). In reality, systemic safety principles have been successfully implemented only by a few – primarily the wealthiest – countries. As a discipline, traffic psychology can support the shift towards a systemic approach by clarifying the principles of human error and behaviour, as well as its limitations.

The second aspect concerns how traffic psychologists will approach assessment and rehabilitation in the coming decades. While playing a key role in the traditional view of road safety, these areas are just some among many – and not the priority ones – within the systemic approach. Decades of road safety research have made it clear that the traditional approach has significant limitations and so do tools such as assessments or rehabilitation (Shi et al., 2023). In all fairness, one can carry out assessments and offer rehabilitation indefinitely if the system continues to generate new clients. Both instruments help to improve the current safety situation and mitigate errors and weaknesses of the traditional system, but they do not have the potential to eliminate crashes, which is what the new systemic approach strives for. As long as the system allows for errors and violations, people will commit them, and assessments and rehabilitation will remain a never-ending task.

The future of rehabilitation will therefore depend on governments' ability to implement the systemic approach. The more effectively this system is established, the fewer potential participants there will be for such programmes. In the decades ahead, we can thus expect a transformation and a shift away from rehabilitation as a "hard" corrective instrument towards programmes with a preventive purpose. The focus will particularly be on groups such as novice drivers or those who commit defined offences within a specific period. In assessment, there will be a need to collect data on human abilities, skills, and traits; not for the purpose of direct evaluation, but to determine the parameters of a system in which everyone can move safely. This means, for instance, collecting evidence to inform measures designed to prevent drivers from violating the rules, rather than assessing who is likely to pose a risk. In other words: under what circumstances and in what ways should vehicles be equipped with alcohol interlocks or speed limiters? At what age, and through what process, should driving licences be issued? What are the psychological limits regarding fatigue for the purposes of setting rest periods and safety breaks? Traffic psychology will thus continue to address human characteristics, but not with the primary aim of "profiling" participants; rather, the goal will be to use this knowledge to design a safe system.

However, this shift may prove difficult because of the long-standing tradition of conceiving traffic psychology as an expert discipline concerned with risk assessment. As noted in the opening section, the growing number of crashes gave rise to the idea, rooted in the notion of road safety and crashes as accidental events, that a small group of "risky" drivers is responsible for most of them. Although this is true statistically and in practice, it spawned what became, in effect, a "hunt" for such drivers. (Shaw & Sichel, 1971). Research focused on identifying characteristics of drivers prone to accidents (and on who is to blame for them), and this led to the search for psychological tests as predictors of accident-prone drivers. One of the earliest examples of this approach was described by Forbes (1922). Although the concept of riskiness appears under different names, the term most often encountered in the literature is *accident-proneness*. This phrase (introduced by Farmer in 1925) is derived from a once-popular theory of drivers prone to accidents which emerged in the 1920s with the development of psychotechnics (see, e.g. Burnham, 2009) but was later heavily criticised and lost scientific credibility (Reason, 1990). Nevertheless, it repeatedly resurfaces in traffic psychology discourse as an attractive and simple explanation of road crashes (e.g. Herzberg, 2009). One reason for this is the fragmentation of traffic psychology and, consequently, the limited extent to which the common evidence base is shared. The 100-year-old perspective of psychological assessment of fitness to drive as a tool for selecting risky drivers has changed, and today – in the context of the systemic approach – is considered both ineffective and unethical.

Accident-proneness, however, is not the only concept that traffic psychology will need to reconsider in the near future. The same applies, for example, to the notion of vulnerable road users (VRUs). In the light of the paradigm shift, pedestrians and cyclists should be seen as those who determine the parameters of the system. If they are jeopardised, injured, or killed, it is not due to their own vulnerability, but rather the result of a poorly designed (vulnerable) system. That is why, for example, in the Netherlands – following the implementation of the Safe System approach – the risk of dying in a road crash is the same for pedestrians as it is for vehicle occupants (Shi et al., 2021). Another concept in need of revision is the distinction between *accident* and *crash*. Traffic psychologists should revise the terminology and the ways in which key events in transport are discussed and labelled. The word *accident* implies randomness and suggests a low level of influence or responsibility for the occurrence (see, e.g. Loimer & Guarnieri, 1996; Tortosa et al., 2003).

Technological challenges

Since the 1990s, with the advancement of ITS, ADAS, automation, and digitalisation, traffic psychology has increasingly explored human–machine interaction and the influence of technology on behaviour. As the technological orientation of transport is expected to predominate in the coming decades, traffic psychology will likewise need to pay more attention to these issues. This technological boom is primarily spurred by the effort to mitigate the negative impacts of road transport and deal with the capacity exhaustion of existing infrastructure. Traffic psychology will thus need to address both the technical and human aspects of using such technologies.

In technological terms, the most significant trend will concern the issues associated with *automated vehicles*. Given their anticipated widespread adoption, key challenges will involve psychological aspects such as the automation paradox, takeover scenarios, intercultural adaptation, and motion sickness. The latter represents a particularly serious challenge, as it has the potential to exclude a substantial portion of the population from using automated transport (Zámečník et al., 2022). At the same time, the widespread use of information and communication technologies is expected to have a major influence on traffic behaviour. Traffic psychology will thus need to deal with issues such as cognitive overload, distraction, and the saturation of the public space with advertising messages.

Regarding the "human" aspects of technology use, there will be ethical concerns pertaining to automation and the large-scale deployment of ITS. The so-called "trolley problem", that is, the question of who should be given safety priority in transport, often features in such discussions. Although largely a theoretical issue that is already resolved in many countries, it continues to attract attention because of its controversial nature. More pressing practical challenges associated with automated vehicles will relate to constant surveillance (or the sense thereof), intrusions into privacy, the perception of technology as something that takes away personal freedom (or employment), and, last but not least, the reduced need for physical exercise. Traffic psychology will therefore need to develop strategies, especially through social marketing, to help people balance mobility and other needs while maintaining quality of life and health. This is particularly relevant in terms of physical activity: at the beginning of the 20th century, the average daily walking distance per person was around six kilometres, but by the start of the 21st century, it had dropped to less than one kilometre. As transport becomes more accessible, efficient, and automated, there are fewer practical reasons for walking. Yet walking remains important source of exercise, especially in the light of rising obesity rates; even a simple walk to and from work can fulfil the daily requirement for physical activity necessary to maintain physical health. Traffic psychology will thus also need to look for ways of helping people overcome their preference for comfort (offered by technology) as the key motivation behind transport and mobility behaviour.

Societal challenges

In terms of societal challenges, two new aspects will be of paramount significance for traffic psychology: the systemic (the Safe System approach to road safety) and the environmental (transport's response to the climate crisis). Global climate change and efforts to mitigate it exert pressure on the better carbon efficiency of transport, that is, reducing emissions. For the time being, the transport sector has responded primarily through technological means –developing, producing, and deploying more efficient and environmentally friendly technologies. While this development and implementation of new technologies opens up new possibilities and challenges for traffic psychology, there are significant misgivings about whether technology alone can truly deliver the required outcomes. Existing evidence from traffic psychology provides no definitive answers, and the subject deserves serious attention. History (i.e. past traffic behaviour) shows that increasing the efficiency of a resource reduces the cost of using it, which in turn increases demand and cancels out the original efficiency gains – a phenomenon known as Jevons' paradox or the rebound effect. This has long been evident in transport: since the 1960s, it has become clear that more roads often lead to increased congestion and longer travel times as a result of drivers' self-interested behaviour (Braess' paradox). Similarly, induced demand shows that new lanes or roads often increase road capacity consumption without reducing travel times (more roads = more traffic) and that the added value of new infrastructure may disappear within weeks or months (the Lewis-Mogridge position). In other words, every time more efficient production or systems have been introduced in transport, the efficiency gains have been offset by increased consumption, with no net savings in key resources or commodities. In transport, this tendency to behavioural change can be seen not only in relation to new infrastructure, but more broadly in mobility. Marchetti's constant and Zahavi's hypothesis reveal that more efficient technologies lead to longer travel distances rather than (time) savings. This gives rise to a form of "mobility consumption homeostasis", where means of transport are used to maximise mobility rather than to maintain current mobility levels at lower cost (in terms of time, emissions, money, etc.). It is therefore reasonable to assume that more efficient (in terms of emissions) technologies may lead to more frequent or intensive use. This will be a key element in the success (or lack thereof) of transport's technological response to the climate crisis and a major challenge for traffic psychology, which will need to describe, explain, and help steer user behaviour in the desired direction.

Electromobility offers a good example. Its promotion has not been based on the principles of social marketing in transport, and the pressure to adopt it has often proved counterproductive. Psychological research shows that travel behaviour is habitual and that some changes are best allowed to occur naturally, even if they may take longer. Consumer surveys (e.g. Hiep, 2024) suggest that the vast majority of users who purchase an electric vehicle do not return to combustion engines. There is thus no need to coerce consumers and face a backlash in return.

As with environmental issues, traffic psychology also plays a crucial role in addressing systemic aspects. About a decade after the Safe System approach was introduced, studies evaluating its limitations began to emerge (e.g. Elvik et al., 2009), because systems theory alone cannot completely eliminate road crashes. People adapt to the traffic system, grow accustomed to its structure, and change their behaviour over time. Additionally, as risk homeostasis theory shows, the safer the system becomes, the more likely it is to encourage risk-taking behaviour. An ultimately safe system would therefore need not only to be safe in objective terms but also to feel less safe to users, so as not to trigger compensatory risk-taking mechanisms or habitual behaviour "disconnected" from real-time conditions. Moreover, the systemic approach relies heavily on technical and engineering measures, which are extremely costly, and, therefore, developing countries, and even some developed ones, "cannot afford" them. In the coming decades, traffic psychology will thus need to explore behavioural limits within the Safe System approach, as well as identifying economically viable ways of shaping road users' behaviour.

References

Barjonet, P. E., & Tortosa, F. (2001). Transport psychology in Europe: A historical approach. In G. Underwood (Ed.), *Traffic psychology today* (pp. 13–29). Elsevier.

Baumgarten, F. (1928). *Die Berufs eignungsprüfungen: Theorie und Praxis*. R. Oldenbourg.

Baumgarten, F. (1949). *Progress of psychotechnics – fortschritte der psychotechnik 1939–1945*. A. Francke.

Belin, M. Å., Tillgren, P., & Vedung, E. (2012). Vision zero – a road safety policy innovation. *International Journal of Injury Control and Safety Promotion*, *19*(2), 171–179. https://doi.org/10.1080/17457300.2011.635213

Burnham, J. (2009). *Accident prone: A history of technology, psychology, and misfits of the machine age*. University of Chicago Press.

Elvik, R., Høye, A., Vaa, T., & Sørensen, M. (2009). *The handbook of road safety measures* (2nd ed.). Emerald Insight.

Farmer, E. (1925). *The method of grouping by differential tests in relation to accident proneness. Annual report* (pp. 43–45). Industrial Fatigue Research Board.

Forbes, T. W. (1922). Age-performance relationships among accident-repeater automobile drivers. *Journal of Consulting Psychology*, 2, 143–148.

Guttman, N. (2014). *Communication, public discourse, and road safety campaigns: Persuading people to be safer*. Routledge.

Haddon, W., Jr. (1968). The changing approach to the epidemiology, prevention, and amelioration of trauma: The transition to approaches etiologically rather than descriptively based. *American Journal of Public Health and the Nation's Health*, *58*(8), 1431–1438. https://doi.org/10.2105/AJPH.58.8.1431

Haddon, W., Jr. (1972). A logical framework for categorizing highway safety phenomena and activity. *Journal of Trauma*, *12*(3), 193–207. https://doi.org/10.1097/00005373-197203000-00002

Hagenzieker, M. P., Commandeur, J. J., & Bijleveld, F. D. (2014). The history of road safety research: A quantitative approach. *Transportation Research Part F: Traffic Psychology and Behaviour*, *25*, 150–162. https://doi.org/10.1016/j.trf.2013.10.004

Herzberg, P. Y. (2009). Beyond "accident-proneness": Using five-factor model prototypes to predict driving behavior. *Journal of Research in Personality*, *43*(6), 1096–1100. https://doi.org/10.1016/j.jrp.2009.08.008

Hiep, E. (2024, 10. December). Global EV driver survey: 92% of EV drivers say they'll never go back*!* Global EV Drivers Alliance. https://globalevalliance.com/world-wide-ev-drivers-survey-92-of-ev-drivers-will-never-go-back/

Hyder, A. A., Hoe, C., Hijar, M., & Peden, M. (2022). The political and social contexts of global road safety: Challenges for the next decade. *The Lancet*. https://doi.org/10.1016/S0140-6736(22)00917-5

Loimer, H., & Guarnieri, M. (1996). Accidents and acts of God: A history of the terms. *American Journal of Public Health*, *86*(1), 101–107. https://doi.org/10.2105/AJPH.86.1.101

McAndrews, C. (2013). Road safety as a shared responsibility and a public problem in Swedish road safety policy. *Science, Technology, & Human Values*, *38*(6), 749–772. https://doi.org/10.1177/0162243913493675

Mira, L. E. (1924). Psicotecnia. *Enciclopedia espasa* (pp. 1509–1516).

Münsterberg, H. (1912). *Psychologie und Wirtschaftsleben: Ein beitrag zur angewandten experimental-psychologie*. Barth.

Organisation for Economic Co-operation and Development (OECD). (1997). *Road safety principles and models: Review of descriptive, predictive, risk and accident consequence models* (OCDE/GD(97)153). https://one.oecd.org/document/OCDE/GD(97)153/En/pdf

Reason, J. (1990). *Accident proneness: The history of an idea*. Cambridge University Press.

Rothengatter, T. (1997). Psychological aspects of road user behaviour. *Applied Psychology: An International Review*, *46*(3), 223–234. https://doi.org/10.1111/j.1464-0597.1997.tb01227.x

Shaw, L., & Sichel, H. S. (1971). *Accident proneness: Research in the occurrence, causation, and prevention of road accidents*. Pergamon Press.

Shi, G., Methoxha, V., Atkinson-Palombo, C., & Garrick, N. W. (2021). Sustainable safety in The Netherlands: Creating a road environment where people on foot and on bikes are as safe as people in cars. *Transportation Research Record: Journal of the Transportation Research Board*, *2675*(11), 792–803. https://doi.org/10.1177/03611981211019736

Shi, Q., Li, Z., Xu, Y., Yan, T., & Chen, M. (2023). Dynamic scenario simulations of sustainable rural and towns development in China: The case of Wujiang district. *Sustainability*, *15*(19), 14254. https://doi.org/10.3390/su151914254

Tortosa, F., Barjonet, P., Civera, C., & Montoro, L. (2003). A history of European traffic and transport psychology. *Anuario de Psicología*, *34*(3), 401–416.

van Drunen, P. C. (1997). Psychotechnics. In A. R. Luria & J. H. Wozniak (Eds.), *A pictorial history of psychology* (pp. 480–484). Quintessence Publishing Co. Inc.

Wilde, G. J. S. (1982). The theory of risk homeostasis: Implications for safety and health. *Risk Analysis*, 2(4), 209–225. https://doi.org/10.1111/j.1539-6924.1982.tb01384.x

World Health Organization (WHO). (2017). *Save LIVES: A road safety technical package*. www.who.int/publications/i/item/save-lives-a-road-safety-technical-package

World Resources Institute (WRI). (2018). *Sustainable and safe: A vision and guidance for zero road deaths*. www.wri.org/publication/sustainable-and-safe-vision-and-guidance-zero-road-deaths

Zámečník, P., Havlíčková, D., & Vondráčková, L. (2022). Psychosociální aspekty automatizované dopravy. In D. Černý, O. Vaculín, & P. Zámečník (Eds.), *Automatizované řízení vozidel a autonomní doprava: Technické a humanitní perspektivy* (pp. 181–200). Academia.

Index

Note: Page numbers in *italics* indicate figures.